Smart Agriculture

Radhakrishnan Edayileveetil Krishnankutty
Ajay Kumar
Kallamparambil Balakrishnan Soni
Aswani Ravi
Editors

Smart Agriculture

Artificial Intelligence for Sustainable Food Systems

Springer

Editors
Radhakrishnan Edayileveetil Krishnankutty
School of Biosciences
Mahatma Gandhi University
Kottayam, Kerala, India

Ajay Kumar
Amity Institute of Biotechnology
Amity University
Noida, Uttar Pradesh, India

Kallamparambil Balakrishnan Soni
Department of Molecular Biology and Biotechnology
Kerala Agricultural University
Thiruvananthapuram, Kerala, India

Aswani Ravi
Department of Bioscience
Rajagiri College of Social Sciences (Autonomous)
Kochi, Kerala, India

ISBN 978-981-95-8784-1 ISBN 978-981-95-8785-8 (eBook)
https://doi.org/10.1007/978-981-95-8785-8

This Springer imprint is published by the registered company Springer Nature Singapore Pte Ltd.
The registered company address is: 152 Beach Road, #21-01/04 Gateway East, Singapore 189721, Singapore

Preface

Agriculture stands at a decisive crossroads. The growing global population, changing climate, and increasing scarcity of arable land demand transformational solutions to safeguard food security and environmental well-being. Traditional farming models alone can no longer meet these accelerated challenges. In this context, the integration of Artificial Intelligence (AI) with agriculture emerges not merely as an innovation but as a vital pathway toward sustainability and resilience.

The present book, *Smart and Sustainable Agriculture through Artificial Intelligence*, explores how advanced computational intelligence ranging from machine learning and computer vision to robotics, big data analytics, and the Internet of Things is redefining the agricultural ecosystem. These technologies enable precise resource utilization, proactive crop and livestock management, early detection of diseases and pests, climate-adaptive farming practices, and data-driven decision support systems. They empower farmers and agribusinesses to enhance productivity while minimizing environmental footprints.

This book brings together multidisciplinary insights from research, technological development, and practical field implementation. It highlights success stories, challenges, and future opportunities for AI-enabled farming systems, placing equal emphasis on inclusivity, ethics, and accessibility ensuring that technological progress benefits farmers of all scales across the globe.

As digital transformation reshapes rural landscapes, collaboration between scientists, engineers, policymakers, educators, and farmers becomes crucial. We hope this book will inspire readers to envision and contribute to the future of agriculture one where intelligence and sustainability grow hand in hand, fostering a healthy planet and a nourished society.

Kottayam, Kerala, India — Radhakrishnan Edayileveetil Krishnankutty
Noida, Uttar Pradesh, India — Ajay Kumar
Thiruvananthapuram, Kerala, India — Kallamparambil Balakrishnan Soni
Kochi, Kerala, India — Aswani Ravi

Contents

1 **AI Tools, Techniques, and Technologies in Agriculture** 1
Geetha Radhakrishnan and R. Rakhi

2 **Opportunities for AI Interventions at Various Stages of Agricultural Practices** 29
R. Rakhi, R. Geetha, and S. Sarada

3 **Use of AI in Crop Health Monitoring** 41
Ersin Elbasi, Greeshma Varghese, Yehia Ibrahim Alzoubi, Ahmet E. Topcu, and Aymen I. Zreikat

4 **Sustainable Development Goals (SDG) based AI integrated Irrigation Systems** 57
Prem Rajak, Suchandra Bhattacharya, Abhratanu Ganguly, and Satadal Adhikary

5 **Scope of AI in Future Farming** 73
Ersin Elbasi, Elda Cina, Nour Moustafa, Aymen I. Zreikat, Ahmed Shdefat, and Ahmet E. Topcu

6 **Strategies for Strengthening Biosecurity in Indian Agriculture** 91
Chetan Chauhan, Shanta Kumari, and Taru Singh

7 **Plant Disease Management Through Artificial Intelligence Tools** 115
Govindh A. Nair, Arsha Kamal, and N. S. Radhika

8 **Microbiomes in Weed and Postharvest Management** 139
P. Sukrith, Gholamreza Salehi Jouzani, and Radhakrishnan Edayileveetil Krishnankutty

9 **Unlocking the Potential of Algal-Mediated Nanoparticles in Plant Science: Synthesis, Challenges, and Applications** 157
Arya Gupta, Jyotsna Bora, Niharika V. Rao, Monika Joshi, and Subhasha Nigam

10 **Integrated Metagenomic Insight into Microbial Dye Degradation** . . . 181
Bishwarup Sarkar and Sougata Ghosh

11 Current Trends in Agriculture Research Based on AI and Machine Learning 197
Saumya Dwivedi, R. Rajesh, and Manickavelu Alagu

12 Applications of AI in Future Farming 227
Aparna G. Kumar, B. Gayathri, and Radhakrishnan Edayileveetil Krishnankutty

13 Genotyping Advancements for Mitigating Biotic Stress-Induced Damage in Wheat 267
Chandranandani Negi, Mansi Singh, Neeraj Kumar Vasistha, Dharmendra Singh, Krishan Kumar, and Pritesh Vyas

About the Editors

Radhakrishnan Edayileveetil Krishnankutty School of Biosciences, Mahatma Gandhi University, Kottayam, Kerala, India. radhakrishnanek@mgu.ac.in

Dr. Radhakrishnan E. K. is an Associate Professor at the School of Biosciences, Mahatma Gandhi University, Kottayam, and holds several honorary directorships including Business Innovation and Incubation Centre (BIIC), Inter University Centre for Organic Farming and Sustainable Agriculture (IUCOFSA), Dr. N. Radhakrishnan International Centre for Medical Innovation (NRICMI), and Industry Academia Interface Centre (IAIC). He earned his PhD from RGCB, Thiruvananthapuram (2008), and pursued postdoctoral research at The University of Tokyo (2010). With 156 publications, 55 book chapters, and 10 edited books, his research spans plant–microbe interactions, microbial nanotechnology, and antimicrobial materials. He has guided 14 PhD scholars, filed four patents, and launched "Nanopower." Recognized among the top 2% scientists globally (Stanford–Elsevier, 2022–2023), he has received several awards including the Kairali Gaveshana Puraskaram (2022).

Ajay Kumar Amity Institute of Biotechnology, Amity University, Noida, India. ajaykumar_bhu@yahoo.com

Dr. Ajay Kumar is an Assistant Professor at the Amity Institute of Biotechnology, Amity University, Noida, India. He served as a Visiting Scientist (2018–2022) at the Agricultural Research Organization, Volcani Center, Israel, and earned his PhD in Plant–Microbe Interactions from Banaras Hindu University, Varanasi. He has authored over 235 scientific contributions, including research articles, reviews, and book chapters, and has edited more than 48 books with leading publishers such as Elsevier, Springer, CRC Press, and Wiley. His research focuses on plant–microbe interactions, microbial biocontrol, postharvest fruit management, and microbial endophytes. He also serves as Associate Editor for *Frontiers in Microbiology* and *BMC Microbiology*.

Kallamparambil Balakrishnan Soni Department of Molecular Biology and Biotechnology, College of Agriculture, Vellayani, Thiruvananthapuram, Kerala, India. soni.kb@kau.in

Dr. K. B. Soni is Professor of Biotechnology at the Department of Molecular Biology and Biotechnology, College of Agriculture, Vellayani, Thiruvananthapuram. She earned her PhD in Biochemistry from Amala Cancer Research Centre, Thrissur, under Mahatma Gandhi University, Kerala, and completed her postdoctoral studies at the Centre for Genetic Engineering, Indian Institute of Science, Bangalore, in 1995. A recipient of the prestigious Dr. K.S. Krishnan Award from the Department of Atomic Energy, BARC, Mumbai, she has authored numerous research and review articles and book chapters. Her research focuses on biotic and abiotic stress tolerance, siRNA-mediated virus resistance in banana, and metagenomic exploration of novel enzymes.

Aswani Ravi Rajagiri College of Social Sciences (Autonomous), Kalamassery, Cochin, Kerala, India. aswaniraveendran92@gmail.com

Dr. Aswani R. is a Post-Doctoral Fellow at the Department of Bioscience, Rajagiri College of Social Sciences, Cochin, and a former Chief Minister's NavaKerala Post-Doctoral Fellow at the College of Agriculture, Kerala Agricultural University, Thiruvananthapuram. She earned her PhD in Microbiology from the School of Biosciences, Mahatma Gandhi University, Kottayam, and holds an MPhil in Biosciences and an MSc in Microbiology. She has authored several research papers, reviews, and book chapters, and co-edited books such as *Biocontrol Mechanisms of Endophytic Microorganisms* and *The Potential of Microbes for a Circular Economy* (Elsevier). Her research focuses on biocontrol mechanisms, plant defense induction, and biopolymer development from the phytomicrobiome.

AI Tools, Techniques, and Technologies in Agriculture

1

Geetha Radhakrishnan and R. Rakhi

Abstract

Agriculture is under pressure from climate change, limited resources, and increasing food demand. Artificial intelligence (AI) offers practical solutions by improving precision, efficiency, and sustainability in farming. This chapter reviews key AI tools such as machine learning, deep learning, and computer vision, supported by technologies, including sensors, IoT, GIS, GPS, drones, big data, and ICT. Applications cover the full agricultural value chain, from land use planning and resource management to pest detection, yield forecasting, robotics, postharvest handling, and market integration. Case studies show progress in crop monitoring, disease diagnosis, irrigation management, and autonomous machinery. Hybrid models that combine process-based simulations with AI are also discussed for better predictions. Challenges such as high costs, weak infrastructure, poor data quality, and ethical issues persist. The review concludes that AI can drive sustainable and climate-smart agriculture if accessibility, scalability, and farmer-oriented adoption are ensured.

Keywords

Artificial intelligence · Machine learning · Deep learning · Computer vision · Internet of Things (IoT) · Sensors · Precision agriculture · Robotics · Big data · Smart farming · Decision support systems · Climate-smart agriculture

G. Radhakrishnan (✉)
RARS(SZ), Kerala Agricultural University, Vellayani, Thiruvananthapuram, Kerala, India
e-mail: geetha.ramesh@kau.in

R. Rakhi
Department of Plantation, Spices, Medicinal and Aromatic Crops, College of Agriculture, Vellayani, Kerala Agricultural University, Thiruvananthapuram, Kerala, India
e-mail: rakhi.r@kau.in

R. E. Krishnankutty et al. (eds.), *Smart Agriculture*,
https://doi.org/10.1007/978-981-95-8785-8_1

1 Introduction

Agriculture has always been the backbone of human civilization, supplying food, fiber, and raw materials. In India, it plays a vital role in sustainable economic development, accounting for a significant share of GDP and employing a large proportion of the population. Despite record food grain production, the sector faces numerous challenges, including population growth, resource competition, climate change, labor shortages, and environmental degradation. To address these challenges, modern approaches such as smart farming and artificial intelligence (AI)-driven agriculture are being increasingly adopted. Smart farming integrates advanced technologies, viz. sensors, IoT, UAVs, robotics, remote sensing, big data analytics, mobile apps, and cloud computing to optimize farming operations. AI complements these technologies by enabling tools and techniques such as machine learning, deep learning, computer vision, and decision support frameworks that enhance efficiency, precision, and sustainability.

Artificial intelligence (AI), particularly machine learning (ML) and deep learning (DL), has brought about significant changes across various fields (Mishra 2024). AI helps farmers use resources more efficiently. It aids in yield improvement, pest management, and decision-making (Nautiyal et al. 2025). The rise of AI-enabled precision agriculture also supports sustainable and climate-smart farming. Studies highlight the use of AI tools such as ML, DL, the Internet of Things (IoT), and decision support systems (DSS). These are used in precision irrigation, pest control, yield optimization, and decision-making (Benos et al. 2021).

The flowchart (Fig. 1.1) illustrates the movement of agricultural data through a structured pipeline. The workflow of AI applications begins with data acquisition, where information is gathered from diverse sources such as IoT sensors, drones, satellites, weather stations, and soil testing devices. These inputs include critical parameters like soil moisture, temperature, pH, rainfall, and crop images, which form the foundation for intelligent decision-making. Once collected, the information moves to the data processing stage. Here, raw datasets are cleaned, integrated, and analyzed using advanced techniques such as machine learning (ML) and deep learning (DL). This step transforms large and often unstructured data into actionable insights, for example, by extracting features from crop images or developing predictive models for irrigation requirements.

The processed data is then applied across multiple agricultural applications, supporting tasks such as pest and disease detection, irrigation scheduling, yield forecasting, soil nutrient management, and even the operation of autonomous farm machinery. These applications demonstrate how AI transforms complex datasets into real-world tools that directly support farmers. The process leads to measurable outcomes, that include improvised crop yields, reduced input costs, efficient resource use, enhanced decision-making, and progress toward sustainable farming practices.

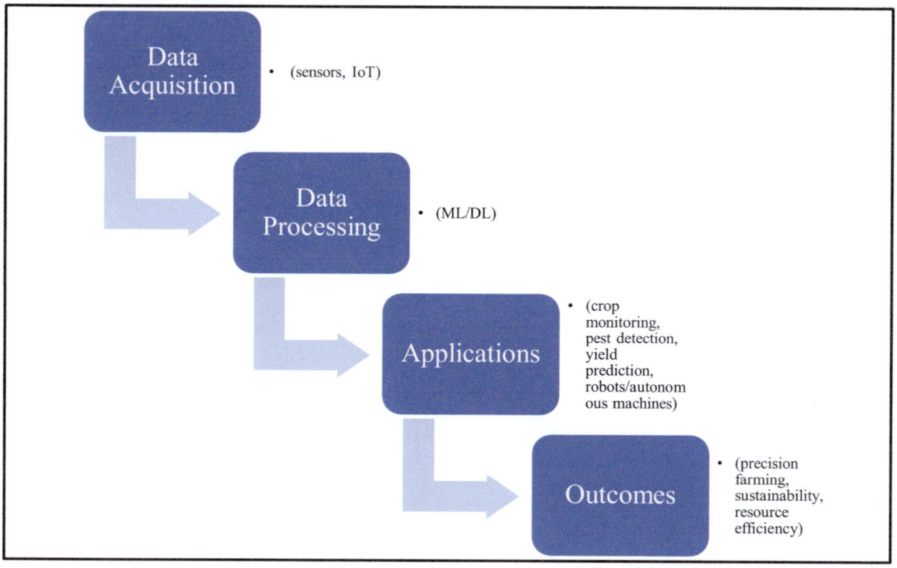

Fig. 1.1 Overview of AI tool categories in agriculture

2 Supporting Technologies for AI in Agriculture (Fig. 1.2)

2.1 Sensors

A sensor is an electrotechnical device that measures physical or chemical quantities from the environment and converts them into electrical signals that can be read by instruments (Pivoto et al. 2018). In agriculture, such devices provide data for farmers to use to monitor crops and ensure practices that are adapted to the environmental conditions. It is estimated that by 2050, sensors will be in use across more than 500 million farms worldwide (Triantafyllou et al. 2019). Sensors are used in weather stations, drones, harvesters, weeders, and robots. They measure variables such as temperature, humidity, light, pressure, soil moisture, other soil properties, airflow, carbon dioxide, and plant water stress. These sensors provide data to AI and IoT systems for real-time decision-making (Sheikh Mansoor et al. 2025; Dhanaraju et al. 2022; Choudhary et al. 2025). Sensors have been tested in several contexts. Gómez et al. (2006) evaluated an electronic nose for tomato ripening. Fekete et al. (2018) compared sensory evaluation with an electronic tongue to assess taste differences in grafted and nongrafted watermelons. Peng et al. (2021) developed a portable molecular imprinting sensor to detect pesticide residues in mango, cowpea, and water samples.

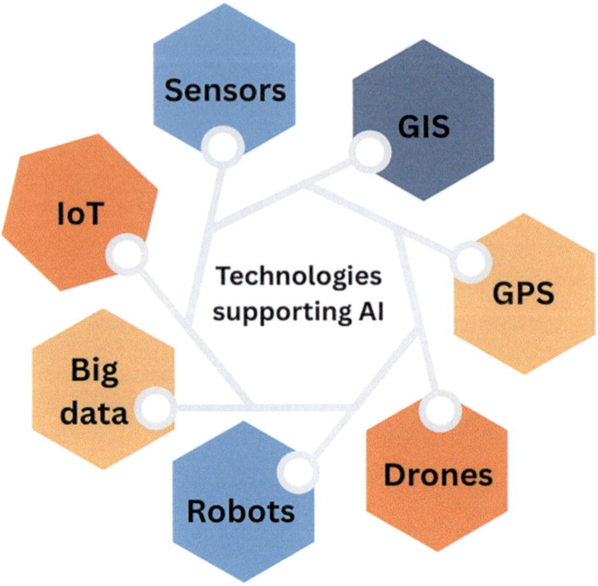

Fig. 1.2 Technologies supporting AI

Sensor type	Principle/description	Applications in agriculture	References
Temperature sensors	Measure heat using thermocouples, thermistors, or infrared devices	Greenhouse monitoring, soil and air temperature tracking, climate control	Mansoor et al. (2025)
Proximity sensors	Detect presence/absence of objects without contact (inductive, capacitive, ultrasonic)	Automation in farm machinery, seed placement, monitoring equipment position	Liu et al. (2025)
Pressure sensors	Convert fluid/gas pressure into electrical signals	Irrigation systems, soil compaction studies, monitoring of hydraulic systems	Aarif et al. (2025)
Light sensors	Detect and measure light intensity (photoresistors, photodiodes, phototransistors)	Crop growth monitoring, greenhouse light control, photosynthesis studies	Austria et al. (2023)
Image sensors	Convert light into digital signals	Remote sensing, UAV imaging, plant disease detection, phenotyping	Shahi et al. (2023)
Optical sensors	Passive non-contact sensors that measure reflected/ absorbed light	Soil property estimation (organic matter, clay, moisture), fruit maturity assessment	Aarif et al. (2025)
Location sensors (GPS)	Determine precise geographic position	Precision farming, mapping fields, equipment tracking, livestock monitoring	Shahi et al. (2023)
Servo sensors	Provide continuous feedback in servo motors for real-time adjustments	Automated irrigation, robotic arms for planting and harvesting	Li et al. (2023)

Sensor type	Principle/description	Applications in agriculture	References
Infrared (IR) sensors	Detect emitted or reflected IR radiation	Temperature monitoring, motion detection (pests, intruders), crop canopy analysis	Al-Najadi et al. (2025)
Relay sensors	Switch states based on electrical signals	Automated irrigation pumps, greenhouse climate control systems	Kunt (2025)
Level sensors	Measure fluid or solid levels (capacitance, ultrasonic, vibrating, float switch)	Irrigation tanks, grain silos, fertilizer reservoirs, pond water management	Xu et al. (2022)
Motion sensors	Detect movement using PIR, microwave, or ultrasonic principles	Monitoring animal/pest activity, irrigation automation, intruder detection, soil moisture tomography	Ruban et al. (2025)
Chemical sensors	Detect chemical substances (electrochemical, semiconductor)	Soil nutrient analysis, gas leak detection, pesticide residue monitoring	Hossain et al. (2024)
MEMS sensors	Miniaturized devices integrating mechanical and electronic components	Accelerometers for drones, gyroscopes for autonomous robots, pressure monitoring in irrigation systems	Sonmezocak (2025)
Capacitive sensors	Detect changes in capacitance between conductive plates	Touch-based interfaces, soil moisture monitoring, level sensing, fertilizer management	Xu et al. (2022)

2.2 Geographical Information System (GIS)

A geographic information system (GIS) is a computer-based tool that stores, integrates, analyzes, and visualizes spatially referenced data (Mennecke and Crossland 1996). According to Ojo and Ilunga (2018), GIS databases can hold detailed information on field characteristics, such as soil type, nutrient status, chemical inputs, and topography, and help establish how these factors interact to influence crop growth. Abdel Rahman et al. (2025) used AHP-GIS techniques to create detailed soil and land suitability maps in arid and semi-arid regions. Their model integrates soil, climate, and agro-practices to guide sustainable land use. Zhai et al. (2025) applied land evaluation and site assessment (LESA) method with ArcGIS at the watershed scale in China. They mapped agricultural capability, including both soil attributes and socioenvironmental factors. Both proprietary and open-source GIS platforms are in use. ArcGIS, MapInfo, GeoMedia, and ArcView are common commercial systems, while QGIS and GRASS GIS are widely adopted open-access tools. GIS has been applied to evaluate site suitability and agro-climatic potential (Falasca et al. 2014). These tools support various GIS platforms. Commercial software (e.g., ArcGIS) is widely used. Open-source options like QGIS and GRASS GIS are also popular because they are more accessible in many regions. GIS is applied for site suitability, yield estimation, soil fertility mapping, agro-climatic analyses, and policy support.

2.3 Global Positioning Systems (GPS)

The global positioning system (GPS) is a satellite navigation system. It provides precise location data, latitude, longitude, and elevation, by measuring signals from multiple satellites (Lang 1992). GPS paired with real-time kinematic (RTK) and IoT sensors now offer high accuracy under field conditions (Zhao et al. 2024). Farmers use GPS-guided machinery for planting, spraying, and harvesting with minimal error (Getahun et al. 2024).

GIS and GPS are often integrated in agricultural research. Hussain et al. (2018) used a combination of GIS and GPS to monitor the population of applied GPS and GIS to monitor *Pieris brassicae* (cabbage butterfly) in India. Their georeferenced maps demonstrated how spatial data can support pest forecasting and targeted management. Similarly, Slaughter et al. (2012) developed and field-tested an automated intra-row weed control system in tomato fields. This system used GPS-generated crop maps to guide mechanical weed knives for the precise removal of weeds.

2.4 Drones

Drones are unmanned aerial vehicles that fly using GPS and carry cameras and sensors. They collect spatial, spectral, and temporal data. These data help track crop health, detect disease early, and forecast yield. Recent work shows UAV-hyperspectral imaging can detect Potato Virus Y in early season using visible, red-edge, and near-infrared bands (Nesar et al. 2025). Another study used UAV multispectral imagery and plant height data to predict potato tuber yield parameters like fresh tuber weight and number of tubers with high accuracy (Tatsumi and Usami 2024). In vegetable crops, Canicattì and Vallone (2024) found that drones improve management by helping assess plant vigor, map weeds, and support precision pest control. Researchers in Italy used multispectral UAV images to monitor crop cycle and map crop vigor with different nitrogen fertilization levels (Messina et al. 2021). UAV-based multispectral and texture-data model forecast yield and market quality of Vidalia sweet onions, producing useful predictions of yield and market class across different days before harvest (Barbosa Júnior et al. 2024).

2.5 Internet of Things

The Internet of Things (IoT) is central to modern farming. It connects sensors, devices, and platforms to gather real-time data from fields. These systems help farmers monitor soil moisture, temperature, humidity, pH, and other key variables. (Sheikh Mansoor et al. 2025). Smart fertigation is one emerging application. An IoT-based fertigation system (Adamo et al. 2025) adjusts water and fertilizer delivery in greenhouses. It bases its decisions on short-term predictions and sensor feedback. IoT-based irrigation and fertigation has been reviewed by Miller et al. (2025). There is also research on connectivity technologies. A comparative study of

LPWAN, hybrid LPWAN-5G, and cellular networks shows how combining them can reduce IoT connectivity costs while improving coverage in remote areas. IoT enables precision farming, reduces resource waste, and improves yield stability. But many small farms still struggle with infrastructure, cost, and technical skills.

2.6 Big Data

In agriculture, data are generated from multiple sources, that is, IoT sensors, satellite imagery, drones, weather stations, and farm machinery. This large and complex dataset is known as big data. Big data cannot be processed using traditional methods. To extract meaningful insights from this varied and voluminous data, advanced analytics was implemented (Zaborowicz and Frankowski 2025; Wolfert et al. 2017).

The importance of big data in agriculture lies in its ability to improve decision-making at every stage of the farming cycle. Analysis of weather patterns, soil characteristics, and crop performance assists optimized decision-making regarding planting schedules, fertilizer application, and irrigation. Hussein et al. (2025) report that big data analytics enable site-specific crop management, reduce the input costs, and minimize environmental impacts.

Big data also plays a role in risk management. Predictive models built on large datasets can forecast pest outbreaks, disease spread, or drought events. This allows farmers to take preventive measures.

Big data also supports sustainability goals. By monitoring soil moisture, nutrient cycling, and greenhouse gas emissions, analytics platforms guide resource-efficient practices. Cloud-based frameworks that link crop models with real-time data, such as the system proposed by Akhter et al. (2025), demonstrate how big data can support adaptive responses to environmental stress while maintaining productivity.

2.7 Information and Communication Technology (ICT)

Information and communication technology (ICT) plays a crucial role in modern agriculture. It provides easy access to information, supports decision-making, and facilitates the efficient use of resources. ICT encompasses tools such as mobile phones, Internet platforms, geographic information systems (GIS), global positioning systems (GPS), and decision-support software that link farmers with timely knowledge and services (Aker et al. 2016).

ICT systems facilitate precision agriculture by integrating data from sensors, drones, satellites, and farm machinery into accessible digital platforms. Farmers can monitor soil health, crop status, and weather conditions in real time, enabling site-specific management of inputs such as water, fertilizer, and pesticides (Mittal and Mehar 2016). Mobile-based advisory systems and ICT-driven platforms have significantly improved smallholder access to market information, weather forecasts, and pest/disease advisories, contributing to better risk management and increased productivity.

ICT also supports digital extension services. Cloud-based dashboards and mobile applications deliver customized advisories on irrigation scheduling, nutrient application, and crop protection. In 2024, Devi et al. underscore the improved farming practices through ICT-enabled extension activities. ICT-based e-learning tools, online databases, and decision-support systems contribute to the dissemination of knowledge and capacity building among farmers, researchers, and extension workers.

The integration of ICT with AI and IoT further enhances its potential. ICT platforms act as the backbone for data exchange between devices and users, enabling innovations such as AI-driven predictive analytics, IoT-enabled smart irrigation, and GIS-based site-specific planning (Miller et al. 2025). This convergence of ICT with emerging digital technologies is driving agriculture toward higher efficiency, sustainability, and inclusivity.

3 AI Tools and Techniques

AI tools encompass machine learning (ML), deep learning (DL), and computer vision approaches that support tasks such as yield prediction, pest and disease detection, soil monitoring, and livestock management. ML algorithms like decision trees, random forest, SVM, and gradient boosting have been widely applied in agriculture. DL and computer vision models such as CNNs, ResNet, and YOLO are crucial for image-based applications in crop monitoring and pest detection (Fig. 1.3).

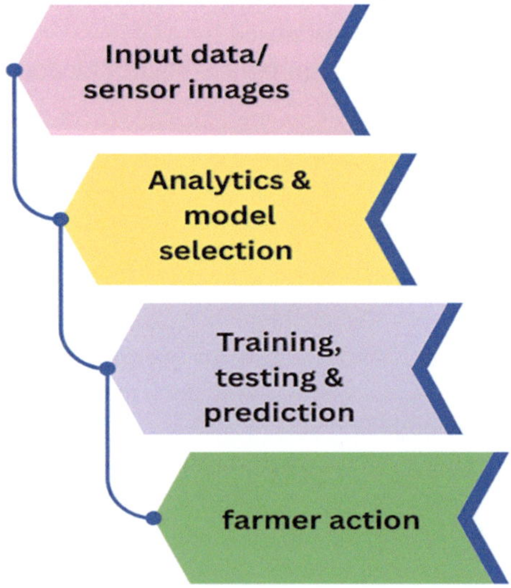

Fig. 1.3 Technologies supporting AI

3.1 Machine Learning (ML) in Agriculture

Machine learning (ML) has become a core technology in agriculture due to its ability to process large and complex datasets. ML algorithms learn from historical and real-time data to make predictions and support decision-making. The supervised models, viz. decision trees, random forests, gradient boosting, and support vector machines (SVMs), have been widely used for crop yield prediction, soil fertility assessment, and disease detection (Liakos et al. 2018). Artificial neural networks (ANNs) are also increasingly applied in livestock monitoring, crop classification, and precision irrigation. A systematic review by Benos et al. (2021) shows that ANNs outperform traditional statistical methods for tasks involving high-dimensional and nonlinear data. ML is also integral to sustainable resource management.

ML in sustainable agriculture has evolved from understanding weather data to integrating advanced technologies like the Internet of Things, remote sensing, and smart farming. Research highlights the need for the expansion of predominant concepts, like deep learning and smart farming, to explore new applications that will help maximize the benefits of ML in agricultural sustainability (Botero-Valencia et al. 2025).

3.2 Deep Learning (DL) in Agriculture

Deep learning (DL), a subfield of ML, uses multilayered neural networks to capture complex spatial and temporal patterns. Convolutional neural networks (CNNs) are especially effective for image-based applications, such as plant disease detection, weed identification, and fruit maturity classification (Kamilaris and Prenafeta-Boldú 2018).

Recent advances include the use of architectures like U-Net for lesion-level segmentation and Mask R-CNN for object detection in crop health monitoring. Vision transformers (ViTs) are also emerging, offering improved handling of long-range dependencies in agricultural imagery (Upadhyay et al. 2025).

3.3 Computer Vision (CV) in Agriculture

Computer vision (CV) enables the automatic interpretation of images and videos, supporting precision agriculture. CV systems analyze RGB, multispectral, and hyperspectral imagery to monitor crop growth, assess nutrient deficiencies, detect pests and diseases, and guide robotic operations (Cao et al. 2025). Image acquisition, image stitching and photogrammetry, image analysis, decision-making, treatment, and planning are some of the tasks performed by CV (Ghazal et al. 2024). CV is also key to automation: drones and ground robots equipped with cameras and CV algorithms enable weed mapping, fruit harvesting, and real-time crop condition monitoring (Vijayakumar et al. 2025).

4 AI-Powered Applications in Agriculture

By integrating AI with IoT, remote sensing, robotics, and big data analytics, agriculture is moving toward a new era of smart farming. The practical applications of AI now span soil and crop monitoring, livestock management, postharvest processing, and supply chain optimization. The applications of AI in agriculture enable farmers to make precise, real-time decisions based on sensor data, imagery, and predictive models.

4.1 AI across the Agricultural Value Chain

Artificial intelligence (AI) is integrated across multiple stages of the agricultural value chain—from planning to market.

Figure 1.4 is a flowchart illustrating the sequential connection between planning, production, protection, postharvest, and marketing, which are the key agricultural stages.

The flowchart highlights the contribution of artificial intelligence (AI) in decision-making across the agricultural value chain. The process begins with land use and crop planning, where machine learning techniques combine information on soil, climate, and topography to guide crop selection and optimize land allocation. It then moves into resource management, where real-time data from IoT sensors, interpreted through AI algorithms, supports precise scheduling of irrigation, fertilizer use, and pesticide application—improving efficiency while reducing waste.

During the production stage, technologies such as drones, satellite imagery, and predictive models enable continuous monitoring of crops and livestock, ensuring that problems are detected early and interventions are timely. Protection follows as a critical step, with computer vision and deep learning tools identifying pests, diseases, and weeds at an early stage, which allows targeted solutions such as robotic weeding or precision spraying. As crops mature, yield prediction models integrate weather, soil, and remote sensing data to estimate harvests in advance, providing valuable input for both farmers and policymakers.

Once harvesting is completed, AI applications extend into the postharvest and logistics phase, where they support grading, sorting, and storage as well as optimize transport systems to reduce spoilage and maintain product quality. The market and decision support systems employ predictive analytics to forecast demand, anticipate price fluctuations, and link farmers with buyers. These stages demonstrate a connected pipeline for AI, from planning and production to market integration. This ultimately enhances productivity, conserves resources, and supports sustainable agricultural practices.

Table 1.1 summarizes how artificial intelligence (AI) tools are deployed at different stages of the agricultural value chain. In the land use and crop planning stage, machine learning (ML) models combined with GIS techniques are applied to classify land according to its suitability for different crops, enabling more informed planning decisions. For resource management, IoT sensors integrated with ML

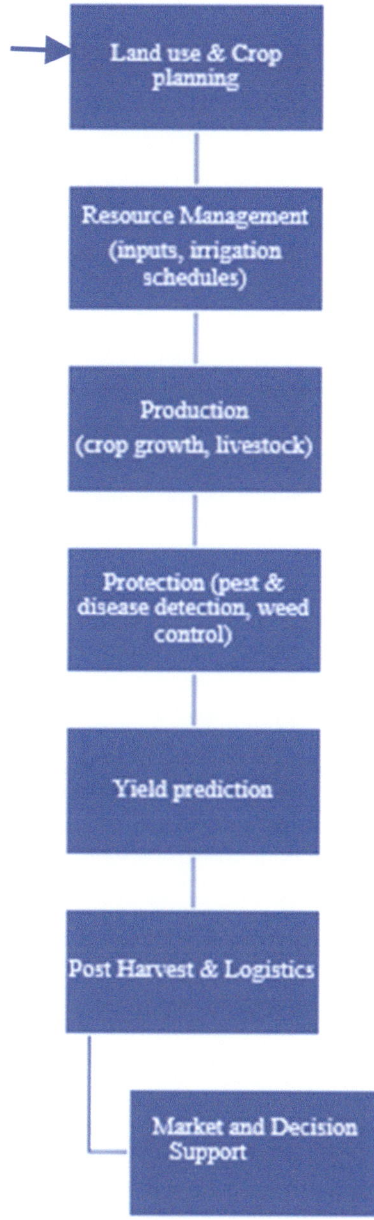

Fig. 1.4 AI contributions in decision-making

algorithms provide real-time data on soil and environmental conditions, which is then analyzed to optimize irrigation scheduling and ensure efficient use of water and inputs.

In the area of disease detection, computer vision and deep learning models, such as CNNs and YOLO variants, are trained alongside traditional classifiers like SVMs and random forests to recognize symptoms of crop diseases from images with high

Table 1.1 AI tools deployed in agriculture

Stage	AI tool/technique	Example application	References
Land use and crop planning	ML-based suitability modeling, GIS + AI	Balanced bagging classifier used to determine the suitability levels of land use	Agrawal et al. (2025)
Resource management	IoT sensors + ML for irrigation scheduling	An ensemble of decision tree and random forest models, trained on the great time dataset, analyzes soil and environmental parameters. The IoT-enabled framework then applies real-time sensor data and ML to optimize irrigation scheduling	Roy et al. (2025)
Disease detection	Computer vision + CNN/YOLO models	Support vector machines (SVMs), random forest (RF), k-nearest neighbor (KNN), and deep models like VGG16, ResNet50, and DenseNet121 were used for disease detection in crops	Ngugi et al. (2024)
Weed control	DL-based image recognition, robotics	The robot user laser technology in combination with AI algorithms to target weeds with lasers, reducing herbicide application	B. T. et al. (2025)
Yield prediction	Remote sensing (satellite, drone) + ML Predictive analysis	Random forest and gradient boosting algorithms applied to drone-based, multispectral images improved rice yield predictions. Beehive clustering	Venugopal et al. (2021)
Livestock management	AI vision, ML anomaly detection	CNN-based video analytics used to monitor cattle behavior, detecting early signs of lameness and disease	Neethirajan (2024)
Logistics and market	Predictive analytics, DSS, reinforcement learning	AI-based price forecasting models (e.g., LSTM networks) help predict market demand for perishable vegetables	Ananthara et al. (2013)
Climate resilience	AI-driven early warning systems	ML incorporating geospatial and meteorological foundation models for EWS, crisis management	Reichstein et al. (2025)

accuracy. Similarly, weed control has advanced through robotics and deep learning, where machines equipped with AI-driven laser systems can target weeds selectively, reducing reliance on herbicides. Remote sensing technologies such as drones and satellites are combined with ML models like random forests and gradient boosting to improve accuracy in estimating/predicting crop yields before harvest.

AI applications also extend to livestock management, where video analytics powered by convolutional neural networks (CNNs) are used to monitor animal behavior and detect early signs of stress, lameness, or disease, allowing timely interventions. Additional methods such as predictive analytics and clustering techniques, contribute to yield estimation by identifying patterns in complex agricultural datasets. In the logistics and market stage, decision support systems (DSS) and predictive models, including LSTM neural networks, help forecast demand and prices for perishable crops, enabling better marketing strategies. AI supports climate resilience through early warning systems that integrate meteorological and geospatial data, providing timely alerts and crisis management strategies to mitigate the impact of climate-related risks.

4.2 Machine Learning Techniques in Agriculture

Machine learning (ML) is a branch of artificial intelligence that enables computers to learn patterns and make predictions or decisions without being explicitly programmed. Its strength lies in handling complex, nonlinear, and high-dimensional datasets, which are common in agriculture due to the interaction of biological, environmental, and management variables. ML techniques can be broadly categorized into supervised learning, unsupervised learning, and reinforcement learning, each suited to different types of problems. In supervised learning, algorithms such as decision trees, support vector machines (SVMs), and ensemble methods like random forests and gradient boosting learn from labeled data to perform classification or regression tasks (Tamayo-Vera et al. 2024). Unsupervised learning methods, such as clustering and dimensionality reduction, are used to uncover hidden structures in data, for example, in soil-type classification or grouping of crop phenotypes (Bougiouklis et al. 2025). Reinforcement learning (RL), though less widely adopted in agriculture so far, is gaining traction in robotics and autonomous systems, where agents learn to perform optimal actions through trial and error in dynamic environments. Goldenits et al. (2024) show that RL is gaining attention for dynamic environment tasks. Deep learning (DL), a subfield of ML, employs multilayered artificial neural networks (ANNs) and architectures such as convolutional neural networks (CNNs) to capture complex spatial and temporal features, making them highly effective for image recognition, time-series forecasting, and sensor fusion. These techniques have found widespread use in agriculture. Girones et al. (2025) discuss how ANNs model complex non-linear interactions in agriculture and outperform simpler methods in tasks like the prediction of yield and the modeling of crop responses.

A systematic review of ML applications between 2018 and 2020, conducted under PRISMA guidelines, concluded that ANNs consistently outperformed traditional statistical models in managing nonlinear datasets and were especially effective for crop classification, disease detection, and livestock monitoring (Benos et al. 2021). Earlier surveys also reported the versatility of supervised learning algorithms, particularly decision trees, SVMs, random forests, and gradient boosting, in tasks such as yield prediction, soil health assessment, and pest and disease identification (Liakos et al. 2018).

Waqas et al. (2025) have delivered higher spatial resolution and predictive accuracy than the traditional survey methods in weed detection and yield forecasting by using UAV-mounted multispectral sensors and drone imagery analyzed with random forest and gradient boosting models.

In combination with data and images, ML techniques ranging from classical supervised algorithms to advanced neural networks are becoming indispensable in agriculture. They support innovations that make farming more sustainable and resilient, and help it cope with climate change and food security challenges.

4.3 Computer Vision and Deep Learning in Crop Monitoring, Pest Detection, and Yield Prediction

Computer vision and deep learning (DL) have emerged as central technologies for agricultural automation. These are two areas critical to ensuring food security and minimizing yield losses. They enable machines to analyze visual data at high levels of accuracy and scale. Deep learning architectures, viz. convolutional neural networks (CNNs) and their extensions, form the backbone of modern image-based agricultural applications (El Sakka et al. 2025).

Recent advances in object detection and segmentation models have greatly improved crop stress and weed monitoring. A review by Silva et al. (2024), in *Remote Sensing,* highlights that YOLOv8 variants, Mask R-CNN, and U-Net, when used for weed detection and segmentation from UAV imagery, achieve very high mAP, precision, and recall. Daraghmi et al. (2025) report that CNN-based models, with object detection frameworks like YOLOv8, Faster R-CNN, and RetinaNet variants, are utilized in the detection of plant stress in aerial images of potato crops. YOLO, Faster R-CNN, and semantic segmentation networks such as U-Net are now widely used for plant stress detection, weed classification, and yield estimation tasks. These models excel at recognizing complex spatial patterns in RGB, multispectral, and hyperspectral imagery. This enables early detection of physiological changes that are difficult to identify with the naked eye.

Deep learning has demonstrated strong performance in pest and disease detection tasks. Using the large IP102 dataset that contains 39,990 images across 82 pest species, Venkateswara and Padmanabhan (2025) developed CNN-based classifiers with autoencoder-based augmentation, confirming the potential of CNNs in pest monitoring. For plant disease detection, semantic segmentation approaches have proven particularly effective at the lesion level. Singh et al. (2025) employed U-Net to segment leaf disease lesions, enabling accurate localization and quantification of severity. Similarly, Polly and Devi (2024) applied U-Net and CNN-based segmentation to leaf disease classification, achieving high accuracy in identifying and quantifying damaged leaf regions, further demonstrating the utility of DL in precision plant health monitoring.

A comprehensive review by Cao et al. (2025) outlines how deep learning-based computer vision enables core tasks, such as disease monitoring and weed detection in crop growth management, citing rapid adoption of models like CNN to automate image-based assessments in real-world farming.

In agricultural computer vision, domain variability refers to the differences in conditions under which images are captured, such as variations in lighting (bright sunlight versus cloudy skies), background complexity (bare soil compared to dense canopy), crop variety (differences in leaf shape or color within the same species), and seasonal or geographical factors (where the same disease appears differently across regions). Such variability often causes deep learning models trained in one context, for example, rice leaf blight in India, to perform poorly when applied in another location, such as Vietnam. To address this challenge, researchers employ domain adaptation techniques that enhance model robustness and generalizability. These include adversarial training, which enables models to ignore irrelevant differences; unsupervised alignment, which leverages unlabeled target-domain data to align feature distributions; and synthetic data generation using GANs to expand limited training sets. Collectively, these strategies improve the reliability of computer vision systems across diverse agricultural environments.

Recent research emphasizes the importance of enhancing the robustness of computer vision models in agriculture under variable field conditions. Hu et al. (2025) provide a comprehensive review of domain adaptation in agricultural image analysis, highlighting how adversarial training and unsupervised alignment techniques address challenges such as lighting variability, background complexity, and crop diversity to improve model generalization across regions and seasons. Shoaib et al. (2025) show that modern architectures for classification, detection, and segmentation routinely surpass traditional image-processing methods. Also, Xiang et al. (2025) demonstrate a practical approach to domain adaptation with their DODA framework, which leverages generated data to adapt object detectors for dynamic agricultural environments, thereby supporting the transition of deep learning models from experimental to field-ready applications.

4.4 Emerging Architectures: ViTs and GANs

Vision transformers (ViTs) and generative models are beginning to complement CNNs in agricultural domains. Upadhyay et al. (2025) highlight the role of computer vision (CV) in crop disease control. He emphasizes that vision transformers offer improved capacity for capturing long-range dependencies in plant imagery.

Generative adversarial networks (GANs) are increasingly applied in agriculture to address the scarcity of labeled datasets. Olaniyi et al. (2022) and Lu et al. (2022) show the effectiveness in augmenting image datasets for tasks such as plant health, weed, and fruit recognition under limited training conditions. Recent advances such as a dual discriminator GAN demonstrate the generation of fine-detail synthetic crop disease images, significantly improving deep learning model performance (Wang et al. 2025). Conditional GANs like Pix2PixHD have been used to expand small root image datasets, with segmentation networks such as SegNet then applied for precise root analysis (Thesma and Mohammadpour 2022).

4.5 Remote Sensing, IoT, and Yield Prediction

Remote sensing has become a vital tool in modern agriculture, enabling large-scale monitoring of crops and field conditions that are critical for yield forecasting. Recent studies emphasize the increasing use of hyperspectral and UAV-based (drone) imagery in agricultural monitoring, supporting continuous assessment of vegetation status, soil characteristics, and crop growth patterns (Zhang et al. 2024; Matese et al. 2024). Reviews further highlight that advanced platforms such as lidar, radar, and multispectral/hyperspectral sensors, are providing rich spatial and temporal datasets that feed predictive models of crop growth and stress.

The Internet of Things (IoT) has introduced powerful tools for agriculture through sensor-based monitoring systems. These sensors track key parameters such as soil moisture, temperature, humidity, nutrient levels, and canopy health, giving farmers timely information to guide irrigation, fertilization, and crop management (Goap et al. 2018). When linked with predictive models, IoT data supports precision farming by improving input efficiency and reducing waste. One example is a smart irrigation system described by Goap et al. (2018), which combines IoT sensors with algorithms to automate water delivery, resulting in notable savings in water and energy use.

Recent studies show that integrating IoT sensors with machine learning (ML) offers significant advantages over traditional rule-based decision support systems in agriculture. By processing real-time data from soil, canopy, and environmental sensors, ML–IoT frameworks enable adaptive, data-driven strategies for managing pests and diseases. These systems provide more accurate predictions and support timely interventions, thereby improving resilience in crop production (Eze et al. 2025). Machine learning serves as the analytical core of such frameworks, converting raw sensor and imagery data into actionable insights. Advances in sensing technologies, when combined with ML, have also been shown to make crop and environmental assessments more affordable and precise, further reinforcing their role in modern precision agriculture (Shawon et al. 2025).

Disease management is an increasingly promising area for systems that fuse IoT sensors and machine learning (ML). A recent study proposed an AI-IoT smart agriculture pivot for detecting and treating plant diseases using real-time leaf images and sensor data, achieving very high classification metrics (Ibrahim et al. 2025). The integration of remote sensing, IoT, and ML is proving powerful for improving both yield prediction and water efficiency. A critical review by Bassine et al. (2025) emphasizes how combining vegetation indices from remote sensing, microclimate and soil data from IoT sensors, and cloud-hosted ML analytics yields more dynamic and context-aware forecasting and irrigation decisions. Eze et al. (2025) report water savings in validated IoT-ML irrigation systems, showing such integrated approaches outperform static decision systems (Eze et al. 2025).

4.6 AI-Powered Robotics and Automation

Artificial intelligence has significantly advanced agricultural robotics, moving them beyond mere mechanization toward autonomous, intelligent systems capable of real-time decision-making (Wakchaure et al. 2023; Spagnuolo et al. 2025). Robots now integrate vision systems, deep learning models, and sensor fusion to perceive their surroundings. They can localize fruits and guide manipulator arms for tasks like selective spraying and harvesting. Vision perception technology using RGB, depth, and multisensor inputs supports fruit-picking robots. This is possible through object recognition and precise end-effector control even under occlusion and variable lighting (Tan et al. 2025). Compact agricultural robots fuse LiDAR and inertial sensors to perform under-canopy crop-row navigation, maintaining path following across diverse field conditions (Baquero Velasquez et al. 2022). More recent research introduces self-supervised neural methods like CROW for row-following in under-canopy plantations; these methods handle variation in plant growth, lighting, and row topology, boosting robustness in unstructured environments (Affonso et al. 2025). This indicates the efficiency and adaptability of agricultural robots with AI.

One of the uses of AI-powered robotics is in fruit harvesting. Robotic systems employ deep learning and computer vision to detect and localize fruits in orchards. They can identify ripe fruits and use manipulator arms for precise picking. They detect fruits with a single-shot multibox detector (SSD) and a stereo camera for 3D localization (Onishi et al. 2019).

Another important application is autonomous weed removal. Laser-weeding robots use AI-vision to distinguish weeds from crops and apply lasers to eliminate weeds, reducing reliance on herbicides (Czymmek et al. 2024).

Postharvest operations also benefit from the use of AI-based robots. Intelligent grading and sorting systems use computer vision to assess fruit quality, identify size, color, or defects, to maintain market-grade consistency.

4.7 Modeling and Decision Support Frameworks

Process-based models (PBMs) simulate the biological, physical, and chemical processes underlying agricultural systems, including crop development, water and nutrient cycling, and phenology. Widely used models such as APSIM, DSSAT, WOFOST, and STICS capture how inputs like fertilizer, irrigation, weather, and soil interact to determine yield. Their main strengths lie in interpretability and their ability to run "what-if" scenario analyses under different climate or management regimes. However, they often require detailed parameterization and calibration and can be computationally demanding (Jones et al. 2017).

Machine learning (ML) models, in contrast, rely on data-driven pattern recognition using algorithms such as decision trees, random forests, gradient boosting, support vector machines, and deep neural networks. They are particularly effective in handling nonlinear, high-dimensional datasets, including those generated from remote sensing and multiple environmental variables. ML models can achieve strong predictive performance when sufficient data are available, but they face challenges such as overfitting, limited interpretability, and reduced reliability when applied outside their training domains (Liakos et al. 2018).

Hybrid models combine PBMs with ML to leverage the strengths of both approaches. For example, Li et al. (2024) integrated APSIM outputs with ensemble ML methods to predict dryland wheat yield in Gansu, China. Their hybrid framework achieved higher accuracy and better generalization than either PBM or ML alone. Similarly, Shi et al. (2025) classified hybrid methods into two categories: DL-informed PBMs, where neural networks refine specific PBM modules, and PBM-informed DL, where mechanistic constraints guide ML predictions. They demonstrated that such hybrids outperform stand-alone models across varied spatial and environmental conditions.

Decision support systems (DSS) build on these models to deliver actionable insights through dashboards, scenario simulations, and real-time updates. Saikai et al. (2023) showed that coupling APSIM outputs with sensor inputs such as soil moisture, leaf area index, and rainfall allowed reinforcement learning agents to optimize daily irrigation schedules. This system outperformed conventional irrigation rules in simulated wheat systems, improving both water use efficiency and profitability.

Emerging trends highlight the future direction of agricultural modeling and DSS. Explainable AI (XAI) methods, including SHAP and LIME, are increasingly integrated into ML and hybrid models to improve interpretability and user trust (Mohiddin et al. 2026). A second challenge is temporal and spatial generalization. Here, the models are trained on historical or localized data, which often fail under new geographies or climate conditions. Hybrid PBM–DL models enhance robustness by integrating mechanistic knowledge from process-based models with the adaptive learning capacity of deep learning (Shi et al. 2025). Advances in sensor and remote sensing fusion, which integrates proximal sensors, UAVs, and satellites, are improving both the timeliness and spatial detail of model inputs. Researchers are also exploring optimization under uncertainty, with reinforcement learning beginning to explicitly address stochastic weather, soil heterogeneity, and pest or disease risk (Saikai et al. 2023). Scalability and accessibility remain crucial. Cloud- and mobile-based DSS platforms are being developed to meet the needs of smallholder farmers, who often face constraints in data availability and computational resources (Ghandour et al. 2023).

4.8 AI for Climate-Smart and Sustainable Agriculture

Artificial intelligence is becoming central to climate-smart agriculture. It helps optimize water use, improves soil nutrient management, and builds resilience to environmental change.

AI-driven irrigation systems can cut water use in water-stressed regions (Oğuztürk 2025). These tools combine IoT sensors, weather data, and weather forecasts to irrigate only when needed. This reduces waste while maintaining crop yield. Ercan Oğuztürk et al. (2025) tested AI-controlled irrigation on geraniums in Turkey. They found that the system used less water than manual irrigation while retaining the plant quality.

Soil analysis is another important area. Machine learning models can predict nutrient levels such as nitrogen, phosphorus, and organic matter. They rely on data from sensors and remote sensing. Farmers then apply fertilizers only where needed. This reduces excess use, lowers costs, and protects soil health (Gunasekaran et al. 2025).

AI also supports agrivoltaics, wherein crops are grown under solar panels. Algorithms help choose crop types, track water stress, and design panel layouts. The aim is to balance crop growth and energy production. Early studies indicate that this approach maintains yields while producing renewable energy and reducing land use conflicts.

A combination of these AI-enabled tools delivers the goals of climate-smart agriculture. They can improve productivity and resilience while reducing greenhouse gas emissions.

5 Challenges, Ethical, and Social Considerations

Artificial intelligence and robotics face several barriers in agriculture. High implementation costs restrict adoption, especially among smallholder farmers (Spagnuolo et al. 2025; Lemay and Boggs 2024). Many robotic systems are built for specific crops, which limits their use in diverse farming contexts (Spagnuolo et al. 2025). Power supply also remains a challenge: continuous autonomous operation demands reliable and efficient energy sources that many systems lack (Bras et al. 2024). Finally, AI models still struggle with robustness in real field conditions—variations in lighting, weather, and crop diversity often degrade performance (Lemay and Boggs 2024; Spagnuolo et al. 2025).

Robotic agriculture is rapidly progressing toward broader adoption across both large and small farms. A recent systematic review reports increased interest in swarm robotics, IoT-enabled platforms, and generative AI. This highlights that robotics will increasingly be used in both industrial and smallholder settings (Spagnuolo et al. 2025). Another study emphasizes the future use of swarms of cooperating robots for coordinated tasks, such as harvesting or weeding over large areas. It also mentions deeper integration with remote sensing and IoT systems for full-field automation (Shahzad et al. 2023). Mobile robotics reviews also designate greater efficiency, adaptability, and reduced hardware cost, to enable farmers to trust model recommendations (Yépez-Ponce et al. 2023).

While AI promises substantial benefits for agricultural efficiency and sustainability, multiple technical, infrastructural, and social barriers remain. One major obstacle is the availability and quality of data. Many farms, especially in developing regions, lack reliable historical records, sufficient sensor coverage, or standardized weather and soil information—limiting the training and generalization of robust AI/

ML models (Cravero et al. 2022). Another issue is the skills gap. Operating and maintaining AI systems demand expertise in data science, model interpretation, and calibration. These skills are often scarce in smallholder communities or in regions without strong agricultural extension support (Atapattu et al. 2024).

Infrastructure limitations pose a significant obstacle to effective AI deployment in agriculture. Miller et al. (2025) identify unreliable Internet connectivity, high infrastructure costs, limited interoperability, and weak data governance as key barriers in applying AI-integrated sensing systems at scale, especially in rural areas. Technical and social hurdles also include insufficient hardware and software capacity, lack of training for users, uncertainties around data ownership and cybersecurity (Miller et al. 2025).

Ethical and social risks are also prominent. Farmer autonomy may be compromised if tools impose practices that clash with local knowledge or customs. These issues are discussed in the same systematic reviews that document infrastructural and governance gaps.

6 Summary and Future Outlook

Artificial intelligence is increasingly viewed as a transformative force in agriculture, with applications spanning production efficiency, sustainability, and resilience. Nautiyal et al. (2025) provide a comprehensive review of how AI, including machine learning, deep learning, IoT, and decision support systems, is being deployed to improve agricultural practices and enhance the quality of produce. Their study highlights how these tools contribute to more efficient use of resources and better management of operations. Complementing this, Pandey and Mishra (2024) emphasize the role of AI in enabling farms to adapt to climate variability by integrating predictive tools and remote sensing technologies that anticipate environmental stresses such as droughts and extreme temperatures. These studies highlight the dual potential of AI in enhancing the quality and quantity of the produce. It also equips farming systems with the tools to withstand emerging climate challenges.

The impact of AI in agriculture is expected to increase substantially as the supporting infrastructure, data sharing policies, and equitable access improve. According to Ali et al. (2025), advances in connectivity, more accessible sensors, and open-data initiatives are enabling smaller farms to adopt precision-farming practices. These tools facilitate real-time monitoring of crops, early disease detection, and efficient use of inputs.

Similarly, Kumari et al. (2025) argue that tightening policies around emissions, water use, and land sustainability will make AI tools—especially those combining remote sensing, variable-rate technologies, and machine learning—central to meeting climate goals without sacrificing food security.

References

Aarif MKO, Alam A, Hotak Y (2025) Smart sensor technologies shaping the future of precision agriculture: recent advances and future outlooks. J Sens. https://doi.org/10.1155/js/2460098

Abdel Rahman MAE, Yossif TMH, Metwaly MM (2025) Enhancing land suitability assessment through integration of AHP and GIS-based for efficient agricultural planning in arid regions. Sci Rep 15:31370. https://doi.org/10.1038/s41598-025-14051-7

Adamo T, Caivano D, Colizzi L, Dimauro G, Guerriero E (2025) Optimization of irrigation and fertigation in smart agriculture: an IoT-based micro-services framework. Smart Agric Technol 11:100885. https://doi.org/10.1016/j.atech.2025.100885

Affonso F, Tommaselli FAG, Capezzuto G et al (2025) CROW: a self-supervised crop row navigation algorithm for agricultural fields. J Intell Robot Syst 111:28. https://doi.org/10.1007/s10846-025-02219-2

Agrawal N, Govil H, Kumar T (2025) Agricultural land suitability classification and crop suggestion using machine learning and spatial multicriteria decision analysis in semi-arid ecosystem. Environ Dev Sustain 27:13689–13726. https://doi.org/10.1007/s10668-023-04440-1

Aker JC, Ghosh I, Burrell J (2016) The promise (and pitfalls) of ICT for agriculture initiatives. Agric Econ 47(S1):35–48. https://doi.org/10.1111/agec.12301

Akhter S, Honda K, Aida K, Ines AVM (2025) An integrated (Crop model, cloud and big data analytic) framework to support agriculture activity monitoring system. ArXai. arXiv:2411.12303. https://doi.org/10.48550/arXiv.2411.12303

Ali Z, Muhammad A, Lee N, Waqar M, Lee SW (2025) Artificial intelligence for sustainable agriculture: a comprehensive review of AI-driven technologies in crop production. Sustainability 17(5):2281. https://doi.org/10.3390/su17052281

Al-Najadi R, Al-Mulla Y, Al-Abri I et al (2025) Effectiveness of drone-based thermal sensors in optimizing controlled environment agriculture performance under arid conditions. Sci Rep 15:9042. https://doi.org/10.1038/s41598-025-94432-0

Ananthara MG, Arunkumar T, Hemavathy R (2013) CRY – An improved crop yield prediction model using bee hive clustering approach for agricultural data sets. In: 2013 international conference on pattern recognition, informatics and Mobile engineering, Salem, India, pp 473–478. https://doi.org/10.1109/ICPRIME.2013.6496717

Atapattu AJ, Perera LK, Nuwarapaksha TD, Udumann SS, Dissanayaka NS (2024) Challenges in achieving artificial intelligence in agriculture. In: Chouhan SS, Saxena A, Singh UP, Jain S (eds) Artificial intelligence techniques in smart agriculture. Springer, Singapore. https://doi.org/10.1007/978-981-97-5878-4_2

Austria ACH, Fabros JS, Sumilang KRG, JBernardino J, Doctor AC (2023) Development of IoT smart greenhouse system for hydroponic gardens. ArXiv. arXiv:2305.01189

Baquero AE, Hisano VVA, Valverde HM, Narenthiran GAV., Marcelo S, Girish B, Chowdhary (2022) Multi-sensor fusion based robust row following for compact agricultural robots field robotics 21291–1319. https://doi.org/10.55417/fr.2022043

B. T, V. A S, S. S D and S. Kumar G (2025) Weed management using laser technique. In: 2025 3rd international conference on artificial intelligence and machine learning applications theme: healthcare and internet of things (AIMLA), India, pp 1–5. https://doi.org/10.1109/AIMLA63829.2025.11040797

Barbosa Júnior MR, Sales L, Garcia dos Santos R, Vargas BS (2024) Forecasting yield and market classes of vidalia sweet onions: a UAV-based multispectral and texture data-driven approach. https://doi.org/10.2139/ssrn.5049296

Bassine FZ, Epule TE, Kechchour A, Chehbouni A (2025) Recent applications of machine learning, remote sensing, and IoT approaches in yield prediction: a critical review. arXiv, arXiv:2306.04566

Benos L, Tagarakis AC, Dolias G, Berruto R, Kateris D, Bochtis D (2021) Machine learning in agriculture: a comprehensive updated review. Sensors (Basel) 21(11):3758. https://doi.org/10.3390/s21113758

Botero-Valencia J, García-Pineda V, Valencia-Arias A, Valencia J, Reyes-Vera E, Mejia-Herrera M, Hernández-García R (2025) Machine learning in sustainable agriculture: systematic review and research perspectives. Agriculture 15(4):377. https://doi.org/10.3390/agriculture15040377

Bougiouklis JN, Barouchas PE, Petropoulos P, Tsesmelis DE, Moustakas N (2025) Precision soil sampling strategy for the delineation of management zones in olive cultivation using unsupervised machine learning methods. Sci Rep 15:8253. https://doi.org/10.1038/s41598-025-89395-1

Bras A, Montanaro A, Pierre C, Pradel M, Laconte J (2024) Toward a better understanding of robot energy consumption in Agroecological applications. arXiv:2410.07697

Canicattì M, Vallone M (2024) Drones in vegetable crops: a systematic literature review. Smart Agric Technol 7:100396. https://doi.org/10.1016/j.atech.2024.100396

Cao Z, Sun S, Bao X (2025) A review of computer vision and deep learning applications in crop growth management. Appl Sci 15(15):8438. https://doi.org/10.3390/app15158438

Choudhary V, Guha P, Pau G, Mishra S (2025) An overview of smart agriculture using internet of things (IoT) and web services. Environ Sustain Indicators 26:100607. https://doi.org/10.1016/j.indic.2025.100607

Cravero A, Pardo S, Sepúlveda S, Muñoz L (2022) Challenges to use machine learning in agricultural big data: a systematic literature review. Agronomy 12(3):748. https://doi.org/10.3390/agronomy12030748

Czymmek V, Völckner J, Hussmann S (2024) Development and evaluation of a laser system for autonomous weeding robots. AgriEngineering 6(4):4425–4441. https://doi.org/10.3390/agriengineering6040251

Daraghmi Y-A, Naser W, Daraghmi EY, Fouchal H (2025) Drone-assisted plant stress detection using deep learning: a comparative study of YOLOv8, RetinaNet, and faster R-CNN. AgriEngineering 7(8):257. https://doi.org/10.3390/agriengineering7080257

Devi T, Sharma K, Sharma N, Sethi D, Devi N, Ankita K (2024) Role of ICT in bridging the knowledge gap in rural agriculture. Int J Agric Ext Soc Dev 8(SP1):20–24

Dhanaraju M, Chenniappan P, Ramalingam K, Pazhanivelan S, Kaliaperumal R (2022) Smart farming: Internet of Things (IoT)-based sustainable agriculture. Agriculture 12(10):1745

El Sakka M, Ivanovici M, Chaari L, Mothe J (2025) A review of CNN applications in smart agriculture using multimodal data. Sensors (Basel) 25(2):472. https://doi.org/10.3390/s25020472. PMID: 39860841; PMCID: PMC11768470.

Ercan Oğuztürk G, Murat C, Yurtseven M, Oğuztürk T (2025) The effects of AI-supported autonomous irrigation systems on water efficiency and plant quality: a case study of geranium psilostemon Ledeb. Plants (Basel) 14(5):770. https://doi.org/10.3390/plants14050770

Eze VHU, Eze EC, Alaneme GU et al (2025) Integrating IoT sensors and machine learning for sustainable precision agroecology: enhancing crop resilience and resource efficiency through data-driven strategies, challenges, and future prospects. Discover Agric 3:83. https://doi.org/10.1007/s44279-025-00247-y

Falasca S, Pitta S, Bernabé MA (2014) Agroclimatic zoning for Amaranth (Amaranthus caudatus) using a geographical information system in Argentina. J Geogr Environ Earth Sci Int 1(1):9–23. https://doi.org/10.9734/JGEESI/2014/15198

Fekete D, Balázs G, Bőhm V, Várvölgyi E, Kappel N (2018) Sensory evaluation and electronic tongue for sensing grafted and non-grafted watermelon taste attributes. Acta Aliment 47(4):487–494. https://doi.org/10.1556/066.2018.47.4.12

Getahun S, Kefale H, Gelaye Y (2024) Application of precision agriculture technologies for sustainable crop production and environmental sustainability: a systematic review. Sci World J 2024:2126734. https://doi.org/10.1155/2024/2126734

Ghandour O, El Kafhali S Hanini M (2023) Computing resources scalability performance analysis in cloud computing data center. J Grid Computing 21:61. https://doi.org/10.1007/s10723-023-09696-5

Ghazal S, Munir A, Qureshi WS (2024) Computer vision in smart agriculture and precision farming: techniques and applications. Artif Intell Agric 13:64–83. https://doi.org/10.1016/j.aiia.2024.06.004

Girones SC, Munera S, Sober MM, Blasco J, Cubero S, Sanchis JG (2025) Artificial neural networks in agriculture, the core of artificial intelligence: what, when, and why. Comput Electron Agric 230:109938. https://doi.org/10.1016/j.compag.2025.109938
Goap A, Sharma D, Shukla AK, RamaKrishna C (2018) An IoT-based smart irrigation management system using Machine learning and open source technologies. Comput Electron Agric 155:41–49. https://doi.org/10.1016/j.compag.2018.09.040
Goldenits G, Mallinger K, Raubitzek S, Neubauer T (2024) Current applications and potential future directions of reinforcement learning-based Digital Twins in agriculture. Smart Agric Technol 8:100512. https://doi.org/10.1016/j.atech.2024.100512
Gómez AH, García-Pérez T, García A (2006) Monitoring volatile compounds during tomato ripening using a portable electronic nose. Postharvest Biol Technol 40(3):294–303
Gunasekaran K, Karmel A, Sreevardhan P, SamiKannu R (2025) Real-time soil fertility analysis, crop prediction, and insights using machine learning and deep learning algorithms. Frontiers 5:1652058. https://doi.org/10.3389/fsoil.2025.1652058
Hossain MI, Khaleque MA, Ali MR, Bacchu MS, Hossain MS, Shahed SMF, Aly MAS, Md. Khan ZH (2024) Development of electrochemical sensors for quick detection of environmental (soil and water) NPK ions. RSC Adv 14:9137–9158. https://doi.org/10.1039/D4RA00034J
Hu X, Chen S, Duan Q, Ahn CK, Shang H, Zhang D (2025) Domain adaptation for big data in agricultural image analysis: a comprehensive review. arXiv:2506.05972. https://doi.org/10.48550/arXiv.2506.05972
Hussain B, War AR, Pfeiffer DG (2018) Mapping foliage damage index and monitoring of Pieris brassicae by using GIS-GPS technology on cole crops. J Entomol Zool Stud 64(3):933–938
Hussein JB, Workneh TS, Kassim A, Ntsowe K, Melesse SF, El-Mesery HS (2025) A review on the impact of big data analytics in transforming agricultural practices, food processing, and preservation strategies. Appl Food Res 5(2):101234. https://doi.org/10.1016/j.afres.2025.101234
Ibrahim AS, Mohsen S, Selim IM et al (2025) AI-IoT-based smart agriculture pivot for plant disease detection and treatment. Sci Rep 15:16576. https://doi.org/10.1038/s41598-025-98454-6
Jones JW, Antle JM, Basso B, Boote KJ, Conant RT, Foster I, Godfray HCJ, Herrero M, Howitt RE, Janssen S, Keating BA, Munoz-Carpena R, Porter CH, Rosenzweig C, Wheeler TR (2017) Toward a new generation of agricultural system data, models, and knowledge products: state of agricultural systems science. Agric Syst 155:269–288. https://doi.org/10.1016/j.agsy.2016.09.021
Kamilaris A, Prenafeta-Boldú FX (2018) Deep learning in agriculture: a survey. Comput Electron Agric 147:70–90. https://doi.org/10.1016/j.compag.2018.02.016
Kumar RR, Sharma SK, Goswami S, Singh GP, Singh R, Singh K, Pathak H, Rai RD (2013) Characterization of differentially expressed stress-associated proteins in starch granule development under heat stress in wheat (Triticum aestivum L.). Indian J Biochem Biophys 50(2):126–138
Kumari K, Mirzakhani Nafchi A, Mirzaee S, Abdalla A (2025) AI-driven future farming: achieving climate-smart and sustainable agriculture. Agric Eng 7(3):89
Kunt YE (2025) Development of a smart autonomous irrigation system using IoT and AI. ArXiv:arXiv:2506.11835
Lang L (1992) GPS+ GIS+ remote sensing: an overview. In: Earth observation magazine, pp 23–26
Lemay MA, Boggs J (2024) Determinants of adoption of automation and robotics technology in the agriculture sector – a mixed methods, narrative, interpretive knowledge synthesis. PLOS Sustain Transform 3(11):e0000110
Li Y-R, Lien W-Y, Huang Z-H, Chen C-T (2023) Hybrid visual servo control of a robotic manipulator for cherry tomato harvesting. Actuators 12(6):253. https://doi.org/10.3390/act12060253
Li Z, Nie Z, Li G (2024) Integrating crop modeling and machine learning for the improved prediction of dryland wheat yield. Agronomy 14(4):777. https://doi.org/10.3390/agronomy14040777
Liakos KG, Busato P, Moshou D, Pearson S, Bochtis D (2018) Machine learning in agriculture: a review. Sensors 18(8):2674. https://doi.org/10.3390/s18082674

Liu W, Zhou J, Zhang T, Zhang P, Yao M, Li J, Sun Z, Ma G, Chen X, Hu J (2025) Key technologies in intelligent seeding machinery for cereals: recent advances and future perspectives. Agriculture 15(1):8. https://doi.org/10.3390/agriculture15010008

Lu Y, Chen D, Olaniyi E, Huang Y (2022) Generative adversarial networks (GANs) for image augmentation in agriculture: a systematic review. Comput Electron Agric 200:107208. https://doi.org/10.1016/j.compag.2022.107208

Mansoor S, Iqbal S, Popescu SM, Kim SL, Chung YS, Baek J-H (2025) Integration of smart sensors and IOT in precision agriculture: trends, challenges and future prospectives. Front Plant Sci 16:1587869. https://doi.org/10.3389/fpls.2025.1587869

Matese A, Czarnecki JMP, Samiappan S, Moorhead R (2024) Are unmanned aerial vehicle-based hyperspectral imaging and machine learning advancing crop science? Trends Plant Sci 29(2):196–209. https://doi.org/10.1016/j.tplants.2023.09.001

Mathenge M, Sonneveld BGJS, Broerse JEW (2022) Application of GIS in agriculture in promoting evidence-informed decision making for improving agriculture sustainability: a systematic review. Sustainability 14(16):9974. https://doi.org/10.3390/su14169974

Mennecke BE, Crossland MD (1996) Geographic information systems: applications and research opportunities for information systems researchers. In: Proceedings of HICSS-29: 29th Hawaii international conference on system sciences; 3–6, January 1996, Hawaii, vol 3, pp 537–546

Messina G, Praticò S, Badagliacca G, Di Fazio S, Monti M, Modica G (2021) Monitoring onion crop "Cipolla Rossa di Tropea Calabria IGP" growth and yield response to varying nitrogen fertilizer application rates using UAV imagery. Drones 5(3):61. https://doi.org/10.3390/drones5030061

Miller T, Mikiciuk G, Durlik I, Mikiciuk M, Łobodzińska A, Śnieg M (2025) The IoT and AI in agriculture: the time is now-a systematic review of smart sensing technologies. Sensors 25(12):3583. https://doi.org/10.3390/s25123583

Mishra A (2024) A comprehensive review of artificial intelligence and machine learning: concepts, trends, and applications. Int J Sci Res Sci Technol 11(5):126–142. https://doi.org/10.32628/IJSRST2411587

Mittal S, Mehar M (2016) Socio-economic factors affecting adoption of modern information and communication technology by farmers in India: analysis using multivariate probit model. J Agric Educ Ext 22(2):199–212. https://doi.org/10.1080/1389224X.2014.997255

Mohiddin SK, Sharmila S, Rao BM, Varalakshmi M (2026) Explainable AI models for transparent and trustworthy decision-making in precision agriculture. In: Dutta PK, Singh R, Haghi AK, Hamad AA (eds) Harnessing AI to reshape the future of agriculture. Springer, Cham. https://doi.org/10.1007/978-3-032-12118-9_10

Nautiyal M, Joshi S, Hussain I, Rawat H, Joshi A, Saini A, Kapoor R, Verma H, Nautiyal A, Chikara A, Ahmad W, Kumar S (2025) Revolutionizing agriculture: a comprehensive review on artificial intelligence applications in enhancing properties of agricultural produce. Food Chem X, 29:102748. https://doi.org/10.1016/j.fochx.2025.102748

Nautiyal M, Joshi S, Hussain I, Rawat H, Joshi A, Saini A, Kapoor R, Verma H, Nautiyal A, Chikara A, Ahmad W, Kumar S (2025) Revolutionizing agriculture: a comprehensive review on artificial intelligence applications in enhancing properties of agricultural produce. Food Chem: X 29:102748. https://doi.org/10.1016/j.fochx.2025.102748

Neethirajan S (2024) Artificial intelligence and sensor innovations: enhancing livestock welfare with a human-centric approach. Hum-Centric Intell Syst 4:77–92. https://doi.org/10.1007/s44230-023-00050-2

Nesar SB, Nugent PW, Zidack NK, Whitaker BM (2025) Unmanned aerial vehicle-based hyperspectral imaging for potato virus Y detection: machine learning insights. Remote Sens 17(10):1735. https://doi.org/10.3390/rs17101735

Ngugi HN, Akinyelu AA, Ezugwu AE (2024) Machine learning and deep learning for crop disease diagnosis: performance analysis and review. Agronomy 14(12):3001. https://doi.org/10.3390/agronomy14123001

Onishi Y, Yoshida T, Kurita H, Fukao T, Arihara H, Iwai A (2019) An automated fruit harvesting robot by using deep learning. ROBOMECH Journal 6(13). https://doi.org/10.1186/s40648-019-0141-2

Oğuztürk GE (2025) AI-driven irrigation systems for sustainable water management: a systematic review and meta-analytical insights. Smart Agric Technol 11:100982. https://doi.org/10.1016/j.atech.2025.100982

Ojo OI, Ilunga F (2018) Geospatial analysis for irrigated land assessment, modeling and mapping. In: Rustamov RB, Hasanova S, Zeynalova MH (eds) Multi-purposeful application of geospatial data, vol 9. IntechOpen, London, pp 65–84

Olaniyi E, Chen D, Lu Y, Huang Y (2022) Generative adversarial networks for image augmentation in agriculture: a systematic review. ArXiv. https://doi.org/10.48550/arXiv.2204.04707

Pandey DK, Mishra R (2024) Towards sustainable agriculture: harnessing AI for global food security. Artif Intell Agric 12:72–84. https://doi.org/10.1016/j.aiia.2024.04.003

Peng S, Yang S, Zhang X, Jia J, Chen Q, Lian Y, Wang A, Zeng B, Yang H, Li J, Dan J (2021) Analysis of imidacloprid residues in mango, cowpea and water samples based on portable molecular imprinting sensors. PLoS One 16(9):e0257042

Pivoto D, Waquil PD, Talamini E, Finocchio CPS, Dalla Corte VF, de Vargas Mores G (2018) Scientific development of smart farming technologies and their application in Brazil. Inf Process Agric 5(1):21–32. https://doi.org/10.1016/j.inpa.2017.12.002

Polly R, Devi EA (2024) Semantic segmentation for plant leaf disease classification and damage detection: a deep learning approach. Smart Agric Technol 9:100526. https://doi.org/10.1016/j.atech.2024.100526

Rafi MSM, Behjati M, Rafsanjani AS (2025) Reliable and cost-efficient IoT connectivity for smart agriculture: a comparative study of LPWAN, 5G, and hybrid connectivity models. arXiV:arXiv.2503.11162. https://doi.org/10.48550/arXiv.2503.11162

Reichstein M, Benson V, Blunk J et al (2025) Early warning of complex climate risk with integrated artificial intelligence. Nat Commun 16:2564. https://doi.org/10.1038/s41467-025-57640-w

Roy BPU, Sattar KNA, Elngar AA (2025) A smart irrigation system using the IoT and advanced machine learning model. J Smart Internet Things 2024(2):13–25. https://doi.org/10.2478/jsiot-2024-0009

Ruban M, Mani GN, Samynathan R, Prabhakaran M, Revathy M (2025) Intelligent motion based animal repulsion system for agriculture. J Artif Intell Capsule Netw 7(2):175–190. https://doi.org/10.36548/jaicn.2025.2.006

Saikai Y, Peake A, Chenu K (2023) Deep reinforcement learning for irrigation scheduling using high-dimensional sensor feedback. PLOS Water 2(9):e0000169

Shahi TB, Xu C-Y, Neupane A, Guo W (2023) Recent advances in crop disease detection using UAV and deep learning techniques. Remote Sens 15(9):2450. https://doi.org/10.3390/rs15092450

Shahzad MM, Saeed Z, Akhtar A, Munawar H, Yousaf MH, Baloach NK, Hussain F (2023) A review of swarm robotics in a nut Shell. Drones 7(4):269. https://doi.org/10.3390/drones7040269

Shawon SM, Ema FB, Mahi AK, Niha FL, Zubair HT (2025) Crop yield prediction using machine learning: an extensive and systematic literature review. Smart Agric Technol 10:100718. https://doi.org/10.1016/j.atech.2024.100718

Shi Y, Han L, Zhang X, Sobeih T, Gaiser T, Thuy NH, Behrend D, Srivastava AK, Halder K, Ewert F (2025) Deep learning meets process-based models: a hybrid approach to agricultural challenges. arXiv. https://doi.org/10.48550/arXiv.2504.16141

Shoaib M, Sadeghi-Niaraki A, Ali F, Hussain I, Khalid S (2025) Leveraging deep learning for plant disease and pest detection: a comprehensive review and future directions. Front Plant Sci 16:1538163. https://doi.org/10.3389/fpls.2025.1538163

Silva JAOS, Siqueira VSD, Mesquita M, Vale LSR, Marques TDNB, Silva JLBD et al (2024) Deep learning for weed detection and segmentation in agricultural crops using images captured by unmanned aerial vehicle. Remote Sens 16(23):4394

Singh G, Al-Huqail AA, Almogren A, Kaur S, Joshi K, Singh A, Bharany S, Hussen S, Rehman AU (2025) Enhanced leaf disease segmentation using U-net architecture for precision agriculture:

a deep learning approach. Food Science and Nutrition 13(7):e70594. https://doi.org/10.1002/fsn3.70594

Slaughter DC, Pérez Ruiz M, Fathallah F, Upadhyaya S, Gliever CJ, Miller B (2012) GPS-based intra-row weed control system: performance and labor savings. In: Rovira F (ed) Automation technology for off-road equipment, 11–12, July, 2012, Spain

Sonmezocak T (2025) Intelligent UAV health monitoring: detecting propeller and structural faults with MEMS-based vibration. Eng Sci Technol Int J 69:102130. https://doi.org/10.1016/j.jestch.2025.102130

Soussi A, Zero E, Sacile R, Trinchero D, Fossa M (2024) Smart sensors and smart data for precision agriculture: a review. Sensors 24(8):2647. https://doi.org/10.3390/s24082647

Spagnuolo M, Todde G, Caria M, Furnitto N, Schillaci G, Failla S (2025) Agricultural robotics: a technical review addressing challenges in sustainable crop production. Robotics 14(2):9. https://doi.org/10.3390/robotics14020009

Tamayo-Vera D, Wang X, Mesbah M (2024) A review of machine learning techniques in agroclimatic studies. Agriculture 14(3):481. https://doi.org/10.3390/agriculture14030481

Tan Y, Liu X, Zhang J, Wang Y, Hu Y (2025) A review of research on fruit and vegetable picking robots based on deep learning. Sensors 25(12):3677. https://doi.org/10.3390/s25123677

Tatsumi K, Usami T (2024) Plant-level prediction of potato yield using machine learning and unmanned aerial vehicle (UAV) multispectral imagery. Discover Appl Sci 6:649. https://doi.org/10.1007/s42452-024-06362-7

Thesma V, Mohammadpour VJ (2022) Plant root phenotyping using deep conditional GANs and binary semantic segmentation. Sensors (Basel) 23(1):309. https://doi.org/10.3390/s23010309

Triantafyllou A, Tsouros DC, Sarigiannidis P, Bibi S (2019) An architecture model for smart farming. In: 15th international conference on distributed computing in sensor systems (DCOSS), 29–31 May 2019, Santorini Island, Greece, pp 385–392

Upadhyay A, Chandel NS, Chakraborty SK (2024) Disease control measures using vision-enabled agricultural robotics. In: Chouhan SS, Singh UP, Jain S (eds) Applications of computer vision and drone Technology in Agriculture 4.0. Springer, Singapore. https://doi.org/10.1007/978-981-99-8684-2_10

Upadhyay A, Chandel NS, Singh KP et al (2025) Deep learning and computer vision in plant disease detection: a comprehensive review of techniques, models, and trends in precision agriculture. Artif Intell Rev 58:92. https://doi.org/10.1007/s10462-024-11100-x

Venkateswara S, Padmanabhan J (2025) Deep learning based agricultural pest monitoring and classification. Sci Rep 15:8684. https://doi.org/10.1038/s41598-025-92659-5

Venugopal A, Aparna S, Mani J, Mathew R, Williams V (2021) Crop yield prediction using machine learning algorithms. Int J Eng Res Technol 9(13):87–91. https://doi.org/10.17577/IJERTCONV9IS13019

Vijayakumar S, Shanmugapriya P, Saravanane P, Ramesh T, Murugaiyan V, Ilakkiya S (2025) Precision weed control using unmanned aerial vehicles and robots: assessing feasibility, bottlenecks, and recommendations for scaling. NDT 3(2):10. https://doi.org/10.3390/ndt3020010

Wakchaure M, Patle BK, Mahindrakar AK (2023) Application of AI techniques and robotics in agriculture: a review. Artif Intell Life Sci 3:100057. https://doi.org/10.1016/j.ailsci.2023.100057

Wang C, Xia Y, Xia L et al (2025) Dual discriminator GAN-based synthetic crop disease image generation for precise crop disease identification. Plant Methods 21:46. https://doi.org/10.1186/s13007-025-01361-0

Waqas M, Naseem A, Humphries UW, Hlaing PT, Dechpichai P, Wangwongchai A (2025) Applications of machine learning and deep learning in agriculture: a comprehensive review. Green Technol Sustainability 3(3):100199. https://doi.org/10.1016/j.grets.2025.100199

Wolfert S, Ge L, Verdouw C, Bogaardt MJ (2017) Big data in smart farming—a review. Agric Syst 153:69–80. https://doi.org/10.1016/j.agsy.2017.01.023

Xu L, Tang Y, Zhang C, Liu F, Chen J, Xuan W, Jin H, Ye Z, Cao Z, Li Y, Wang X, Dong S, Luo J (2022) Fully self-powered instantaneous wireless liquid level sensor system based on triboelectric nanogenerator. Nano Res 15:5425–5434. https://doi.org/10.1007/s12274-022-4125-9

Xiang S, Blok PM, Burridge J, Wang H, Guo W (2025) DODA: adapting object detectors to dynamic agricultural environments in real-time with diffusion, arxiv- computer vision and pattern recognition (cs.CV). https://doi.org/10.48550/arXiv.2403.18334

Yépez-Ponce DF, Salcedo JV, Rosero-Montalvo PD, Sanchis J (2023) Mobile robotics in smart farming: current trends and applications. Front Artif Intell 6:1213330. https://doi.org/10.3389/frai.2023.1213330

Zaccarin AM, Peddada N, Machold M, Turner KT, Olsson RH (2025) Inductively coupled capacitive soil moisture sensors printed on a biodegradable substrate: characterization and long-term testing. Smart Agr Technol. 11:101052. https://doi.org/10.1016/j.atech.2025.101052

Zaborowicz M, Frankowski J (2025) Big data analytics and machine learning for smart agriculture. Agriculture 15(7):757. https://doi.org/10.3390/agriculture15070757

Zhai X, Yi M, An Z et al (2025) Evaluating land capability using site assessment and land evaluation model in GIS in a watershed scale. Sci Rep 15:17767. https://doi.org/10.1038/s41598-025-95586-7

Zhang Y, Wang Y, Hao H, Li Z, Long Y, Zhang X, Xia C (2024) Unmanned Aerial Vehicle (UAV) hyperspectral imagery mining to identify new spectral indices for predicting the field-scale yield of spring maize. Sustainability 16(24):10916. https://doi.org/10.3390/su162410916

Zhao B, McDermott RL, Erickson GE, Xiong Y (2024) Technical note: assessing GPS sensor accuracy using real-time kinematic device for livestock tracking. J Anim Sci 102:skae253. https://doi.org/10.1093/jas/skae253

Opportunities for AI Interventions at Various Stages of Agricultural Practices

2

R. Rakhi, R. Geetha, and S. Sarada

Abstract

Artificial intelligence is a transformative force in modern agriculture, enabling significant advancements in key operational areas. This chapter provides a review of the opportunities for AI intervention at critical stages of agricultural practices. It examines specific applications, such as automated pest diagnosis, the prediction of optimal sowing and harvesting timings, and agricultural price forecasting. The chapter's central contribution is the systematic categorization of these AI technologies according to the critical stages of agricultural practice at which they are deployed, providing a structured framework for understanding their integrative potential in enhancing productivity and decision-making.

Keywords

Artificial intelligence · Precision agriculture · Machine learning · Sustainable farming

---

R. Rakhi (✉)
Department of Plantation, Spices, Medicinal and Aromatic Crops, College of Agriculture, Vellayani, Kerala Agricultural University, Thiruvananthapuram, Kerala, India
e-mail: rakhi.r@kau.in

R. Geetha
Regional Agricultural Research Station (Southern Zone), Kerala Agricultural University, Vellayani, Thiruvananthapuram, Kerala, India
e-mail: geetha.ramesh@kau.in

S. Sarada
Department of Vegetable Science, College of Agriculture, Vellayani, Kerala Agricultural University, Thiruvananthapuram, Kerala, India
e-mail: sarada.s@kau.in

R. E. Krishnankutty et al. (eds.), *Smart Agriculture*,
https://doi.org/10.1007/978-981-95-8785-8_2

1 Introduction

The agricultural sector constitutes a critically important component of the global economy, and its contribution to India's gross domestic product (GDP) is substantially higher than the global average (Economic Survey 2018). Even though 52.3% of India's arable land produces food for the majority of its population (World Bank Open Data 2020), the agriculture sector is facing numerous challenges, such as population growth, land degradation, and pollution of natural resources (Gupta et al. 2020). The situation is overwhelmed by greenhouse gas emissions, climate change, natural calamities, etc. (Shukla et al. 2019).

The integration of innovative technologies and practices facilitates substantial enhancements in crop yield per unit of arable land (Matson et al. 1997). In this context, artificial intelligence (AI) has established itself as a transformative discipline, leveraging digital tools to boost agricultural productivity and concurrently mitigate its environmental footprint. Rapid growth in digitized data and major advancements in information and communication technology (ICT), coupled with artificial intelligence (AI), can create intelligent farming solutions.

2 Artificial Intelligence

The idea of machines having their own intelligence was first proposed by Alan Turing in 1950. The field of artificial intelligence (AI), a term first coined by John McCarthy at the 1956 Dartmouth Conference (van Assen et al. 2022), is a branch of computer science dedicated to developing intelligent systems capable of reasoning, learning, and perceiving. It involves creating algorithms that emulate human cognitive functions to perform complex tasks (Pannu 2015). The application of this technology is rapidly expanding, with the AI sector in Indian agriculture, for instance, projected to grow at 22.5% annually to reach a market value of $2.6 billion by 2025 (Gkikas et al. 2023). Implementation of AI in the agriculture sector provides data-driven suggestions and insights to agricultural stakeholders to optimize farm management, improving precision, profitability, and overall yield (Bhardwaj et al. 2022).

The study of artificial intelligence is crucial due to its quick technological improvements and usefulness in problem-solving (Sharma et al. 2022). The integration of AI technologies facilitates analysis of agricultural data, empowering growers to make informed decisions regarding crop selection, real-time monitoring, and harvesting time to maximize economic returns. Furthermore, AI-driven systems minimize operational errors and enhance precision in farm management, allowing for timely interventions. By mitigating agricultural risks and improving yield stability, AI-powered agriculture is critical for ensuring future food and nutritional security. AI-driven automation presents a viable solution to address the growing shortage of skilled labor in the agricultural sector. To facilitate the adoption of innovative technologies in agriculture, including artificial intelligence, a financial allocation of INR 2422.7 crores was allotted to the States by the Government of India during 2020–2022 (Sharma et al. 2023).

3 Different Approaches in Artificial Intelligence

Artificial intelligence can be implemented by various approaches like machine learning (ML), natural language processing (NLP), fuzzy logic system (FL), and expert system.

Artificial intelligence (AI) encompasses a suite of technologies, including machine learning (ML) and deep learning (DL), which enable systems to autonomously solve problems and achieve specified objectives without direct human intervention. The field of ML, a cornerstone of modern AI, was formally named by Arthur Samuel in 1959 to describe the capacity of computers to learn from experience absent explicit programming for each task (Samuel 1959). ML algorithms operate by constructing predictive models from sample training data, a capability that has seen growing application in agricultural crop management for yield prediction and disease detection (Liakos et al. 2018). A more advanced subset of ML, deep learning, utilizes multilayered artificial neural networks (ANNs) to tackle complex computational problems through representation learning. These ANNs are computing architectures inspired by the biological neural networks of the human brain, processing information via interconnected nodes (Agatonovic-Kustrin and Beresford 2000).

A fundamental distinction often drawn between artificial neural networks and the natural brains is based upon their core operational characteristics. ANNs frequently demonstrate static and symbolic processing, whereas the natural brains are defined by their dynamic, plastic, and analog nature (Bengio et al. 2015). Unlike traditional algorithmic approaches that require explicit programming, ANNs learn complex, nonlinear relationships directly from data through a process of training. Despite this architectural difference, a significant strength of ANNs lies in their capacity for parallel distributed computation, which enables powerful predictive and forecasting capabilities.

Another approach is fuzzy logic (FL), which was formalized by Lotfi Zadeh. It is a reasoning methodology that approximates human decision-making processes. Unlike classical binary logic, FL operates on a continuum of truth values, enabling the representation of concepts with inherent vagueness, such as "possibly yes" or "partially true." Within computational linguistics, FL provides a foundational framework for natural language processing, a subset spanning artificial intelligence, computer science, and linguistics. NLP is concerned with enabling machines to process, analyze, and generate human language data.

4 Applications of Artificial Intelligence in Agriculture

Artificial intelligence (AI) has the potential to revolutionize core agricultural practices, including soil and water management, modeling of input requirements, and crop management, through enhanced yield forecasting. Various applications include automating farming activities, identifying potential pest and disease outbreaks, managing crop quality, monitoring biotic and abiotic factors, and utilizing machine vision systems to analyze phenotypes and various crop growth processes.

4.1 Crop Selection

AI models use geospatial data, climate history, and market trends to recommend suitable crops for each location (Akpoti et al. 2019). ML algorithms predict crop viability under changing climates to help farmers make resilient choices.

4.2 Crop Breeding

AI accelerates the breeding of superior crop varieties. Machine learning algorithms facilitate the analysis of complex genomic datasets to pinpoint genetic markers linked to agriculturally desirable traits, including drought tolerance, enhanced pest resistance, and superior nutritional value (Montesinos-López et al. 2021). This greatly shortens the traditional breeding cycle, facilitating the development of climate-resilient crops.

4.3 Soil Testing and Soil Health Monitoring

Machine learning techniques provide a powerful computational framework for the accurate prediction and characterization of critical soil properties, such as temperature and moisture content, thereby enabling data-driven strategies for enhanced soil management.

4.4 Crop Management

AI-driven systems automate and optimize the application of water, fertilizers, and pesticides. Autonomous tractors and robotic harvesters leverage integrated technologies, including computer vision, sensor fusion, and reinforcement learning to navigate fields, perform essential tasks like plowing and seeding, and identify and harvest ripe produce. AI algorithms create prescription maps that guide agricultural machinery to apply fertilizers and pesticides at variable rates across a field, reducing input use, lowering costs, and minimizing environmental runoff (Mulla 2013). Unmanned aerial vehicles (UAVs) vision-based acquisition technology can provide quick and efficient data acquisition. Drones facilitate comprehensive field analysis and continuous crop monitoring by capturing aerial imagery, and they can also provide real-time weather alerts.

In agriculture, several systems serve as integrated management tools for disseminating location-specific advice. They are targeted at narrow, well-defined areas like crop production to aid in critical decisions pertaining to irrigation, fertilization, and pest control. "PRITHVI" is an expert system based on fuzzy logic developed for soybean cultivation in Rajasthan (Prakash et al. 2013). The PRITHVI system operates as a decision-support tool, featuring dedicated modules for critical agricultural management functions such as irrigation scheduling, fertilization, and pest control.

Its architecture exemplifies a knowledge-based system (KBS), a class of computer programs engineered to emulate human expert reasoning within a specialized domain. A parallel development in this field is an AI-powered sowing application, created by the International Crops Research Institute for the Semiarid Tropics (ICRISAT) in partnership with Microsoft (IDC 2017). Microsoft, in collaboration with ICRISAT, developed an Al sowing app and gives sowing advice for crops like groundnut, ragı, maize, cotton, and rice.

4.5 Smart Irrigation

Automated irrigation systems, managed by ML algorithms, leverage a data fusion approach from inputs such as in situ soil moisture sensors, prognostic weather data, and evapotranspiration calculations. This results in the spatiotemporally precise application of water, a key strategy for enhancing water-use efficiency and conservation efforts in agriculture (Goap et al. 2018). Lorvanleuang and Zhao (2018) developed an intelligent system that can analyze regional rainfall patterns to forecast weather and geography-specific irrigation needs, minimizing water waste and enhancing crop yield. This can be integrated with an automated irrigation implementation, which utilizes soil moisture sensors to accurately predict and deliver the precise quantity of water required, thereby saving significant time and resources (Choudhary et al. 2019). Similarly, in hydrology, these models have proven highly effective for forecasting critical water resource variables, offering improved accuracy over conventional methods (Maier and Dandy 2000).

4.6 Weed Management

Earlier approaches to species-level identification include machine vision strategies. Sogaard (2005) presented an automated method for classifying weed species using an active shape model (ASM) algorithm, achieving classification accuracies ranging from 65% to 90%. Beyond standard imaging, advanced remote sensing (RS) techniques offer enhanced capabilities for crop monitoring. The integration of hyperspectral imaging and 3D laser scanning facilitates the detection of both crop diseases and weed infestations through the analysis of spectral and structural data. A synergistic approach, combining remotely sensed data with sophisticated image classification algorithms, weather information, and ancillary geospatial data, can significantly improve the accuracy of distinguishing weeds from crops. This precision enables site-specific weed management (SSWM), confining herbicide application only to areas requiring treatment, which reduces chemical usage and environmental impact. The development of integrated weed detection systems utilizing both unmanned aerial vehicles (UAVs) and unmanned ground vehicles (UGVs) has been documented by Esposito et al. (2021).

Deep learning methodologies can be used for automated weed detection in agricultural scenarios (Bah et al. 2018). A fully automatic learning method using convolutional neural network was developed for automatic detection of the crop lines and weeds, thereby constructing robust models for image-based classification. DL-based computer vision systems mounted on sprayer booms or autonomous robots can distinguish between crops and weeds in real time. This enables mechanical weed removal or the precise spraying of herbicides only on the weeds, reducing herbicide use by up to 90% (Partel et al. 2019).

4.7 Soil and Crop Health Monitoring

The deployment of artificial intelligence systems for the automated identification of agricultural pests and diseases has expanded significantly, underpinned by advances in machine learning (ML). Architectures, including artificial neural networks (ANNs), convolutional neural networks (CNNs), and recurrent neural networks (RNNs), are employed to detect intricate patterns and relationships within multimodal agricultural datasets. The performance of these models is contingent upon high-fidelity input data, which is frequently supplied by advanced remote sensing platforms. Hyperspectral imaging and 3D laser scanning, in particular, are foundational to generating precise crop metrics across extensive agricultural landscapes, enabling continuous monitoring over thousands of acres.

A significant advancement in this domain is the emergence of portable, palm-sized snapshot hyperspectral imagers, which greatly enhance the accessibility and flexibility of spectral data acquisition (Misra et al. 2022). Selvaraj et al. (2020) reported a deep learning CNN model to detect banana disease and pest symptoms. Sanga et al. (2020) developed a deep learning-based tool for banana plant disease detection, evaluating five distinct convolutional neural network (CNN) architectures. This framework was subsequently deployed as a mobile application, providing an interface for farmers to obtain diagnoses by submitting images of infected leaves through a smartphone.

AGREX (Agricultural Expert System) is a user-friendly computer-based package developed by Ganesan (2006). This system was designed to provide farmers with precise, timely information and advisory services, including diagnoses of plant diseases and pests alongside recommendations for prevention and control measures. FARMWAT is another specialized decision-support system for precision irrigation and RICEDIAG for automated rice disease diagnosis and nutrient management. Such systems integrate sensor data with predictive models to provide farmers with actionable recommendations (Sriram and Philip 2016). Computer vision models analyze imagery captured by drones, satellites, or ground-based robots to monitor crop health.

Modern agricultural systems leverage advanced spectral analysis, such as the calculation of the normalized difference vegetation index (NDVI), to identify crop abiotic and biotic stresses. These include nutrient deficits, water stress, and early-stage disease infections, often long before such issues become apparent through visual inspection (Weiss et al. 2020). This capability is further enhanced by integrating artificial intelligence (AI) with remote sensing data. For instance, Kerkech et al. (2020) demonstrated the efficacy of a deep learning model for semantic segmentation on multispectral imagery, achieving precise spatial detection and mapping of vine disease outbreaks. This represents a significant advancement toward fully automated, high-resolution crop monitoring systems central to precision agriculture.

Beyond disease detection, AI algorithms are increasingly applied to predictive modeling of pest population dynamics, enabling proactive pest control strategies. Satellite remote sensing provides large-scale surveillance of crop health, offering early warnings for potential pest infestations. Furthermore, ground-based technologies, such as AI-powered acoustic sensors and automated image traps, facilitate the accurate identification of specific pest species. These systems can generate real-time alerts, which are critical for implementing timely integrated pest management (IPM) interventions (Aslam et al. 2024). An example of such an application is the "Pest Risk Prediction App," developed by Microsoft in partnership with United Phosphorous Limited, which utilizes weather data to forecast pest risks.

4.8 Precision Farming

Precision farming refers to a method of managing farm resources through the use of IoT and ICT. Precision agriculture techniques can acquire images of the complete agricultural field with a very high spatial resolution at a reasonable cost.

The convergence of artificial intelligence (AI) with precision farming has established a distinct, data-driven discipline focused on enhancing productivity. Central to this evolution is the advancement of farm management systems, which are transitioning from basic monitoring tools into sophisticated AI platforms. By leveraging machine learning to analyze heterogeneous sensor data, these systems provide actionable recommendations and data-driven insights. This capability facilitates informed decision-making and operational actions, thereby optimizing agricultural output.

Smart greenhouses promote action to maintain the most beneficial conditions for crop growth and yield (Alreshidi 2019). Researchers have utilized ANNs to develop robust vision systems capable of discriminating between crops and weeds in precision farming (Gliever and Slaughter 2001). Research by Garg et al. (2021) architected a multimodal framework for precision farming that integrates smart irrigation, fertilizer dosage recommendation, and employs deep learning for crop disease detection alongside machine learning models for damage prediction. This approach is complemented by IoT-based environmental monitoring systems, such as the wireless sensor network (WSN) developed by Shinde and Siddiqui (2018) for greenhouse management, which utilized sensors for temperature, humidity, soil moisture, and light intensity to optimize conditions for crops like tomato and brinjal. India's largest precision agriculture farm, "Simply Fresh," located in Hyderabad, serves as a prominent case study in the integration of artificial intelligence (AI) within controlled-environment agriculture. Its infrastructure is defined by an AI-assisted multilayered retractable roof system, a continuous automated process from seeding to harvest, a network of environmental sensors, and a high-pressure fogging system for climate modulation.

4.9 Crop Harvesting

Substantial research is currently directed toward harnessing progress in artificial intelligence, particularly deep learning, to enhance the automation of modern harvesting systems. The integration of artificial intelligence (AI) into agricultural robotics presents a transformative opportunity to automate harvesting operations, a domain still predominantly reliant on manual labor. AI-enabled harvesting robots may reduce dependency on human labor by an estimated four workers per acre, as reported by Panpatte (2018). The development of automated fruit harvesting robots is considered a viable solution to the challenges of labor scarcity and consistency. A critical component of such automated systems is the accurate detection of fruit maturity. Color, spectral, or thermal cameras can be used for the detection of fruit maturity (Bulanon et al. 2011; Linker et al. 2012). In a specific application addressing labor-intensive tasks, Yuan et al. (2016) developed an automated robotic system designed to perform phytohormone application for the pollination of tomato flowers within inclined trellising systems. The integration of robotic systems into vegetable production is increasingly recognized as a pivotal innovation for enhancing both the sustainability and efficiency of agricultural operations (Lowenberg-DeBoer et al. 2020).

4.10 Postharvest Monitoring

Maximizing yield involves minimizing losses after harvest. Computer vision systems on processing lines can sort fruits, vegetables, and grains based on size, color, shape, and external defects with superhuman speed and accuracy (Bhargava and Bansal 2021). More advanced systems using hyperspectral imaging can even detect internal defects that are not externally visible, ensuring higher quality produce reaches the market.

The development of artificial intelligence (AI)-based systems for fruit monitoring and grading has been profoundly enhanced by the adoption of deep convolutional neural networks (DCNNs), a class of algorithms renowned for their efficacy in object detection and classification tasks (Kamilaris and Prenafeta-Boldú 2018). A key advantage of DCNN-based approaches over conventional techniques is their superior performance in managing complex visual challenges inherent to agricultural settings, including occlusion, and variations in the orientation and position of objects (LeCun et al. 2015).

4.11 Supply Chain Management

Artificial intelligence significantly enhances the efficiency of postharvest agricultural processes, including sorting, logistics, and market linkages. Within supply chain management, AI-driven planning and optimization tools facilitate advanced demand forecasting. This capability can yield substantial cost savings for producers and mitigates information asymmetry for buyers, thereby improving overall supply chain efficiency and reducing spoilage (Oliveira et al. 2017). Furthermore, AI-powered predictive models analyze market prices by processing vast datasets that encompass historical price trends, weather patterns, and geopolitical events. These models provide accurate demand–supply information and demand predictions, enabling farmers to make informed decisions on the timing and location of sales for optimal price realization (Giri et al. 2020). An empirical example of this application is the advanced price forecasting system developed by IBM for the Karnataka Agricultural Prices Commission (KAPC), which predicts market price trends with a lead time of a fortnight.

AI-powered predictive models forecast demand, optimize inventory levels, and plan the most efficient transportation routes. By analyzing factors like shelf life, transportation conditions, and market demand, these systems reduce food waste and improve profitability (Bechtsis et al. 2018). Blockchain technology, often integrated with AI, can enhance traceability from farm to fork, providing consumers with transparent product histories. Leveraging historical yield data, satellite imagery from previous seasons and weather forecasts, ML models can generate highly accurate pre-season yield predictions (Van Klompenburg et al. 2020). These forecasts are crucial for supply chain planning, commodity pricing, and national food security policies.

5 AI Initiatives

The National Institution for Transforming India (NITI Aayog) has acknowledged the strategic significance of artificial intelligence by launching the National Program on AI. This initiative is designed to steer research and development in alignment with the National Strategy for Artificial Intelligence, focusing on new and emerging technologies. The agricultural sector has witnessed a significant influx of start-ups specializing in artificial intelligence (AI), each targeting specific challenges to enhance efficiency and sustainability. Notable examples include Blue River Technology, which focuses on AI-driven precision weed control; Harvest CROO Robotics, which develops automated solutions for crop harvesting; PEAT, which employs machine vision for diagnosing plant pests and soil defects; Sky Squirrel Technologies Inc., which utilizes drone-based imaging and computer vision for detailed crop analysis; and "aWhere," which leverages satellite data for weather prediction and crop sustainability monitoring (Kumar 2019). These innovations represent a growing trend toward data-driven, intelligent farming systems.

6 Conclusions

The incorporation of artificial intelligence into agricultural systems is not merely an incremental improvement but a radical transformation for generations to come. The arrival of artificial intelligence in agriculture is far more than just another new tool;

it is a radical transformation. By putting data-driven insights at the forefront of decision-making, AI is fundamentally changing how we farm. We are moving toward a future of cyber-physical agricultural systems, where AI acts as intelligent hub, constantly sensing, learning from, and fine-tuning the intricate relationships between plant health, soil conditions, local weather, and global market dynamics. It will also require a collaborative push from scientists, industry leaders, and governments to overcome the hurdles to adoption. AI could truly unlock a new age for agriculture, one that is not only more productive but also deeply sustainable, guaranteeing food security for the future.

Acknowledgments We thank Kerala Agricultural University for providing the necessary facilities, resources, support, and guidance that contributed to the successful completion of this chapter.

Disclaimer The views and conclusions expressed in this chapter are solely those of the authors and do not necessarily represent the views of their affiliated institutions. The authors are responsible for the accuracy and completeness of the information provided, but do not accept any liability for any direct or indirect losses resulting from the use of this content.

Declarations

Competing Interests The authors declare no conflict of interest with respect to the research, authorship and/or publication of this chapter.

Statement of Usage of Generative AI During the preparation of this work, the authors used ChatGPT (a generative AI-assisted language model) to rephrase text drafted by the authors to enhance the readability of the chapter. After using this tool, the authors reviewed and edited the content as needed and take full responsibility for the content of the chapter.

References

Agatonovic-Kustrin S, Beresford R (2000) Basic concepts of artificial neural network (ANN) modeling and its application in pharmaceutical research. J Pharm Biomed Anal 22(5):717–727

Akpoti K, Kabo-bah AT, Zwart SJ (2019) Agricultural land suitability analysis: state-of-the-art and outlooks for integration of climate change analysis. Agric Syst 173:172–208

Alreshidi E (2019) Smart sustainable agriculture (SSA) solution underpinned by internet of things (IoT) and artificial intelligence (AI). arXiv preprint arXiv:1906.03106

Aslam A, Ijaz MU, Aslam MT, Chattha MU, Khan I, Gulzar MZ, Ali F (2024) Monitoring and detection of insect pests using smart trap technologies. In: Revolutionizing pest management for sustainable agriculture. IGI Global, pp 443–468

Bah MD, Hafiane A, Canals R (2018) Deep learning with unsupervised data labelling for weed detection in line crops in UAV images. Remote Sens 10:1690

Bechtsis D, Tsolakis N, Vlachos D, Srai JS (2018) Intelligent autonomous vehicles in digital supply chains: a framework for integrating innovations towards sustainable value networks. J Clean Prod 181:60–71

Bengio Y, Lee D, Bornschein J, Mesnard T, Lin Z (2015) Towards biologically plausible deep learning. arXiv 1502: 04156

Bhardwaj A, Kishore S, Pande DK (2022) Artificial intelligence in biological sciences. Life 12:1430

Bhargava A, Bansal A (2021) Fruits and vegetables quality evaluation using computer vision: a review. J King Saud Univ Comput Inf Sci 33(3):243–257

Bulanon DM, Burks TF, Alchanatis V (2011) Visible and thermal images for fruit detection. In: Encyclopedia of agrophysics. Springer, Dordrecht, pp 944–954

Choudhary S, Gaurav V, Singh A, Agarwal S (2019) Autonomous crop irrigation system using artificial intelligence. Int J Eng Adv Technol 8(5):46–51

Economic Survey (2018) Ministry of Finance, & Government of India. (2018). Economic Survey 2017–18 (Volume I and Volume II). Oxford University Press

Esposito M, Crimaldi M, Cirillo V, Sarghini F, Maggio A (2021) Drone and sensor technology for sustainable weed management: a review. Chem Biol Technol Agric 8(1):18

Ganesan V (2006) Decision support system "Crop-9-DSS for identified crops". Proc World Acad Sci Eng Tech 12:1307–6884

Garg S, Pundir P, Jindal H, Saini H, Garg S (2021) Towards a multimodal system for precision agriculture using IoT and machine learning. In: Conference proceedings of 12th international computing communication and networking technologies (ICCCNT); 6–8. IIT, Kharagpur, pp 1–7

Giri A, Saxena DRR, Saini P, Rawte DS (2020) Role of artificial intelligence in advancement of agriculture. Int J Chem Stud 8(2):375–380

Gkikas DC, Theodoridis PK, Gkikas MC (2023) Artificial intelligence (AI) use for e-governance in agriculture: exploring the bioeconomy landscape. In: Recent advances in data and algorithms for e-government. Springer International Publishing, Cham, pp 141–172

Gliever C, Slaughter DC (2001) Crop verses weed recognition with artificial neural networks. ASAE paper 01-3104: 1–21

Goap A, Sharma D, Shukla AK, Krishna CR (2018) An IoT based smart irrigation management system using machine learning and open-source technologies. Comput Electron Agric 155:41–49

Gupta M, Abdelsalam M, Khorsandroo S, Mittal S (2020) Security and privacy in smart farming: challenges and opportunities. IEEE Access 8:34564–34584

IDC (2017). http://idc.icrisat.org/idc/index.php/2017/01/

Kamilaris A, Prenafeta-Boldú FX (2018) Deep learning in agriculture: a survey. Comput Electron Agric 147:70–90

Kerkech M, Hafiane A, Canals R (2020) Vine disease detection in UAV multispectral images using optimized image registration and deep learning segmentation approach. Comp Elect Agri 174:105446

Kumar S (2019) Transforming Indian agriculture through digital platforms: status, support systems & strategies. J Busin Res

LeCun Y, Bengio Y, Hinton G (2015) Deep learning. Nature 521(7553):436–444

Liakos KG, Busato P, Moshou D, Pearson S, Bochtis D (2018) Machine learning in agriculture: a review. Sensors 18(8):2674

Linker R, Cohen O, Naor A (2012) Determination of the number of green apples in RGB images recorded in orchards. Comput Electron Agric 81:45–57

Lorvanleuang S, Zhao Y (2018) Automatic irrigation system using android. OA Lib J 5(4):1–7

Lowenberg-DeBoer J, Huang IY, Grigoriadis V, Blackmore S (2020) Economics of robots and automation in field crop production. Precis Agric 21(3):278–299

Maier HR, Dandy GC (2000) Neural networks for the prediction and forecasting of water resources variables: a review of modelling issues and applications. Environ Model Soft 15:101–124

Matson PA, Parton WJ, Power AG, Swift MJ (1997) Agricultural intensification and ecosystem properties. Science 277(5325):504–509

Misra NN, Dixit Y, Al-Mallahi A, Bhullar MS, Upadhyay R, Martynenko A (2022) IoT, big data, and artificial intelligence in agriculture and food industry. IEEE Internet Things J 9–9:6305–6324

Montesinos-López OA, Montesinos-López A, Pérez-Rodríguez P, Barrón-López JA, Martini JW, Fajardo-Flores SB, Crossa J (2021) A review of deep learning applications for genomic selection. BMC Genomics 22(1):19

Mulla DJ (2013) Twenty-five years of remote sensing in precision agriculture: key advances and remaining knowledge gaps. Biosyst Eng 114(4):358–371

Oliveira N, Cortez P, Areal N (2017) The impact of microblogging data for stock market prediction: using twitter to predict returns, volatility, trading volume and survey sentiment indices. Expert Syst Appl 73:125–144

Pannu A (2015) Artificial intelligence and its application in different areas. Artif Intell 4(10):79–84

Panpatte DG (2018) Artificial intelligence in agriculture: an emerging era of research. Anand Agricultural University, pp 1–8

Partel V, Kakarla SC, Ampatzidis Y (2019) Development and evaluation of a low-cost and smart technology for precision weed management utilizing artificial intelligence. Comput Electron Agric 157:339–350

Prakash C, Rathor AS, Thakur GS (2013) Fuzzy based agriculture expert system for Soyabean. Intl Conf Comput Sci:1–13

Samuel AL (1959) Some studies in machine learning using the game of checkers. IBM Journal of research and development 3(3):210–229

Sanga S, Mero V, Machuve D, Mwanganda D (2020) Mobile-based deep learning models for banana diseases detection. arXiv preprint arXiv:2004.03718

Selvaraj MG, Vergara A, Montenegro F, Ruiz HA, Safari N, Raymaekers D, Blomme G (2020) Detection of banana plants and their major diseases through aerial images and machine learning methods: a case study in DR Congo and Republic of Benin. ISPRS J Photogramm Remote Sens 169:110–124

Sharma A, Georgi M, Tregubenko M, Tselykh A, Tselykh A (2022) Enabling smart agriculture by implementing artificial intelligence and embedded sensing. Comput Ind Eng 165:107936

Sharma R, Rajput N, Bisen P (2023) Artificial intelligence-concept and applications in Indian agriculture. In: Modern agronomy-theory and practices. p 91

Shinde D, Siddiqui N (2018) IoT based environment change monitoring & controlling in greenhouse using WSN. In: International conference on information, communication, engineering and technology (ICICET); 29–31, August, 2018, Zeal College of Engineering and Research, Pune, pp 1–5

Shukla PR, Skeg J, Buendia EC, Masson-Delmotte V, Portner HO, Roberts DC, Malley J (2019) Climate change and land: an IPCC special report on climate change, desertification, land degradation, sustainable land management, food security and greenhouse gas fluxes in terrestrial ecosystems

Sogaard HT (2005) Weed classification by active shape models. Biosyst Eng 91(3):271–281

Sriram N, Philip H (2016) Expert system for decision support in agriculture. TNAU Agritech

van Assen M, Muscogiuri E, Tessarin G, De Cecco CN (2022) Artificial intelligence: a century-old story. In: Artificial intelligence in cardiothoracic imaging. Springer International Publishing, Cham, pp 3–13

Van Klompenburg T, Kassahun A, Catal C (2020) Crop yield prediction using machine learning: a systematic literature review. Comput Electron Agric 177:105709

Weiss M, Jacob F, Duveiller G (2020) Remote sensing for agricultural applications: a meta-review. Remote Sens Environ 236:111402

World Bank (2020) World Bank Open Data. Available: https://www.data. worldbank.org. [11/09/2023]

Yuan T, Zhang S, Sheng X, Wang D, Gong Y, Li W (2016) An autonomous pollination robot for hormone treatment of tomato flower in greenhouse. In: 3rd international conference on systems and informatics (ICSAI) 19–21. pp 108–113

Use of AI in Crop Health Monitoring

3

Ersin Elbasi, Greeshma Varghese, Yehia Ibrahim Alzoubi, Ahmet E. Topcu, and Aymen I. Zreikat

Abstract

Accurate and timely diagnosis of plant diseases is crucial for maintaining agricultural productivity and ensuring global food security. Traditional methods, such as visual inspections, are time-consuming, error-prone, and impractical for large-scale farming operations. This chapter delves into the use of artificial intelligence (AI) technologies, including machine learning (ML), deep learning (DL), and the Internet of Things (IoT), for diagnosing plant diseases. By reviewing recent advancements, it emphasizes the high accuracy of AI models, such as convolutional neural networks (CNNs), in detecting and classifying plant diseases. The integration of IoT with AI has further enabled real-time disease detection and monitoring, offering significant benefits for large-scale agricultural applications. Nevertheless, challenges such as limited datasets, environmental variability, high implementation costs, and data privacy concerns remain. This analysis provides an in-depth overview of these technologies, addressing their benefits, limitations, and prospects, and aims to advance scalable AI-driven solutions for effective plant disease management.

Keywords

Plant diseases · Artificial intelligence · AI tools · Machine learning · Diagnosis · Agriculture

E. Elbasi (✉) · G. Varghese · A. E. Topcu · A. I. Zreikat
College of Engineering and Technology, American University of the Middle East, Egaila, Kuwait
e-mail: ersin.elbasi@aum.edu.kw; greeshma.varghese@aum.edu.kw; ahmet.topcu@aum.edu.kw; Aymen.Zreikat@aum.edu.kw

Y. I. Alzoubi
College of Business Administration, American University of the Middle East, Egaila, Kuwait
e-mail: yehia.alzoubi@aum.edu.kw

R. E. Krishnankutty et al. (eds.), *Smart Agriculture*,
https://doi.org/10.1007/978-981-95-8785-8_3

1 Introduction

The early and accurate diagnosis of plant diseases is crucial for maintaining agricultural productivity and ensuring food security. Traditional methods of identifying plant diseases often rely on visual inspection by experts, which can be time consuming, labor intensive, and prone to human error. As a result, researchers have turned to artificial intelligence (AI) and other information technologies to develop automated and precise tools for detecting plant diseases. AI-based solutions, particularly those leveraging machine learning (ML), Internet of Things (IoT), and deep learning (DL), have demonstrated great potential in identifying plant diseases from visual symptoms, such as leaf spots, blights, and mildew (Mohanty et al. 2016).

One of the primary approaches in this domain involves using convolutional neural networks (CNNs), which are specialized in image classification tasks. CNNs have been employed in various studies to detect and classify plant diseases with high accuracy. For example, in a study by Ferentinos (2018), a deep learning-based model was trained on a large dataset of plant images and achieved an average accuracy of 99.53% across different crop species and diseases. This underscores the robustness and generalizability of deep learning models in complex agricultural settings.

In addition to CNNs, other AI tools like support vector machines (SVMs), decision trees, and random forest algorithms have been utilized in plant disease diagnosis. These models can be combined with feature extraction techniques to capture subtle variations in plant disease symptoms that are often indistinguishable to the human eye (Barbedo 2018). Furthermore, the integration of AI with Internet of Things (IoT) devices, such as smart cameras, sensors, and robots, enables real-time monitoring of plant health, providing farmers with timely and actionable insights (Pantazi et al. 2016).

The use of AI in plant disease diagnosis not only reduces the dependency on expert knowledge but also offers scalability and precision, making it an invaluable tool for modern agriculture. As research progresses, there is a growing interest in developing more sophisticated models that can address challenges such as variations in lighting, occlusions, and complex backgrounds (Lu et al. 2017a, b). There are several reasons for using AI in plant disease recognition as follows:

- AI enables early identification of plant diseases, reducing the risk of widespread outbreaks.
- Automated systems streamline the monitoring process, saving time and labor.
- AI tools can effectively manage extensive agricultural operations.
- Machine learning algorithms minimize human error by analyzing data precisely.
- AI optimizes the use of pesticides and fertilizers, supporting sustainable agriculture.

This chapter discusses several key sections, including the literature review, the evolution of plant disease diagnosis, and AI tools for plant disease detection such as

IoT, computer vision, and machine learning. Additionally, it examines case studies and explores the challenges and limitations in this field.

2 Background and Literature Review

The agricultural industry is crucial to a nation's economy, food security, and the general welfare of its people. The farming industry employs several workers and plays a role in a nation's export revenue. Nevertheless, agriculture encounters several obstacles, such as climate change and plant diseases; specifically, plant diseases can lead to significant decreases in crop yields and quality, negatively impacting farmers' livelihoods and the nation's food security.

For example, Egypt is one of the primary nations that must prioritize the detection of plant diseases to ensure a successful crop yield, given that the country predominantly depends on agriculture for its revenue. Additionally, other nations depend on the fruits shipped out of Egypt like oranges, since Egypt is well known as one of the biggest orange exporters globally (Ibrahim and Elghamrawy 2023). Thus, plant disease detection is crucial to preventing crop losses and ensuring healthy crop yields for food. However, farmers are not well informed or use advanced disease detection methods, which are more precise and quicker than traditional methods like visual inspection by agriculture experts. This old approach relies on human error-prone "experience and skills" to identify diseases accurately. On the other hand, artificial intelligence (AI) tools and machine learning techniques play an important role in identifying plant illnesses because they apply information most frequently and provide excellent methods for disease diagnosis (Munawar et al. 2024).

Diverse studies have discussed this issue from different perspectives. For example, Singh et al. (2018) suggested an integrated and collaborative platform for the automated diagnosis, monitoring, and prediction of diseases. By using a mobile application, farmers can quickly and precisely identify diseases by capturing images of the affected plant parts. The platform facilitates real-time diagnosis through advanced artificial intelligence (AI) algorithms that support Cloud-based image processing. The AI model is continually refined by learning from images uploaded by users and recommendations from experts, which improves its accuracy. In addition, farmers can engage with local experts via the platform. According to the experimental results, the authors reported an accuracy of 95% in the identification of diseases. In Banadda et al. (2023), the primary aim of the research was to provide farmers with vital resources that enhance agricultural productivity through accurate and efficient identification of maize diseases. This is achieved by developing and training a tailored model that employs supervised machine learning techniques alongside transfer learning with established models, such as VGG19 and MobileNet, to refine the diagnostic process for maize diseases. The authors claimed to obtain over 95% accuracy in disease detection even with an increased dataset for predicting early symptoms of maize diseases.

To meet the high demand for maize production in India due to the increase in the population, there was a need to utilize up-to-date AI techniques for early detection and fighting these diseases so that the country's overall maize production could be increased. Regarding fighting maize diseases in India, compared with the work conducted in Banadda et al. (2023), Gupta et al. (2022) claimed to achieve 97.93% accuracy. In this chapter, we have introduced a deep convolutional neural network model that integrates transfer learning techniques, leading to a decrease in the training time required by the artificial intelligence model. To further enhance the accuracy, data augmentation methods have been employed to increase the variability within the dataset used for training the network.

In Coletta et al. (2022), the authors proposed a crowd-sensing framework in which smartphones are handed out to farmers who oversee providing the fundamental sensing activities of the framework. Smartphones are equipped with the Nuru application, which can detect plant diseases at early stages and constitutes the key component of the crowd-sensing framework. By conducting simulations and real-life situations from Kenya, the authors claim that their solution outperforms the current solutions in terms of accuracy and completeness at a lower cost. The research in Anand et al. (2023) utilizes machine-learning algorithms, such as K-nearest neighbors (KNN), logistic regression, random forest (RF), naïve Bayes, decision trees, and support vector machines (SVM), on a varied dataset of crop images. A convolutional neural network (CNN)-based framework integrated with machine learning soil analysis can identify suitable crops, thereby facilitating informed decisions regarding planting. This study introduces an advanced evaluation system that employs machine learning and pretrained models (VGG16, InceptionV3, ResNet152) to classify plant diseases within the datasets. Additionally, the incorporation of the Internet of Things (IoT) enhances the prediction of diseases and yields, which are essential for contemporary agricultural practices. Kumar and Monika (2024) present a thorough approach to predicting leaf diseases through the utilization of convolutional neural networks and transfer learning techniques, implemented in Python using TensorFlow, Keras libraries, and the Flask framework. The workflow includes data augmentation, model development, training, testing, and validation. The dataset comprises images sourced from a public repository, which are augmented to enhance model performance. The proposed model was trained and assessed over 15 epochs, achieving an impressive training accuracy of approximately 95.84%, while the accuracy during testing or validation was approximately 87.0%.

Some other research (Srivastava et al. 2024a, b) has investigated not only the techniques for disease detection but also the possible methods for treatment. The authors proposed a model using the PlantVillage dataset, which contains a large collection of photos of diverse plant species and diseases. The proposed model employs a deep learning technique called convolutional neural network (CNN) and uses Python's Fast.ai package. The strategy consists of many procedures, including data preprocessing, model selection, model training, and outcome analysis. The accuracy obtained using this architecture was 99.68% on the validation set, demonstrating the model's ability to accurately detect disease in plants. Compared to

Srivastava et al. (2024a, b), Biswas and Yadav (2023) also utilized PlantVillage dataset in their investigation by studying various deep learning (DL) algorithms where 54,303 images are categorized by species and disease into 38 categories. The authors claimed that the proposed approach can be used by farmers anywhere in the world for early prediction of plant diseases and in a short time. Other researchers (Rose Anandhi and Sathiamoorthy 2023) have utilized machine learning (ML) and computer vision (CV) techniques to detect diseases related specifically to rice crops. This study employs the deep learning-based automated rice plant disease recognition and classification (DL-ARPDRC) technique. The DL-ARPDRC approach combines various techniques to improve accuracy and diagnostic performance. The first level contains preprocessing stages such as image scaling and the Gaussian filter (GF) to improve image quality. Next, Otsu's threshold-based segmentation was used to separate the lesion from the background, isolating the area of interest. Feature extraction was achieved using the VGG-19 architecture. The collected features were input to an XGBoost classification model to help discriminate between different types of disorders. Through the experimental results and analysis, the authors declared that DL-ARPDRC technique outperformed other approaches in the literature.

3 The Evolution of Plant Disease Diagnosis

The evolution of plant disease diagnosis has followed a fascinating trajectory, from early observational methods to cutting-edge technologies like artificial intelligence (AI). This evolution (Fig. 3.1) shows a clear shift from basic observation to sophisticated, AI-driven diagnostic techniques, enabling faster, more accurate, and large-scale plant disease management. Here are the main milestones in this evolution (Sharma and Sharma 2016; Stackhouse et al. 2020; Wang et al. 2022):

- Early observation and empirical knowledge: In ancient times, farmers relied on visual observations and experience to identify plant diseases. They noticed patterns in plant symptoms like yellowing leaves, wilting, or stunted growth, associating them with poor health or environmental factors.
- Microscopy and germ theory (nineteenth century): The invention of the microscope in the nineteenth century allowed scientists to observe microorganisms, leading to the discovery of plant pathogens like fungi, bacteria, and viruses. The germ theory of disease postulated that specific diseases were caused by specific microorganisms. This theory laid the foundation for understanding how diseases spread and how pathogens such as fungi, bacteria, and viruses infect plants. The application of germ theory to plants led to the discovery that plant diseases were not merely a result of bad weather or poor soil but were caused by infectious agents.
- Plant pathology as a scientific field (early twentieth century): As plant diseases were increasingly understood to be caused by specific microorganisms, the need

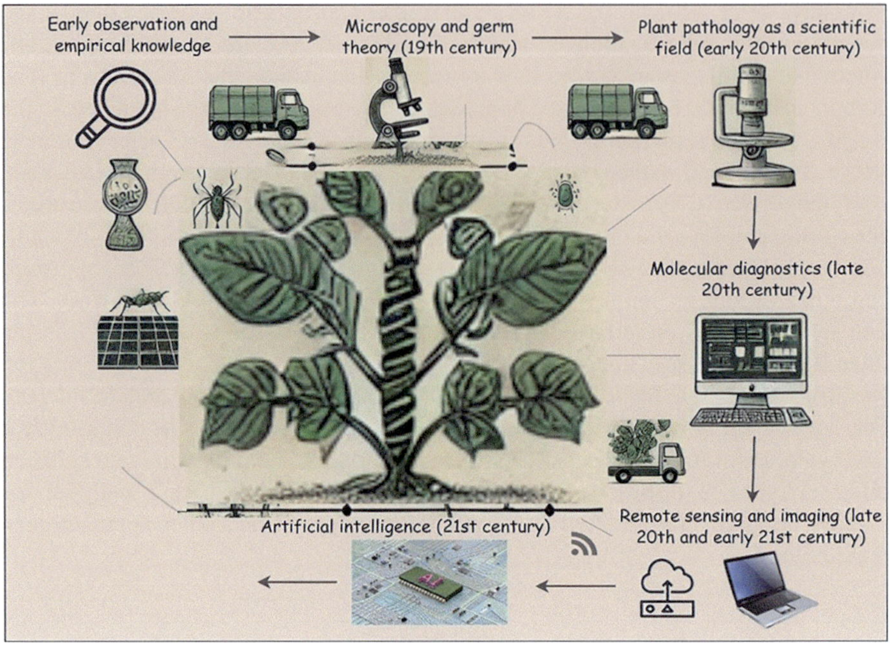

Fig. 3.1 Evaluation of AI in plant leaf diagnosis

to systematically study them became apparent. This led to the establishment of plant pathology departments in universities and research institutions. The establishment of plant pathology as a scientific field created a structured approach to studying and diagnosing plant diseases. This stage set the foundation for later developments in molecular diagnostics and technological advancements, leading to even more precise methods for identifying and managing plant diseases in the following decades. Scientists developed methods like culture-based assays, where pathogens were grown in controlled environments, and staining techniques to observe them more clearly under microscopes.

- Molecular diagnostics (late twentieth century): The advancement of molecular biology revolutionized plant disease diagnosis. The most often used methods are enzyme-linked immunosorbent assay (ELISA), DNA sequencing, quantitative PCR (qPCR), and polymerase chain reaction (PCR). Through the amplification of certain DNA or RNA segments, PCR makes it possible to identify even minuscule amounts of pathogen genetic material. A more accurate diagnosis is possible thanks to the qPCR, an enhanced PCR that not only finds pathogens but also measures their quantity in the sample's DNA or RNA. Through DNA sequencing, diseases' genetic codes are read, allowing for species-level identification. To find out whether certain proteins or antibodies produced by pathogens are present in plant tissues, ELISA is utilized in conjunction with molecular methods. Techniques like polymerase chain reaction (PCR) allowed for the detection of specific pathogen DNA or RNA, enabling faster and more accurate identification of diseases at the molecular level. This was a game-changer for diagnosing diseases that were difficult to detect through traditional methods.

- Remote sensing and imaging (late twentieth and early twenty-first century): With the rise of satellite imaging and remote sensing technologies, researchers began to analyze large-scale patterns of plant health from afar. The most used techniques include multispectral and hyperspectral imaging, thermal imaging, drone and satellite imaging, and unmanned aerial vehicles (UAVs). The multispectral and hyperspectral imaging techniques capture images in various wavelengths of light (visible, infrared, etc.). By analyzing the spectral signatures of plants, researchers can detect physiological changes caused by disease, such as chlorophyll breakdown or water stress, often before symptoms are visible to the naked eye. The thermal imaging cameras detect temperature changes in plants. Diseased plants may have different water usage or transpiration rates, causing variations in temperature that can be detected remotely. The drones equipped with sensors (multispectral, thermal, etc.) provide high-resolution images of crop fields, enabling early detection of disease outbreaks. Satellite imaging offers a broader view, allowing for regional or even global monitoring of plant health. UAVs fitted with various sensors provide precise, on-demand data collection for large agricultural areas. They are used to track disease spread and apply targeted treatments. These methods helped in detecting diseases across vast agricultural areas by observing changes in plant physiology that indicated stress or infection.
- AI and machine learning (twenty-first century): Today, AI and machine learning tools are leading the way in plant disease diagnosis. Common AI-based techniques include convolutional neural networks (CNNs), deep learning models, machine learning algorithms, predictive analytics, and mobile applications. AI-based systems, often powered by deep learning models, can analyze vast amounts of image data to identify disease symptoms at early stages with high accuracy. These tools can be integrated with smartphone apps, drones, or sensors to provide real-time diagnosis in the field. By training AI models on thousands of images of healthy and diseased plants, these systems can quickly and effectively diagnose diseases, making them invaluable for farmers and researchers aiming to protect crops from disease outbreaks.

4 AI Tools for Plant Disease Diagnosis

Plant diseases pose a significant threat to global food security. Fortunately, advancements in artificial intelligence (AI) offer promising solutions for early and accurate disease detection. AI-based tools for plant disease diagnosis can be categorized into three main approaches: image-based, sensor-based, and machine learning-based techniques. Figure 3.2 demonstrates AI tools in plant leaf diagnosis (Elbasi et al. 2023a, 2024b).

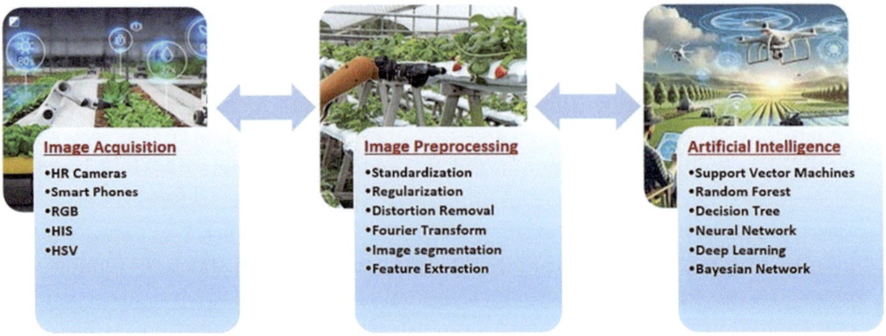

Fig. 3.2 AI tools in plant leaf diagnosis

4.1 Image-Based Diagnosis

Image-based techniques leverage computer vision and machine learning algorithms to analyze digital images of plants. Research into plant disease detection using image processing has provided valuable insights through remote sensing analysis. The study (Shanmugam et al. 2017) was conducted in two phases: In the first phase, the model was trained using datasets of healthy and infected plants, and threshold values were extracted from the images. In the second phase, various analysis tools for crop monitoring were applied, including Canny edge detection and histogram mapping. Regular RGB measurements taken from field samples through remote sensing and compared with the threshold images aided in disease identification.

Content-based image retrieval techniques (Veni et al. 2021) revealed promising results across multiple crops. The content-based image retrieval model uses support vector machine (SVM) for leaf classification and the K-nearest neighbors (KNN) method for disease classification. Key crops such as apples, tomatoes, potatoes, grapes, and corn were used in testing the model.

A combination of image processing and pattern recognition techniques can be employed for plant disease detection (Wang et al. 2012). Color features, shape features, texture features, and their combined features from diseased plant images will be extracted after image processing. Based on the extracted feature data, backpropagation networks are constructed using different function combinations, and these networks are used as classifiers to identify diseases in plants.

Plant disease image recognition method based on Siamese Networks and image processing (Shang et al. 2022) revealed new possibilities in disease recognition. Siamese Networks has two inputs, which are mapped to a new space by two neural networks. The experiments with EMBRAPA dataset show the Siamese Network method can identify the plant diseases with more than 80% accuracy. This approach can use a deep learning framework based on TensorFlow and Keras.

Principal component analysis (PCA) and neural networks were used to implement image recognition of plant diseases based on extracted color features, shape features, texture features, and their combined features from disease images (Wang et al. 2012). Plant disease recognition is carried out using backpropagation (BP) networks, radial basis function (RBF) neural networks, generalized regression neural networks (GRNNs), and probabilistic neural networks (PNNs). By using PCA, the dimensions of the data extracted from plant disease images can be reduced, which reduces the number of neurons in the input layer and increases the speed of the neural networks.

Image-based detection of black gram crop disease (DBCD) (Harika et al. 2023) benefits from integrated analysis methods. The two main modules in this detection method are classification analysis and prediction. The classification analysis consists of five submodules: image preprocessing, training, testing, and performance evaluation. The techniques used include artificial neural networks, convolutional neural networks, decision trees, random forests, and K-nearest neighbor algorithms.

A novel crop disease detection model using multiparametric bioinspired feature representation and ensemble classification methods (Lohi and Bhatt 2023) provides new insights into disease detection. This method employs multiple feature extraction units to extract color, shape, texture, and convolutional features from images. The extracted features are then processed using a genetic algorithm-based model to identify the optimal feature sets. These selected features are classified using an ensemble classification model that includes support vector machines, multilayer perceptron, linear regression, decision trees, and naive Bayes classifiers. The proposed model achieved 99.5% accuracy and is capable of classifying diseases in soybean, squash, potato, pepper, apple, and raspberry crops.

4.2 Sensor-Based Diagnosis

Sensor-based plant disease diagnosis uses a field setup that includes sensors to track key environmental factors such as moisture levels, temperature changes, and leaf color variations. The field deployment of Zigbee networks facilitated extensive data collection (Zhang et al. 2014). This approach uses a big data platform to develop an intelligent warning system. IoT-based data acquisition relies on sensors to gather data, including images of injured plant parts, temperature, humidity, wind speed, CO_2 concentration, rainfall, and soil temperature and moisture levels.

An automated system (Kumar et al. 2023) has been developed to identify crop deterioration using sensors like humidity, wetness, temperature, and plant leaf color. The system utilizes Arduino and various sensors to monitor crop health. Farmers are provided with access to the analyzed data, which helps them track plant health issues.

Through IoT network integration (Manoharan et al. 2024a, b), researchers used field sensors to monitor various environmental factors such as temperature, humidity, soil moisture, and imaging equipment for capturing crop images. The data from these sensors are transmitted to a centralized server, where it is processed to detect the presence of plant diseases using image analysis and machine learning algorithms, such as CNN and SVM.

4.3 Machine Learning-Based Diagnosis

Artificial intelligence (AI) enables rapid and efficient disease diagnosis through predictive models. AI-driven systems leverage image data for machine vision tasks, with convolutional neural networks (CNNs) being one of the most powerful techniques (Elbasi et al. 2023b, 2024a).

A CNN-based classification model (Gupta et al. 2023) for disease detection involves several steps, including data collection, preprocessing, cleaning, and classification. The collected data is used to train the machine learning model to predict the disease. To enhance the model's performance, data augmentation is applied to generate diverse data subsets that may not be present in the original dataset, thereby improving the training process. An approach (Saeed et al. 2021) to enhance the accuracy of the CNN model involves experimenting with a subset of the dataset consisting of diseased leaf images that are difficult for human observers to identify. Additionally, an extra dense layer is added before the fully connected (FC) layer in the CNN model. The CNN-based classification model involves collecting and preprocessing the dataset, training the CNN to identify crop types and detect diseases, and validating the model to obtain accurate results (Kulkarni 2018).

By using specialized datasets (GWHD and LWDC), researchers tracked both individual plant features and broader disease patterns (Etaati et al. 2023). The study also applied object detection techniques to identify and localize diseased plants. A framework that integrates deep learning models with Explainable AI (XAI) (Rakesh and Indiramma 2022) helps to explain how the models make decisions regarding plant health. A key advantage of this approach is its ability to predict disease detection while also offering visual explanations that clarify the model's predictions.

The integration of YOLOv4-Tiny algorithms with multilayer intelligent farming systems enabled detailed monitoring (Qu et al. 2022). A lightweight MobileNetV2 CNN architecture revealed advantages in wheat disease monitoring. The lightweight design showed both improved speed and accuracy compared to existing CNN approaches (Bouskour et al. 2024). Research into YOLOv5 applications for chili crop monitoring encountered real-world challenges, including the lack of an open-source dataset of chili images. Despite these challenges, researchers achieved 75.64% accuracy using standard RGB cameras (Mayalekshmi et al. 2023).

Table 3.1 summarizes various machine learning approaches for plant disease diagnosis from different research studies, including details on the methodology, target crop types, disease types, datasets used, and the accuracy rates of experimental tests.

Table 3.1 Comparison of methodologies in plant leaf disease classification

Contribution	Methodology	Crop type	Disease type	Dataset	Accuracy rate
Gupta et al. (2023)	CNN	Potato	Late blight and early blight	Kaggle	95%
Saeed et al. (2021)	CNN (ResNet152 and InceptionV3)	Rice and corn	Brown spots, hispa, and leaf blast	Kaggle	97.8%
Kulkarni (2018)	CNN (InceptionV3 and MobileNet)	5 types of crops (not mentioned in the chapter)	3 types of diseases (not mentioned in the chapter)	Plantvilla	99.7%
Etaati et al. (2023)	YOLOv4 Object detection and CNN models (VGG19)	Wheat	Wheat head-diseases (Fusarium head blight, loose smut)	Global Wheat Head Detection Data set (GWHD), Large Wheat Disease Classification (LWDC) Dataset	95%
Rakesh and Indiramma (2022)	CNN models (Inception V3 and ResNet-9), Explainable AI (LIME and Grad-CAM)	14 different crop species	24 different crop diseases	Plant Village, New Plant Disease	99.2%
Bouskour et al. (2024)	MobileNetV2 CNN detection algorithm	Wheat leaf	Stripe rust and Septoria	Open-source dataset	99%
Mayalekshmi et al. (2023)	YOLO V5	Chilly leaf	Leaf spot and leaf curl	Images collected from actual field conditions	75.64%

5 Case Studies

The various studies to detect various plant diseases can be categorized into the following sections.

5.1 Image-Based Detection Systems

AI models, especially deep learning techniques, are primarily employed to detect plant diseases via images of leaves or stems. Convolutional neural networks (CNNs) are applied to classify diseases, often achieving accuracy above 90%. These models examine images captured by smartphones or drones to detect disease symptoms before human observation. With 98.49% accuracy, an integrated framework using the CNN model found diseases on tomato leaves (Trivedi et al. 2021). This shows that AI can help with precision agriculture.

5.2 Agriculture-Related Mobile Applications

Another area is that researchers have developed deep learning-powered mobile apps to assist farmers in real-time plant disease diagnosis. These applications utilize computer vision models to identify symptoms from uploaded photos, improving accessibility for small-scale farmers and enhancing early detection (Li et al. 2021).

5.3 Models for Forecasting Disease Outbreaks

AI forecasting models incorporate weather, soil, and crop data to predict disease outbreaks before they occur. These models help pre-emptively apply treatments, reducing pesticide use and minimizing crop losses.

Case Study I Artificial Intelligence: Tool for Application in Phytopathology
This work examines the pivotal role of AI in detecting and managing plant diseases, making substantial contributions to phytopathology. It underscores the role of AI, particularly machine learning, in improving accuracy in disease detection, forecasting, and decision-making within agriculture. This AI model classifies plant diseases. Classification techniques like support vector machines (SVM) and random forests are highly effective in distinguishing healthy plants from those afflicted by disease. They also used deep learning technologies, especially convolutional neural networks (CNNs), which are very good at finding plant diseases through image analysis by finding patterns humans cannot see (González-Rodríguez et al. 2024). By possessing these capabilities, the following outcomes can be attained to establish high standards in predicting plant diseases:

- Disease detection: AI systems can precisely identify diseases from images of leaves, fruits, or stems, surpassing conventional visual inspections.
- Forecasting and mitigation: Models can anticipate outbreaks based on environmental factors, facilitating proactive measures.
- Precision agriculture: AI collaborates with IoT devices for instantaneous monitoring and decision-making, minimizing pesticide application and enhancing yield efficiency.

Key challenges include data quality, as the effective implementation of models is heavily reliant on the availability of high-quality, labeled datasets. Furthermore, model generalization is crucial for ensuring that models sustain robust performance across varied environmental conditions and different crops. AI presents significant advancements in phytopathology by improving disease detection precision, lowering expenses, and fostering sustainable agricultural methodologies. However, challenges such as data availability and model robustness need further attention for wider adoption.

Case Study II Spatial-Spectral Analysis of Hyperspectral Images Using AI
Downy mildew is a devastating disease that affects grapevines; this work aims to use hyperspectral imaging and machine learning to detect it early. Traditionally, diagnosing this disease is laborious and necessitates specialized knowledge. The researchers aimed to develop a noninvasive, automated method for pathogen detection before the appearance of visible symptoms.

Grapevine leaves inoculated with *Plasmopara viticola* strains were observed in a controlled environment from 0 to 9 days post-inoculation (dpi). With hyperspectral images as training data, a support vector machine (SVM) classifier got an average validation accuracy of 96% and a test accuracy of 99%. The SVM detected downy mildew at 2 dpi, two days before visible symptoms appeared at 4 dpi, with no false positives on control leaves. The automated tool yielded superior evaluations of disease severity by encompassing a broader affected area than visual inspection. This study employs hyperspectral imaging for the early detection and spatial monitoring of downy mildew on grapevines, emphasizing its potential to enhance disease management via timely and precise intervention (Lacotte et al. 2022).

6 Challenges and Limitations

Artificial intelligence tools for plant disease diagnosis face several challenges and limitations, including issues with data quality, imbalanced, or unrealistic datasets that reduce model effectiveness in real-world scenarios. Models often achieve high accuracy in controlled tests but struggle to generalize across diverse crops, diseases, and environmental conditions, particularly with lighting variability, complex backgrounds, and mixed infections. Challenges and limitations can be listed as follows:

- Many machine learning models rely on large, high-quality datasets for training. However, datasets for specific crops or diseases may be insufficient or imbalanced.
- Most datasets are collected under controlled conditions and may not represent real-world variability such as lighting, occlusions, or environmental factors (Srivastava et al. 2024a, b).
- While IoT enhances real-time monitoring, integrating multiple sensors and AI tools can be resource-intensive, requiring expertise and substantial financial investment (Manoharan et al. 2024a, b).
- Scaling AI tools for large-scale agricultural processes may face challenges in data transmission, storage, and analysis.
- Differences in lighting, shadows, and complex backgrounds can affect model predictions, especially in image processing and computer vision applications (Lu et al. 2017a, b).
- Deploying AI-based solutions integrated with IoT can be costly, limiting accessibility for small-scale or resource-poor farmers.

- Data privacy and ownership issues arise from crowdsourcing and IoT frameworks.
- IoT-based systems require complex setups, making large-scale implementation difficult in resource-poor regions.

7 Conclusion

The integration of artificial intelligence tools in plant disease diagnosis marks a transformative step toward more efficient and accurate agricultural practices. AI-based approaches, particularly those leveraging machine learning and deep learning, have demonstrated remarkable potential in diagnosing plant diseases with high precision. Furthermore, the combination of AI and IoT has enabled real-time monitoring and predictive capabilities, enhancing decision-making for farmers. Despite these advancements, challenges such as limited datasets, environmental variability, high implementation costs, and a lack of technological literacy remain significant barriers to widespread adoption. Addressing these limitations through improved datasets, cost-effective solutions, and farmer education will be critical in making AI-driven tools more accessible and effective. As research progresses, the development of explainable AI models and ethical data governance frameworks will also play a vital role in ensuring the trustworthiness and scalability of these technologies. Ultimately, the application of AI in plant disease diagnosis offers immense promise for achieving sustainable agriculture and enhancing global food security.

References

Anand SS, Sankaran S, Ravi S, Hariharan S, Kukreja V, Vasantha SV (2023) Revolutionizing agriculture through leaf disease prediction in plants. In: Global Conference on Information Technologies and Communications (GCITC), Bangalore, pp 1–6

Banadda M, Aloysius NKS, Nakazzi OB, Ernest MSO, Marvin G (2023) Explainable artificial intelligence for maize disease diagnostics. In: IEEE Asia-Pacific Conference on Computer Science and Data Engineering (CSDE), vol 2023, Nadi, pp 1–6

Barbedo JGA (2018) Factors influencing the use of deep learning for plant disease recognition. Biosyst Eng 172:84–91

Biswas B, Yadav RK (2023) A review of convolutional neural network-based approaches for disease detection in plants. In: 2023 International Conference on Intelligent Data Communication Technologies and Internet of Things (IDCIoT), Bengaluru, pp 514–518

Bouskour S, Zaggaf MH, Bahatti L (2024) Deep learning recognition of wheat leaf disease using MobileNetV2 model. In: In 2024 4th International Conference on Innovative Research in Applied Science, Engineering and Technology (IRASET). IEEE, pp 1–6

Coletta A, Bartolini N, Maselli G, Kehs A, McCloskey P, Hughes DP (2022) Optimal deployment in crowd sensing for plant disease diagnosis in developing countries. IEEE Internet Things J 9(9):6359–6373

Elbasi E, Mostafa N, AlArnaout Z, Zreikat AI, Cina E, Varghese G, Shdefat A, Topcu AE, Abdelbaki W, Mathew S, Zaki C (2023a) Artificial intelligence technology in the agricultural sector: a systematic literature review. IEEE Access 11:171–202

Elbasi E, Zaki C, Topcu AE, Abdelbaki W, Zreikat AI, Cina E, Shdefat A, Saker L (2023b) Crop prediction model using machine learning algorithms. Appl Sci 13(16):9288

Elbasi E, Mostafa N, Zaki C, AlArnaout Z, Topcu AE, Saker L (2024a) Optimizing agricultural data analysis techniques through AI-powered decision-making processes. Appl Sci 14(17):8018

Elbasi E, Topcu AE, Cina E, Zreikat AI, Shdefat A, Zaki C, Abdelbaki W (2024b) Enhanced plant leaf classification over a large number of classes using machine learning. Appl Sci 14(22)

Etaati S, Khoramdel J, Najafi E (2023) Automated wheat Disease detection using Deep Learning: an object detection and Classification approach. In 2023 11th RSI International Conference on Robotics and Mechatronics (ICRoM). IEEE, pp. 116–121

Ferentinos KP (2018) Deep learning models for plant disease detection and diagnosis. Comput Electron Agric 145:311–318

González-Rodríguez VE, Izquierdo-Bueno I, Cantoral JM, Carbú M, Garrido C (2024) Artificial intelligence: a promising tool for application in phytopathology. Horticulturae 10(3):197

Gupta S, Vishnoi J, Rao AS (2022) Disease detection in maize plant using deep convolutional neural network. In: 7th International Conference on Communication and Electronics Systems (ICCES), vol 2022, Coimbatore, pp 1330–1335

Gupta U, Vijh S, Kumar S, Roy NR, Jodan JS (2023) Potato leaf disease detection using machine learning techniques for precision agriculture. In: 2023 Seventh International Conference on Image Information Processing (ICIIP). IEEE, pp 913–918

Harika S, Sandhyarani G, Sagar D, Reddy GS (2023) Image-based black gram crop disease detection. In 2023 International conference on inventive computation technologies (ICICT). IEEE, pp. 529–533

Ibrahim AO, Elghamrawy S (2023) Plant disease detection using artificial intelligence techniques for agricultural productivity enhancement. Egypt International Telecommunications Conference (ITC-Egypt), Alexandria, pp. 490–496

Kulkarni O (2018) Crop disease detection using deep learning. In: 2018 Fourth International Conference on Computing Communication Control and Automation (ICCUBEA). IEEE, pp 1–4

Kumar HSV, Monika G (2024) Machine learning based model for leaf diseases detection. In: International Conference on Smart Systems for Applications in Electrical Sciences (ICSSES), Tumakuru, pp 1–5

Kumar CV, Saritha K, Reddy MS, Pai HA (2023) IoT-based smart agriculture for the detection of plant decay. In: 7th International Conference on Computing Methodologies and Communication (ICCMC). IEEE, pp 1376–1379

Lacotte V, Peignier S, Raynal M, Demeaux I, Delmotte F, da Silva P (2022) Spatial–spectral analysis of hyperspectral images reveals early detection of Downy Mildew on grapevine leaves. Int J Mol Sci 23(17):10012

Li L, Zhang S, Wang B (2021) Plant disease detection and classification by deep learning—a review. IEEE Access 9:56683–56698

Lohi SA, Bhatt C (2023) Design of a crop disease detection model using multi-parametric bio-inspired feature representation and. In: Ensemble Classification 4th International Conference on Innovative Trends in Information Technology (ICITIIT). IEEE, pp 1–6

Lu Y, Yi S, Zeng D, Liu X (2017a) Challenges in deep learning-based agricultural disease detection systems. Comp Vision Precis Agric 7(1):12–20

Lu Y, Yi S, Zeng N, Liu Y, Zhang Y (2017b) Identification of rice diseases using deep convolutional neural networks. Neurocomputing 267:378–384

Manoharan G, Ali SDNH, Sathe M, Karthik A, Nagpal A, Sidana A (2024a) IoT-based crop disease detection and management system using machine learning algorithms. In: International Conference on Science Technology Engineering and Management (ICSTEM). IEEE, pp 1–5

Manoharan S, Ramesh T, Varghese K (2024b) IoT-based integrated systems for plant disease detection and environmental monitoring. J Smart Agric Syst 18(1):45–60

Mayalekshmi KM, Ranjan A, Machavaram R (2023) In-field chilli crop disease detection using YOLOv5 deep learning technique. In: IEEE 8th International Conference for Convergence in Technology (I2CT). IEEE, pp 1–6

Mohanty SP, Hughes DP, Salathé M (2016) Using deep learning for image-based plant disease detection. Front Plant Sci 7:1419

Munawar S, Rajendiran D, Sabjan K (2024) Plant diseases diagnosis with artificial intelligence (AI), pp 187–193. https://doi.org/10.1007/978-981-99-9621-6_12
Pantazi XE, Moshou D, Alexandridis T, Whetton R, Mouazen AM (2016) Wheat yield prediction using machine learning and advanced sensing techniques. Comput Electron Agric 121:57–65
Qu P, Liu N, Qin Z, Jin T, Fu H, Li Z, Sang P (2022) Deep learning-based detection of plant nutrient deficiency symptom and design of multi-layer greenhouse system. In: IEEE International Conference on Mechatronics and Automation (ICMA), pp 641–645
Rakesh S, Indiramma M (2022) Explainable AI for Crop disease detection. In: 4th International Conference on Advances in Computing, Communication Control and Networking (ICAC3N), pp. 1601–1608
Rose Anandhi DF, Sathiamoorthy S (2023) Deep learning based automated rice plant disease recognition and classification model. In: First International Conference on Advances in Electrical, Electronics and Computational Intelligence (ICAEECI), Tiruchengode, pp 1–6
Saeed Z, Raza A, Qureshi AH, Yousaf MH (2021) A multi-crop disease detection and classification approach using CNN. In: International Conference on Robotics and Automation in Industry (ICRAI), pp 1–6
Shang Z, He J, Wang D (2022) Image recognition of plant diseases based on Siamese Networks China Automation Congress (CAC), pp 5328–5332
Shanmugam L, Adline ALA, Aishwarya N, Krithika G (2017) Disease detection in crops using remote sensing images. In: IEEE International Conference on Technological Innovations in ICT for Agriculture and Rural Development
Singh SM, Hanchate DB (2018) Improving disease prediction by machine learning. Int J Res Eng Technol 5:1542–1548
Sharma P, Sharma S (2016) Paradigm shift in plant disease diagnostics: a journey from conventional diagnostics to nano-diagnostics. In: Kumar P, Gupta V, Tiwari A, Kamle M (eds) Current trends in plant disease diagnostics and management practices. Fungal biology, pp 237–264
Srivastava AK, Chauhan S, Mishra T, Mishra V (2024a) Rapid plant disease detection using AI techniques. In: IEEE International Conference on Computing, Power and Communication Technologies (IC2PCT), Greater Noida, pp 540–545
Srivastava N, Pandey R, Mehta P (2024b) Deep learning techniques for plant disease detection: Case studies and comparative analysis. Plant Health Inform 16(2):78–94
Stackhouse T, Martinez-Espinoza AD, Ali ME (2020) Turfgrass disease diagnosis: past, present, and future. Plants 9(11):1544
Trivedi NK, Gautam V, Anand A, Aljahdali HM, Villar SG, Anand D, Kadry S (2021) Early detection and classification of tomato leaf disease using high-performance deep neural network. Sensors 21(23):7987
Veni S, Anand R, Mohan D, Sreevidya P (2021) Leaf recognition and disease detection using content based image retrieval. In 2021 7th international conference on advanced computing and communication systems (ICACCS). IEEE 1:243–247
Wang H, Li G, Ma Z, Li X (2012) Image recognition of plant diseases based on principal component analysis and neural networks. In: 2012 8th International Conference on Natural Computation (pp 246–251)
Wang YM, Ostendorf B, Gautam D, Habili N, Pagay V (2022) Plant viral disease detection: from molecular diagnosis to optical sensing technology—a multidisciplinary review. Remote Sens 14(7):1542
Zhang S, Chen X, Wang S (2014) Research on the monitoring system of wheat diseases, pests and weeds based on IOT. In: 2014 9th International Conference on Computer Science & Education (pp 981–985)

4 Sustainable Development Goals (SDG) based AI integrated Irrigation Systems

Prem Rajak, Suchandra Bhattacharya, Abhratanu Ganguly, and Satadal Adhikary

Abstract

In the wake of increased agricultural practices owing to global population explosion, the requirement for water usage in agricultural fields has emerged as a matter of global concern. Among the 17 Sustainable Development Goals (SDGs) by the United Nations, the goal 6 emphasizes "clean water and sanitation for all," which voices in favor of judicial water usage during crop production. This can be achieved by implementing IoT-based precision farming by carefully tailoring the management of soil and other parameters to standardize the individualistic design of smart irrigation for each field through the employment of automated irrigation schemes equipped with handy wireless sensor networks and decision support systems. Further, the data acquired from various sensors is conveyed to the cloud servers, which eventually enables a closed-loop control over the water feed to materialize a fully automated irrigation scheme. This review summarizes the scope of smart technologies for monitoring the parameters for smart irrigation and discusses different irrigation systems that can be fine-tuned by incorporating IoT. The architecture of smart systems from a mechanistic perspective, their benefits, and barriers has also been discussed. Such a discourse will be beneficial for future research in this avenue and will also facilitate the lab-to-land transfer of smart technology by agricultural practitioners by providing necessary information.

P. Rajak (✉) · A. Ganguly
Department of Animal Science, Kazi Nazrul University, Asansol, West Bengal, India

S. Bhattacharya
Department of Chemistry, A. B. N. Seal College, Cooch Behar, West Bengal, India

S. Adhikary (✉)
Post Graduate Department of Zoology, A. B. N. Seal College,
Cooch Behar, West Bengal, India

R. E. Krishnankutty et al. (eds.), *Smart Agriculture*,
https://doi.org/10.1007/978-981-95-8785-8_4

Keywords

Agriculture · IoT · Irrigation · Precision farming · Sensors · Sustainable Development Goal

1 Introduction

Attaining global sustainability obliges to ensure universal access to clean water for every individual around the world. Although it is an important objective of Sustainable Development Goal—6 (SDG6) of the United Nations' (UN) Agenda 2030 for worldwide sustainable development, present-day water scarcity is rapidly expanding and affecting an increasing number of industrial, residential, and agricultural water users (Salehi 2022). Climate changes influence water supply, while urbanization and agricultural expansion have caused an upsurge in water demand and pollution, which in turn impedes the safe use of water. A recent study alarmingly predicts that by 2050 an additional 40 million square kilometers of basin area in addition to 3 billion more people might potentially face water scarcity (Wang et al. 2024). The UN 2030 Agenda is a comprehensive plan that has been endorsed by all 193 UN Member States. It comprises 17 Sustainable Development Goals (SDGs) and 169 definite targets, all to be attained by 2030 (Segovia-Hernández et al. 2023). Goal 6 of SDGs emphasizes hygienic water and sanitation for the inhabitants of the world (Ho et al. 2020). The shortage of water has turned out to be a potent challenge for the sustainable expansion of irrigated agriculture in water-depleted areas. Irrigation water demand is a crucial indicator for defining the expanse of water needed in irrigation areas (Li et al. 2020a). Global agricultural activities utilize at least 70% of the total water supply, and this even rises to over 80% in developing nations (Li et al. 2020a). The emergent utilization of IoT devices and technology has led to a substantial change in human life. Kevin Ashton introduced the phrase "Internet of Things" (IoT) in 1999, which has gradually become a flourishing topic with noteworthy social, technical, and economic prospects, which involves the amalgamation of conventional technologies of Internet connectivity and advanced data processing. This integration holds the potential to revolutionize the trajectories of domestic, agricultural, and industrial practices (Mouha 2021). Owing to the recent revolution in computers, smartphones, Internet, and cloud-based technologies, irrigation systems can now be made smarter. Additionally, the invention of wireless sensor networks (WSNs) and Internet of Things (IoT) technologies has further boosted the proficiencies of these systems (García et al. 2020). The Internet of Things (IoT), smartphone tools, and sensors are scientific innovations that provide agricultural practitioners with real-time information on various parameters of their fields, such as soil temperature, optimal water requirements, and current climatic conditions. IoT is linked to the automation of agricultural and farming activities, to increase productivity along with the reduction of environmental burden, and conservation of resources, including the use of IoT-assisted irrigation, helping in water conservation (Obaideen et al. 2022). A smart irrigation system is employed to ensure judicial water usage in agricultural fields, which in turn will

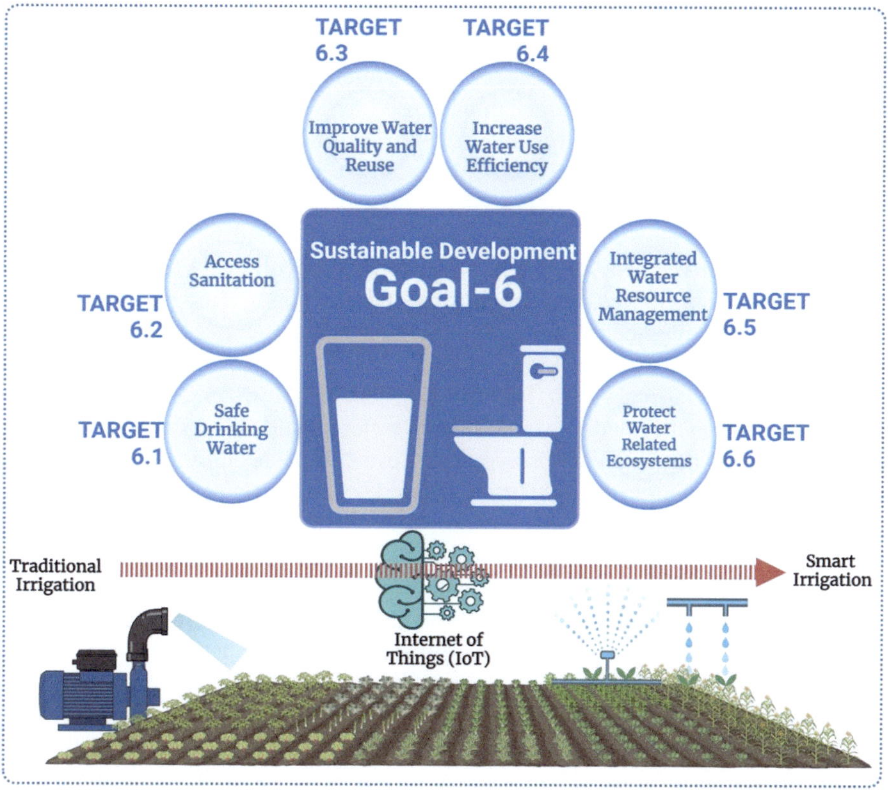

Fig. 4.1 Relationship between Sustainable Development Goal—6 (SDG-6) and Internet of Things (IoT)-enabled smart irrigation. There are several targets to reach SDG-6. To ensure the targets are reached, it is essential to replace traditional irrigation systems with IoT-enabled smart irrigation

lead us to attain SDG-6 (Fig. 4.1). The objective of this chapter is to provide a recent update on the technical transformation from traditional to smart irrigation technologies and justify their impact on crop productivity, crop quality, and water conservation.

2 Prerequisites for Smart Irrigation: Monitoring of Parameters

IoT-based irrigation system requires multiple sets of parameters that should be monitored continuously for better yield. Several environmental parameters like soil moisture, temperatures, precipitation, humidity, and UV index play crucial roles in plant growth and development. Hence, these environmental parameters can be monitored in real time using a number of sensors integrated with the Internet and other accessories to adopt proper measures for improved irrigation (Obaideen et al. 2022). Appropriate measures should be taken promptly to protect the crops from drought, high temperature, and inadequate water availability. Temperature-based sensors can potentially suggest the requirement of irrigation in the agricultural fields depending upon the degree of precipitation and soil moisture. Moreover, moisture level, soil pH, salinity, and soil temperature can be sensed by a number of devices in a large area of agricultural fields (Gałęzewski et al. 2021). This further indicates when and in what amount irrigation is needed for the crops. IoT-based devices are able to detect water content in soil by float switches and notify the farmers through SMS in cell phones when the water content in the soil reduces and falls below a certain level. Low-cost IoT irrigation system integrated with neural network-based intelligence is helpful in remote data monitoring and preventing drought conditions in crop fields and to augment overall yield of crops (Nawandar and Satpute 2019).

Precise monitoring of various parameters is important for sustainability and a more efficient irrigation system. For instance, sensing the water content in soil can help to determine when and in which amount water should be applied in soil. It also helps to reduce wastage and conserve water resources, especially with the event of wireless systems (García et al. 2020). IoT-based irrigation also optimizes the irrigation schedule, thereby reducing the energy consumption and lowering the maintenance cost. The system also enables real-time detection of parameters that enables farmers to resolve the issues promptly and maintain optimum growth conditions. IoT systems facilitate enhanced decision-making in appropriate time. Data analysis on various parameters provides valuable information, enabling the farmers to make informed decisions regarding irrigation, requirements for fertilizer, and pest control. Automation and remote monitoring of irrigation save time and labor and allow farmers to focus on essential tasks (Nikolidakis et al. 2015). Moreover, physical factors like presence of major pollutants in air like PM2.5, PM10, CO, and NOx can also be monitored using IoT devices that affect the crop growth and yield (Dasgupta et al. 2019).

IoT-based irrigation employs a number of automated sensors (García et al. 2020). These devices include sensors for soil moisture, temperature and humidity sensors, water quality sensors, weather forecasting devices, and unmanned aerial devices for aerial sensing. Major IoT platforms and devices employ data analysts platforms, cloud-based sensors for data storage, mobile apps, and web portals. By implementing IoT technologies, farmers and irrigation managers can develop a more efficient, sustainable, and productive irrigation system, ultimately contributing to food security and better yield.

3 Irrigation Systems

3.1 Surface Irrigation

Surface irrigation is the process of applying and distributing water across the soil surface using the gravitational force. Surface irrigation is the predominant method of irrigation and has been used worldwide for many decades with minimal modifications. An efficient surface irrigation system should fill the root zone reservoir equally to reduce plant stress and conserve water as well as nutrients. This can be achieved through the method of flooding the entire field by directing water into narrow channels or by flooding the land surface in strips (borders) until the entire field is uniformly flooded. In regions with low temperature, this irrigation system can protect sensitive fruits and vegetables from damage. Surface irrigation may also promote salt transportation into the roots. Such irritation system is beneficial in enhancing the soil fertility and proper distribution of insecticides (Gamal et al. 2023). Surface irrigation is widely used as it is simple and energy-efficient. However, continuous improvement in surface irrigation is needed to avoid unequal water distribution in soil (Muluneh et al. 2022; Shi et al. 2022). Traditional methods help to penetrate water better into the soil to increase soil fertility. Experiences over the past are essential drivers in improving the surface irrigation methods. In some cases, field-based evaluation of surface irrigation system is difficult, time consuming, and expensive due to a wide variety of agricultural field attributes such as water infiltration rate into the soil, surface texture, water holding capacity, and field slope. Simulation-based models are improved systems that could be cost effective and less time consuming in designing the surface irrigation system (Adamala et al. 2014). Sustainability, data availability, improved irrigation methods, and precision in prediction are essential criteria for selecting simulation models for surface irrigation. Optimizing design for surface irrigation usually requires connecting model elements to crop/orchard productivity. Establishing those relations requires irrigation frequency and timing, water application depth, design variables, crop yield, and irrigation expenses. Surface irrigation design variables include discharge of water, furrow or border length, irrigation cutoff time, distance between borders, and the watered area. These should be connected with the production, costs, and net benefits. Production and irrigation expenses are linked to irrigation quality measurements. Furrow length and irrigation time are the most important design elements that improve the overall surface irrigation.

3.2 Drip Irrigation

Drip irrigation is crucial to overcome water deficit. Controlled and accurate drip irrigation can transport water directly to plant roots. A drip irrigation system delivers water efficiently to the roots of crops. This irrigation method reduces evaporation and runoff, making it the most water-efficient method for irrigation. Modern agriculture uses drip irrigation and organic or inorganic (plastic) mulches. These mulches reduce evaporation, raise soil temperature, and control weeds. It enhances the uniformity in drip irrigation. Any obstruction in this method can possibly hamper the irrigation system and reduce crop yield (Shi et al. 2022). Drip irrigation offers cost-effective and efficient delivery of water to the yard and garden. Drip irrigation is usually common in arid regions. Sprinklers have an efficiency rate of 65–75%, while drip irrigation is far more efficient. In contrast, drip irrigation efficiently employs 90% of water, enabling the maximum use of water by crops. It reduces flow of surface water and evaporation for better availability of water to the pants. Drip irrigation is essential in delivering the water to those sites where it is mostly needed (Mostafa et al. 2024). Drip irrigation is predominantly employed in commercial nursery and farms. In residents, people have the option to utilize drip irrigation systems for their vegetable farming and perennial gardens, and additionally for watering trees and bushes. Drip irrigation entails the placement of tubing equipped with emitters on the ground, adjacent to the plants. The emitters gradually release water into the soil at the root zone. By maintaining moisture levels within an ideal range, the productivity and quality of plants are enhanced (Lamm 2002). Furthermore, drip irrigation effectively mitigates the risk of illness by minimizing direct water contact with the leaves, stems, and fruit of plants. Drip irrigation system enhances the dryness of the rows between plants, facilitating access and minimizing weed proliferation. The system is also very efficient, requires less water, and is cost effective. It also requires less manpower for its application. Drip irrigation augments performance on uneven terrain. It minimizes the loss of water and nutrients by leaching into the area where plant roots can access them.

3.3 Sprinkler Irrigation

Sprinkler irrigation is the process of dispersing water into the atmosphere such that it descends in a manner similar to rainfall. The water discharge by sprays is regulated by the water pressure and flows through a system of pipes, ultimately coming out from little nozzles. Sprinklers should be sized according to design and pressure. The leaching rate of the soil can be used to evenly distribute water throughout the agricultural field (Goap et al. 2018). Many crops can be grown with sprinkler irrigation. These include potato, carrot, onion, garlic, lettuce, and others; oilseeds like peanut, sunflower, and safflower; spices like pepper and cardamom; flowers like carnation and jasmine; and fibers like sisal and cotton (Bortolini and Tolomio 2019). Sprinkler irrigation works on most soils except hard clay (Gamal et al. 2023). It conserves water and provides system mobility. It is ideal for watering oilseeds and vegetables in densely planted regions (Gamal et al. 2023). Sprinkler irrigation systems can be entirely portable, semi-portable, semi-permanent, or totally permanent. Watering sprinklers with lower working pressure uses less energy. However, decreasing the operating pressure and changing the nozzle geometry will affect the hydraulic performance of sprinkler. Different spray characteristics are required to be assessed for better performance. Rectangular sprinkler spacing and operating pressures can be adjusted for circular and noncircular injectors to produce watering similarity coefficients. Round nozzles produce larger droplets than noncircular ones (Chen et al. 2022). However, sprinkler heads are of different types that are employed in various water distribution methods.

4 IoT-Based Approaches for Smart Irrigation

4.1 Communication Technology

The choice of communication technology is crucial for achieving success in IoT-based smart irrigation. There are various wireless systems available for communication between IoT devices, which can be broadly categorized based on their communication range. One technology operates as a node and enables the energy-efficient transmission of small aggregates of data over short distances. Other systems provide the facility to transfer a vast array of data across extended distances, while also requiring high levels of energy consumption (Abagissa et al. 2018). A few widely used communication technologies include Wi-Fi, global system for mobile communication (GSM), long range (LoRa), and message queuing telemetry transport (MQTT). Although Wi-Fi shows several advantages like cost-effectiveness and wide availability, it shows small area coverage and reach (Soto-Garcia et al. 2013). GSM, on the other hand, is a wireless technology that enables long-distance communication. LoRa too has an extensive coverage range, which makes it a highly viable and valuable solution, especially for remote places lacking any connectivity. In contrast, despite MQTT's widespread acceptance because of its minimal power consumption, it is not yet employed for smart irrigation systems on a regular basis (Munir et al. 2019). Finally, ZigBee is a wireless communication technology that is known for its extremely low-power consumption. A standard ZigBee network consists of a coordinator, router, and terminal devices. ZigBee technology has the potential for extensive use in various agricultural applications, including automated irrigation systems (Khriji et al. 2021; Chikankar et al. 2015). A ZigBee technology-based automated irrigation system has also been demonstrated (Chikankar et al. 2015). A summary of different communication systems is provided in Table 4.1.

4.2 Cloud Technology

One of the most important and mostly used storage systems is cloud technology. In the irrigation-based agro-systems, data is gathered and scrutinized using sensors, involving cloud technology. Studies found that data processing occurs within the cloud, which can be retrieved by the user. This technology is implicated in storing large-scale information, which eventually augments efficiency. The introduction of

Table 4.1 Wireless communication system used in smart irrigation practices (Khriji et al. 2021)

Communication protocols	Transmission range	Data rate	Power
Wi-Fi	10–200 m	2–54 mbps	Medium
4G LTE	<15 km	100 mbps	High
Zigbee	10–100 m	20–250 kbps	Low
LoRaWAN	5–10 km	0.3–50 kbps	Medium
BLE	30 m	1–24 mbps	Low

cloud technology has enabled large-scale agro-practitioners, including agricultural bodies to securely store and access information. Regarding its application in designing smart irrigation, cloud technology has been employed to construct algorithms and generate alerts, which have curbed potential hazards that could have infested otherwise (Sales et al. 2015). These alerts render it more practical to modify need-based planning approaches and execute appropriate measures well ahead of damage. In consensus to this, cloud technologies are utilized in irrigation practices to mitigate the prospective risks and thereby render the irrigation system smarter (Fernández-Ahumada et al. 2019). A recent study proposed a system engaging a cloud server, sensors, and a microcontroller to transmit real-time content through network. The collected data is regularly sent to an Internet of Things (IoT) cloud platform. Eventually, data is scrutinized and analyzed to produce results. Following the attainment of definitive soil moisture, the smart irrigation system is automated to water the garden under study. In the study, Arduino Mega 2560 microcontroller has been used in the control unit. Along with that, the sensors transmit data to a cloud channel where they are transformed into graphs and subsequently transferred for additional analysis. Furthermore, the sensors transmit data to a cloud channel, and the data gets converted into graphs for the ease of subsequent analytical study (Nanda et al. 2023).

In a cloud server, machine learning techniques may be employed. Further, information provides a directive to the system in determining if the water flow rate is to be amplified or diminished in order to optimize functional efficiency and minimize water wastage. Integrating Internet of Things (IoT) technologies into smart irrigation systems has the profound prospect to enhance agricultural productivity, with optimized water utilization (Reddy et al. 2024), a step toward SDG-6.

4.3 Big Data Optimization

Agriculture data analytics is an intricate system that involves a large amount of data and climatic parameters. Over the past decade, the field of agriculture has witnessed a revolutionary transformation in the area of big data analytics and machine learning. Big data analytics is a crucial technique used to analyze large amounts of data. Competent management of irrigation water is a formidable challenge in the way of achieving sustainable agricultural goals. This is contingent upon a multitude of climatic parameters pertaining to water, soil, and other allied conditions, which demands a robust modeling approach that is necessary for the accurate estimation of water requirements in agricultural fields (Saggi and Jain 2022). IoT systems generate a substantial array of data due to real-time monitoring of various parameters. Likewise, IoT irrigation systems too play a role in the generation of voluminous data. Even though big data are mostly useful in the analysis and standardize effective irrigation process as a whole, it is essential to prioritize the appropriate administrative control over the information. With the obstacles of managing large volumes of data, it is essential to prioritize sustainable practices in big data management. Several proposals for addressing this issue await on the way, including the elimination of surplus and redundant data, the use of blockchain technology, and the exclusive selection of relevant information (Obaideen et al. 2022).

4.4 Use of Sensors

Many ecological parameters relevant to agriculture, such as temperature, soil moisture, humidity, and light, are captured by radio satellites, sensors, and microcontrollers. Cloud servers, through the utilization of IoT technology, immediately obtain the information from the sensors. Concerned personnel across the globe can obtain them using a network-connected device and plan farming strategies accordingly. The employment of sensor-based applications in present-day farming practices offers cost-effective and environmentally benign solutions. An automated irrigation system applies wireless sensor networks and decision-support technologies to measure environmental factors in an agricultural area from a distant place (Li et al. 2020b). Various terminologies are associated with the irrigation practices that use sensors. These include precision farming, global positioning service (GPS), smart agriculture, variable rate technology (VRT), and site-specific crop management (Srinivasan 2006). The architecture of a sensor-based IoT system consists of the wireless sensor unit (WSU) and wireless information unit (WIU). The sensor units feature various types of sensors to quantify the moisture of the soil along with temperature, whereas the chief microcontroller receives sensor-derived data and eventually optimizes them (Dursun and Ozden 2011). ZigBee is then employed for transmitting information to WIU (Kalaivani et al. 2011). Yet, the on-field installation of sensing stations to establish wired connections with base stations is problematic due to the requirement for labor costs, and maintenance, and is not highly sensible because farm animals damage the wires. Here comes into play wireless networks, which can overcome various concerns and offer a better alternative to orthodox options. Wireless technologies, such as Bluetooth, Wi-Fi, and ZigBee, are utilized in sensor-based IoT approaches (Wall and King 2004) (Table 4.1). It has been reported by scientists that the utilization of sensors in sprinkler irrigation can boost water efficiency by up to 80–90% (Li et al. 2020b).

5 General Architecture of Smart Systems

The fundamental structure of IoT-enabled smart irrigation systems is layered architecture (Munir et al. 2019), primarily comprised of three primary layers. Some other researchers, like Ferrández-Pastor, have contributed to the introduction of a four-layered architecture. Ideally, architectural variation is established based on the layers of architectural units. For example, the levels in a three-layered architecture are perception, network, and application. A new layer called the service layer is inserted between the network and application layers in order to gather and analyze data through technologies like cloud computing (Soto-Garcia et al. 2013). On the other hand, the components of a four-layered architecture are things, edge, communication, and cloud infrastructure. The edge layer facilitates locating key applications, and at the same time, it performs fundamental control procedures. From a general perspective, the lower layer is comprised of actuators and sensor nodes, while the middle layer facilitates data transmission. Finally, the third tier of the architecture is made of cloud databases, services, or applications. Despite the presence of basic layered structures, they can exhibit distinct and diverse features. These specific architectural designs produce differentiation in the diverse components that they are made of. Mostly the designs are frequently organized into functional blocks, in accordance with the specific tasks and functional mechanism that need to be executed (Obaideen et al. 2022).

The design incorporates the deployment of various sensors, including moisture sensors for soil, temperature sensors, and leaf wetness sensors, and such others in agricultural fields (Table 4.2). The data retrieved from these sensors is then compared to pre-established threshold values of crop-specific physical parameters. The data is eventually transmitted through the Arduino Uno processor, which is connected to the data center via a GSM module. Derived information in the data center undergoes data analysis using data mining techniques, such as the Markov model. Finally, the diagnosis report on the physical parameters is sent to the user interface like an Android smartphone where the user can remotely take over the control of the irrigation system by modulating the Arduino to switch the motor pump on or off (Yashaswini et al. 2017) (Fig. 4.2).

In a proposed system for scrutinizing soil moisture and temperature using IoT-based sensor nodes in real time, the architecture consisted of five units: sensing, processing, front-end, actuation, and persistence. The sensing unit consists of many wireless nodes that integrate various sensors, including soil moisture and temperature sensors. On the other hand, the processing unit is made up of a receiver node

Table 4.2 Soil sensors that are utilized in agricultural sectors to control irrigation systems (Li et al. 2020b)

Sensor	Temperature	Moisture	Dielectric/ permittivity	Rain water	Conductivity
107-L temperature sensor	Yes	No	No	No	No
ECRN-100 high-REC rain	No	No	No	Yes	No
ECRN-50 low-REC rain	No	No	No	Yes	No
Tipping bucket rain gage	No	No	No	Yes	No
EC sensor (EC25)	Yes	Yes	No		No
MP406 soil moisture sensor	Yes	Yes	Yes	No	No
Pogo portable soil sensor	Yes	Yes	Yes	Yes	Yes
Hydra probe II soil sensor	Yes	Yes	Yes	Yes	Yes

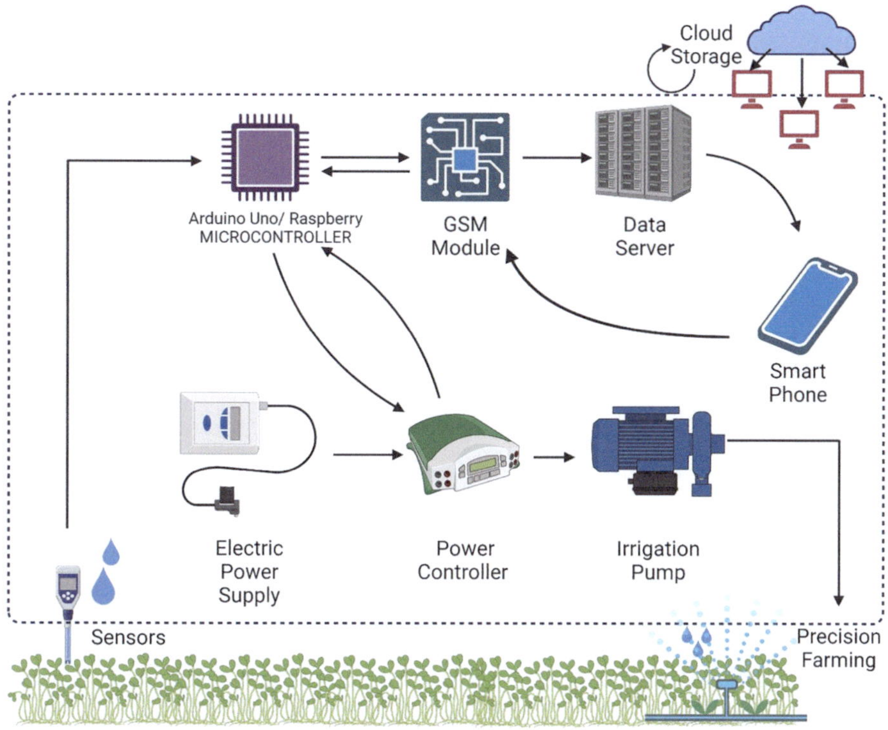

Fig. 4.2 Basic architecture of IoT-enabled smart irrigation system

that is serially linked to a Raspberry Pi. Message queuing telemetry transport (MQTT) communication protocol is generally applied to transmit the sensed data. A mobile application and a graphical web interface enable users to visualize data in real time. HTML/CSS and JavaScript are used to create web interface pages, which are hosted on a Raspberry Pi. Once the soil's moisture levels are above a specific threshold, the user receives a notification through either the mobile or web interface, enabling them to straightaway intervene to set the motor operation and minimize water and energy wastage (Khriji et al. 2021).

6 Benefits of IoT-Based Irrigation System

Deploying IoT smart irrigation systems has various advantages for residential lawns, commercial landscaping, and agricultural applications. IoT smart irrigation systems significantly reduce water consumption and promote water conservation. They rigorously fine-tune watering schedules and durations using real-time data from sensors to meet the water needs of plants. Plants are watered sparingly, as well as necessarily. These systems supply water at exact intervals to maintain appropriate soil moisture levels that promote plant growth. This method promotes the formation of stronger root systems, reduces plant strain, and increases overall plant health and vigor (Chen et al. 2022). IoT smart irrigation systems utilize real-time data to effectively regulate soil moisture levels that are ideal for promoting plant growth. The most recent innovations provide the delivery of water at the appropriate time and in the correct amount. This method encourages the development of more robust root systems, alleviates stress on plants, and enhances overall plant health and vitality. Smart irrigation systems minimize water wastage by employing precise control and automation, eliminating the necessity for human-based watering techniques. Consequently, households can anticipate reduced water expenses, agricultural workers can decrease their labor expenditures, and landscaping businesses can enhance their operational effectiveness. Controlling and overseeing IoT-based smart irrigation systems through smartphone applications or online interfaces offers consumers enhanced flexibility and convenience. Users can access real-time data, modify irrigation schedules, and address problems remotely, provided they have an Internet connection (Morchid et al. 2024). IoT smart irrigation systems in agricultural settings optimize water utilization and enhance crop yield by ensuring that plants receive sufficient water throughout their growth cycle. These technologies enhance production, improve crop quality, and increase profitability by reducing water shortages. Water conservation additionally fosters environmental sustainability by diminishing water runoff and the consequent leaching of fertilizers and pesticides. IoT-based smart irrigation systems mitigate the environmental consequences of irrigation by diminishing soil erosion and safeguarding water resources. Smart irrigation systems utilizing the IoT collect significant data on soil moisture levels, weather patterns, and watering patterns throughout the season. Through meticulous research, this dataset holds the capacity to reveal trends, enabling users to make informed decisions regarding water management and agricultural production, while also enhancing their irrigation practices (Obaideen et al. 2022).

7 Challenges for Smart Irrigation System

One of the difficulties in implementing a smart agriculture system is the process of incorporating sensors and connecting the data collected by these sensors to the analytics that drive automation and response actions. By incorporating data analytics, the agricultural industry may effectively decrease costs and enhance productivity by efficiently managing water, fertilizer, and light, resulting in more yields from the same land area. Smart techniques enable agricultural activities in smaller and more dispersed areas by utilizing remote monitoring, regardless of whether they are conducted indoors or outside. In order to effectively implement a smart agricultural system, it is important to establish a communications network capable of integrating a minimum number of sensors across an extensive expanse of farmland. To do this, it will be necessary to either utilize a third-party network provisioning service or establish a private network that includes access points and uplinks connected to a private backhaul network (Vikranth 2021). This backhaul network will direct all data traffic to centralized monitoring software or an analytics head-end system. The selection of the communications network is crucial because of the network's dependability. The topography of the terrain, as well as the wind and rain, might impact the performance of point-to-point communication technology used in some types of private networks. Mesh networking technology, employed by numerous utility firms, exhibits greater reliability and resilience in the face of network failures. This technology exhibits greater resilience to environmental conditions, but its deployment incurs higher costs (Feng and Mao 2019). Selecting sensor brands and sensor kinds is usually the least challenging aspect of constructing a smart agriculture system. Installing monitoring and automation software packages to enable functions like watering triggers, alerts, and notifications would necessitate software proficiency that may be difficult for certain agricultural specialists. For instance, evaluating the recommended frequency for irrigating a crop and the threshold for soil moisture level that should be reached before initiating a watering regimen can be a difficult task. Integrating software with sensors and agricultural equipment is laborious work that requires information on both seasonal conditions and crops (Dhanaraju et al. 2022). Finally, having a cloud-based network into play, there is always a possibility of hacking and data leakage on the grounds of the evolving darker side of technology.

8 Prospects and Conclusion

In the era of worldwide depletion of natural resources like water, it is essential to implement a combination of weather-based, soil-based, and crop-based monitoring tools in agricultural practices, along with a sound forecasting and control mechanism. It has to be taken into consideration that, unlike environmentally controlled trials in research, real-world agricultural irrigation practices encounter environmental and other uncertainties that require investigation. Future research needs to prioritize the innovation on control and monitoring techniques for irrigation in open-field agro-practices. On the other hand, water scarcity is a worldwide pressing matter and the implementation of a "smarter" irrigation system has become the day's need in the current environmental context. The ramifications of both Internet of Things (IoT) and sensor systems have been momentous in recent times. The Internet of Things (IoT) decreases overall expenditure, and curbs the manual labor, affording the chance to oversee and control the monitoring system from a distance. Wireless sensor networks (WSNs) also play a pivotal role in real-time monitoring for precision farming activities. Still, the IoT-based smart irrigation system has several avenues open for further improvisation. Emphasis should be given to address management and security concerns when executing smart irrigation systems. Ensuring the system security and safeguarding the data is crucial for reducing the dangers associated with digital threats. It is vital to take into account the environmental consequences of the irrigation system and ensure that they correspond to the Sustainable Development Goals to strengthen the three pillars: society, environment, and economy. Additionally, it is crucial to ensure that the operational expenses for IoT-based smart irrigation do not exceed the apparent benefits to sustain this innovative approach in the longer term. Finally, agronomy is still a global platform for employment. Will the encroachment of digitization and automation in irrigation render people living in agricultural practices unemployed in the future? This also remains an open area of serious discourse because we must balance all facets of Sustainable Development Goals since a conquest for clean water and sanitation must not go against the other goals of SDG, for example, conquest for the reduction of inequality. Thus IoT-based smarter irrigation system is going to be a promising trump card requiring judicial implementation to move one step forward toward the Sustainable Development Goals.

Acknowledgments Biorender.com was used for the preparation of figures for this manuscript. The authors acknowledge Swami Vivekananda Merit-cum-Means Scholarship, Department of Higher Education, Government of West Bengal, India, which was awarded to the author, Abhratanu Ganguly (WBP211651743312).

Conflict of Interest None.

References

Abagissa AT, Behura A, Pani SK (2018) IoT based smart agricultural device controlling system. In: 2018 Second International Conference on Inventive Communication and Computational Technologies (ICICCT). IEEE, pp 26–30

Adamala S, Raghuwanshi NS, Mishra A (2014) Development of surface irrigation systems design and evaluation software (SIDES). Comput Electron Agric 100:100–109

Bortolini L, Tolomio M (2019) Influence of irrigation frequency on radicchio (Cichorium Intybus L.) yield. Water 11(12):2473

Chen R, Li H, Wang J, Guo X, Song Z (2022) Comparisons of spray characteristics between non-circular and circular nozzles with rotating sprinklers. Appl Eng Agric 38(1):61–75

Chikankar PB, Mehetre D, Das S (2015) An automatic irrigation system using ZigBee in wireless sensor network. In: 2015 International Conference on Pervasive Computing (ICPC). IEEE, pp 1–5

Dasgupta A, Daruka A, Pandey A, Bose A, Mukherjee S, Saha S (2019) Smart irrigation: IOT-based irrigation monitoring system. In: Proceedings of International Ethical Hacking Conference 2018: eHaCON 2018, Kolkata, India. Springer Singapore, pp 395–403

Dhanaraju M, Chenniappan P, Ramalingam K, Pazhanivelan S, Kaliaperumal R (2022) Smart farming: Internet of Things (IoT)-based sustainable agriculture. Agriculture 12(10):1745

Dursun M, Ozden S (2011) A wireless application of drip irrigation automation supported by soil moisture sensors. Sci Res Essays 6(7):1573–1582

Feng M, Mao S (2019) Dealing with limited backhaul capacity in millimeter-wave systems: a deep reinforcement learning approach. IEEE Commun Mag 57(3):50–55

Fernández-Ahumada LM, Ramírez-Faz J, Torres-Romero M, López-Luque R (2019) Proposal for the design of monitoring and operating irrigation networks based on IoT, cloud computing and free hardware technologies. Sensors 19(10):2318

Gałęzewski L, Jaskulska I, Jaskulski D, Lewandowski A, Szypłowska A, Wilczek A, Szczepańczyk M (2021) Analysis of the need for soil moisture, salinity and temperature sensing in agriculture: a case study in Poland. Sci Rep 11(1):16660

Gamal Y, Soltan A, Said LA, Madian AH, Radwan AG (2023) Smart irrigation systems: overview. IEEE Access

García L, Parra L, Jimenez JM, Lloret J, Lorenz P (2020) IoT-based smart irrigation systems: an overview on the recent trends on sensors and IoT systems for irrigation in precision agriculture. Sensors 20(4):1042

Goap A, Sharma D, Shukla AK, Krishna CR (2018) An IoT based smart irrigation management system using Machine learning and open source technologies. Comput Electron Agric 155:41–49

Ho L, Alonso A, Forio MAE, Vanclooster M, Goethals PL (2020) Water research in support of the Sustainable Development Goal 6: a case study in Belgium. J Clean Prod 277:124082

Kalaivani T, Allirani A, Priya P (2011) A survey on Zigbee based wireless sensor networks in agriculture. In: 3rd International Conference on Trendz in Information Sciences & Computing (TISC2011). IEEE, pp 85–89

Khriji S, El Houssaini D, Kammoun I, Kanoun O (2021) Precision irrigation: an IoT-enabled wireless sensor network for smart irrigation systems. In: Women in precision agriculture: technological breakthroughs, challenges and aspirations for a prosperous and sustainable future, pp 107–129

Lamm FR (2002) Advantages and disadvantages of subsurface drip irrigation. In: International Meeting on Advances in Drip/Micro Irrigation, Puerto de La Cruz, Tenerife, Canary Islands, pp 1–13

Li J, Fei L, Li S, Xue C, Shi Z, Hinkelmann R (2020a) Development of "water-suitable" agriculture based on a statistical analysis of factors affecting irrigation water demand. Sci Total Environ 744:140986

Li W, Awais M, Ru W, Shi W, Ajmal M, Uddin S, Liu C (2020b) Review of sensor network-based irrigation systems using IoT and remote sensing. Adv Meteorol 2020(1):8396164

Morchid A, Alblushi IGM, Khalid HM, El Alami R, Sitaramanan SR, Muyeen SM (2024) High-technology agriculture system to enhance food security: a concept of smart irrigation system using internet of things and cloud computing. J Saudi Soc Agric Sci, 10.1016/j.jssas.2024.02.001

Mostafa EF, Emara EIR, Bakr RH, Abdel Azim KG, Hamed LMM (2024) Enhancing wheat productivity utilizing nano-and bio-fertilizers in water-scarce Sinai, Egypt, under different irrigation systems. Water Sci 38(1):209–224

Mouha RARA (2021) Internet of things (IoT). J Data Anal Inform Process 9(02):77

Muluneh A, Tadesse T, Girma R (2022) Assessing potential land suitable for surface irrigation using GIS and AHP techniques in the Rift Valley Lakes Basin, Ethiopia. Sustai Water Resour Manage 8(2):46

Munir MS, Bajwa IS, Cheema SM (2019) An intelligent and secure smart watering system using fuzzy logic and blockchain. Comput Electr Eng 77:109–119

Nanda I, Singh M, Khatua L (2023) Automated irrigation system using IoT cloud computing. Int J Intell Syst Appl Eng 11(2s):360–365

Nawandar NK, Satpute VR (2019) IoT based low cost and intelligent module for smart irrigation system. Comput Electron Agric 162:979–990

Nikolidakis SA, Kandris D, Vergados DD, Douligeris C (2015) Energy efficient automated control of irrigation in agriculture by using wireless sensor networks. Comput Electron Agric 113:154–163

Obaideen K, Yousef BA, AlMallahi MN, Tan YC, Mahmoud M, Jaber H, Ramadan M (2022) An overview of smart irrigation systems using IoT. Energy Nexus 7:100124

Reddy VS, Harivardhagini S, Sreelakshmi G (2024) IoT and cloud based sustainable smart irrigation system. In: E3S web of conferences, vol 472. EDP Sciences, p 01026

Saggi MK, Jain S (2022) A survey towards decision support system on smart irrigation scheduling using machine learning approaches. Arch Comput Methods Eng 29(6):4455–4478

Salehi M (2022) Global water shortage and potable water safety; today's concern and tomorrow's crisis. Environ Int 158:106936

Sales N, Remédios O, Arsenio A (2015) Wireless sensor and actuator system for smart irrigation on the cloud. In: 2015 IEEE 2nd World Forum on internet of things (WF-IoT). IEEE, pp 693–698

Segovia-Hernández JG, Hernández S, Cossío-Vargas E, Sánchez-Ramírez E (2023) Challenges and opportunities in process intensification to achieve the UN's 2030 agenda: goals 6, 7, 9, 12 and 13. Chem Eng Process:109507

Shi K, Lu T, Zheng W, Zhang X, Zhangzhong L (2022) A review of the category, mechanism, and controlling methods of chemical clogging in drip irrigation system. Agriculture 12(2):202

Soto-Garcia M, Del-Amor-Saavedra P, Martin-Gorriz B, Martínez-Alvarez V (2013) The role of information and communication technologies in the modernisation of water user associations' management. Comput Electron Agric 98:121–130

Srinivasan A (ed) (2006) Handbook of precision agriculture: principles and applications. CRC Press

Vikranth K (2021) An implementation of IoT and data analytics in smart agricultural system–a systematic literature review. Int J Manage Technol Soc Sci 6(1):41–70

Wall RW, King BA (2004) Incorporating plug and play technology into measurement and control systems for irrigation management. In: 2004 ASAE Annual Meeting. American Society of Agricultural and Biological Engineers, Paper number 042189

Wang M, Bodirsky BL, Rijneveld R, Beier F, Bak MP, Batool M, Droppers B, Popp A, van Vliet MTH, Strokal M (2024) A triple increase in global river basins with water scarcity due to future pollution. Nat Commun 15(1):880

Yashaswini LS, Vani HU, Sinchana HN, Kumar N (2017, September) Smart automated irrigation system with disease prediction. In: 2017 IEEE International Conference on Power, Control, Signals and Instrumentation Engineering (ICPCSI). IEEE, pp 422–427

Scope of AI in Future Farming

5

Ersin Elbasi, Elda Cina, Nour Moustafa, Aymen I. Zreikat, Ahmed Shdefat, and Ahmet E. Topcu

Abstract

The integration of artificial intelligence (AI) technologies into agriculture is transforming the sector by addressing critical challenges such as food security, sustainability, and climate resilience. This chapter examines the latest trends and applications of AI in agriculture, including precision farming, autonomous machinery, and disease detection. By leveraging advanced tools such as machine learning, computer vision, and the Internet of Things (IoT), AI enhances decision-making, optimizes resource utilization, and boosts productivity. Moreover, AI-driven innovations, such as automated irrigation systems and predictive analytics, promote sustainable farming practices while minimizing environmental impact. Despite its transformative potential, the widespread adoption of AI faces several challenges, including high implementation costs, data management complexities, and the need for farmer training. The chapter emphasizes the importance of interdisciplinary collaboration and the development of cost-effective, scalable solutions to ensure equitable access to AI technologies. It concludes by envisioning a future where AI plays a central role in achieving global food security and fostering sustainable agricultural systems.

Keywords

Agriculture · Artificial Intelligence · Computer Vision · Machine Learning · Internet of Things · Sustainability

E. Elbasi (✉) · E. Cina · N. Moustafa · A. I. Zreikat · A. Shdefat · A. E. Topcu
College of Engineering and Technology, American University of the Middle East, Egaila, Kuwait
e-mail: ersin.elbasi@aum.edu.kw; Elda.Cina@aum.edu.kw; Nour.Moustafa@aum.edu.kw; Aymen.Zreikat@aum.edu.kw; Ahmed.shdefat@aum.edu.kw; Ahmet.topcu@aum.edu.kw

R. E. Krishnankutty et al. (eds.), *Smart Agriculture*,
https://doi.org/10.1007/978-981-95-8785-8_5

1 Introduction

Over the past decade, AI has emerged as a transformative force across various industries, including agriculture. This transformation is having a profound impact, as new technological advancements are reshaping or even replacing traditionally manual and labor-intensive processes worldwide. The integration of AI technologies is addressing some of the most pressing global challenges, such as population growth, climate change, resource constraints, and sustainability in agriculture. The adoption of precision farming not only boosts productivity but also improves farm management, accelerates decision-making, and reduces carbon emissions and other environmental impacts (Kumar et al. 2024; Rattan et al. 2024; Sivakumar et al. 2023). In this context, AI is set to usher in an unprecedented phase of transformation in the agricultural sector—one that was previously unimaginable.

One of the primary factors of AI adoption in agriculture is the necessity to increase food production while minimizing resource utilization. According to the United Nations Food and Agriculture Organization (FAO), the worldwide population will reach 10 billion by 2050, requiring a 70% rise in food production to satisfy the demand (FAO 2017). So far multiple AI solutions are being developed and tested to assist precision farming. These solutions use IoT technology, networking, and machine learning techniques to assist farmers in enhancing their crop yields, effectively managing resources like water and fertilizers, and monitoring environmental conditions with improved accuracy (Sharma 2021). This guarantees that each input is utilized more efficiently, resulting in sustainable and lucrative farming methods.

- IoT devices—Drones equipped with sensors, capture data on crop health, moisture levels, and soil conditions from the air; ground sensors placed throughout the field and designed to monitor soil moisture, nutrient levels, and temperature in real time.
- Data processing units—including network transmission, cloud computing, and machine learning algorithms.
- Automated irrigation system—This system is shown with pipes and nozzles, automatically adjusting water distribution based on data received from the sensors.

AI solutions are being applied in various activities to improve agriculture. The study by Elbasi et al. lists a considerable number of activities such as crop and soil monitoring, disease diagnosis, automating manual activities through robots, crop yield prediction, predictive insight aiding decision-making, and intelligent spraying (Elbasi et al. 2023a).

Traditionally, identifying crop needs and diseases in crops or infestations requires significant manual labor and expertise. By the utilization of AI tools, however, large volumes of data can be easily processed and analyzed from multiple sources such as satellite imagery, drones, and sensors to detect crop anomalies early. For instance, computer vision and machine learning models can identify patterns in plant health, detecting diseases and pests before they become widespread (Kamilaris and

Prenafeta-Boldú 2018). This early intervention is crucial for preventing yield loss and reducing reliance on chemical pesticides, which can have harmful effects on the environment and human health. Using machine learning models for yield prediction, pest detection, and optimizing agricultural processes can enhance precision agriculture by optimizing inputs, reducing costs, and increasing sustainability in farming practices (Liakos et al. 2018).

An earlier effort of AI to improve agriculture is in the automation of farm processes. Autonomous tractors and other machinery are fast becoming common farm equipment; Utilizing AI algorithms, this machinery can operate continuously, almost independently of human intervention. Such machines use GPS, and real-time sensor data, along with machine learning models to perform tasks such as planting, plowing, or harvesting with high accuracy. AI-driven robots and automated machinery perform tasks such as planting, harvesting, and weeding, which reduces labor costs and increases operational efficiency (Akter et al. 2024; Sharma and Shivandu 2024). According to a report by KBV research, the market size of artificial intelligence has an enormous growth considering 682.1 million in 2018 to a predicted value of 4.9 billion by 2028 (KBV 2023). Another report from Markets and Markets shows that the global market for agricultural robots is experiencing prominent growth from $16.6 billion in 2024 to $51.0 billion by 2029, highlighting the rapid pace of AI adoption in farming machinery (Markets and Markets 2023).

While AI is minimizing manual labor and reducing human errors in agricultural processes, it is accelerating the response time and accuracy in decision-making processes while utilizing the use of big data (Rabhi et al. 2021) and predictive analytics (Sudha and Mohan 2019; Wolfert et al. 2017). Farmers are taking advantage of AI platforms to analyze historical data, weather patterns, soil conditions, crop performance, and other elements to make informed decisions about the type of crops to plant, planting schedules, irrigation management, and harvest timing (Abinaya et al. 2024; Bendre et al. 2016; Chen et al. 2024). These platforms integrate various data sources to provide actionable insights, helping farmers adapt to changing environmental conditions and market demands.

Lastly, predictive analytics play a crucial role in anticipating future challenges, such as climate change effects, weather events, or pest outbreaks, allowing for proactive measures to mitigate risks.

The rest of this chapter continues as follows: Section 2 covers a background and literature review where fundamental and recent studies will show the trend of AI in the agriculture sphere. Section 3 includes the novel trends in agriculture followed by AI techniques that are being deployed nowadays including the Internet of Things, computer vision, and machine learning in Sect. 4. Section 5 will analyze the benefits of AI in agriculture, its challenges, and opportunities for the future. Final remarks are included in Sect. 6.

2 Background and Literature Review

The obstacles to eliminating hunger and food insecurity continue to increase, with the COVID-19 pandemic exposing vulnerabilities in our agri-food systems and disparities in our communities. The Food and Agriculture Organization (FAO) (Food and Agriculture Organization of the United Nations 2024) defines agri-food systems as all the interconnected activities and actors involved in obtaining food from fields to forks (Alexandratos and Bruinsma 2012). This situation becomes increasingly critical as the demand for food increases. FAO predicts that by 2050 there will be approximately 10 billion people, leading to a 70% increase in food demand. Therefore, artificial intelligence (AI) methods used in farming can enhance agricultural operations by boosting the resilience of the food system. Agriculture has been the backbone of human civilization for millennia, providing sustenance and contributing to economic development (Benefits of AI in agriculture 2024). Therefore, innovative ideas are being introduced by AI in every industry, and agriculture is no exception. These innovations are becoming increasingly essential for many global challenges that threaten the sustainability of our food systems, such as climate change, population growth, and resource scarcity. Introducing AI solves many challenges and helps diminish many obstacles to conventional agricultural practices.

The applications of AI in agriculture have been studied from different perspectives [4–13]. In Tetarwal et al. (2024), the authors investigated the problem of predicting crop prices by employing ARIMA modeling within time-series forecasting (TSF) to predict crop prices. The main objectives were to minimize risks associated with price fluctuations and provide practical decision support for farmers, thereby aiding in optimal crop choices and planting times. This chapter emphasizes the crucial role of precise price forecasts for effective risk management by integrating these insights into decision-making processes. Through these processes, farmers can navigate market uncertainties and optimize their agricultural practices by applying these insights to their decision-making processes. Holzinger et al. (2024) investigated the advancements in the application of human-centered AI, focusing on their application in agricultural processes. These technological advancements enhance crop yields, minimize production, reduce labor and resource wastage, waste, and optimize the farm-to-consumer supply chain. The potential of this discussion is to explore how AI can enhance human decision-making, thereby fostering decision-making in the agri-food sector for a sustainable, efficient, and resilient agri-food sector. In addition, a framework for Agriculture 5.0 is presented in this study. The idea of Agriculture 5.0 is to extend Agriculture 4.0, focusing on more advanced objectives such as a stable climate, healthy plants and soils, environmentally friendly cultivation systems, sustainable plant products, and satisfying customers.

The recent growth in artificial intelligence technology allows farmers to deploy autonomous farming machinery equipment and make better predictions of the future based on present and past conditions to minimize informed decisions about crop disease care. Therefore, in Qazi et al. (2022), the authors conducted a survey to offer detailed tutorials on insights into the available advancements in the field of smart agriculture systems using Internet of Things (IoT) technologies and artificial intelligence techniques. Besides, they provided a critical review of these two available technologies and challenges in AI, their widespread deployment: critical evaluation. Additionally, they conducted an in-depth discussion about the future trends, including both technological and social, when smart agriculture systems will be widely adopted by farmers globally.

According to Bertoglio et al. (2021), some authors in the literature have claimed that we are witnessing a digital agricultural revolution (DAR) that will promote sustainable farming. Therefore, in this study, we conducted a literature review to seek evidence of the ongoing DAR process, as well as to clarify its origins, meaning, and future directions. The authors used bibliometric analysis to conduct an impartial and reproducible literature review. They studied 4995 articles by querying the Web of Science database from 2012 to 2019, and they analyzed the dataset to answer three specific research questions: what is the scope of DAR-related terminology? What are the essential publications and the most prominent journals, institutes, and countries? What are the major research streams and new topics? Afterward, the authors grouped the selected publications into five categories based on the given keywords: climate-smart agriculture (CSA), site-specific management (SSM), remote sensing (RS), IoT, and AI. It is understood from this study that the authors could not provide strong evidence of how DAR promotes sustainable farming. Therefore, further investigations are required to justify this.

AI and the Internet of Things have had devastating effects on almost every industry in the world, particularly agriculture. Therefore, there are many new opportunities because of the significant changes being made to current farming practices. In this regard, and to address the issues related to food and agriculture for smart and sustainable agriculture, Anwarul et al. (2022) suggested an IoT- and AI-assisted framework. The proposed framework is organized into four operational layers: monitoring device placement, implementation, data transfer, and processing. The proposed system helps improve crop output by monitoring soil moisture, air, weather, temperature, and pest control without human intervention. Additionally, the authors compared their proposed scheme with existing techniques for agriculture automation systems and demonstrated the effectiveness of their scheme. In Sugunadevi et al. (2023), the authors provide an overview of the use of the Internet of Things in agriculture. The conducted overview and results demonstrate how farming has progressed from agriculture to smart agriculture, which effectively uses many sensors to increase the output and quality of agricultural products. Case studies that include India as an example demonstrate how technologies affect agriculture and, ultimately, food production. Related to this topic, the research in Patil et al. (2024) reviews various machine learning methods for plant leaf disease detection. The evolution of approaches, difficulties faced, and state-of-the-art methods in this area are the main topics of this study. Plant disease identification using machine learning has potential for early diagnosis and efficient disease management. This review summarizes the research results, identifies important developments, and suggests possible directions for further study to support continuous efforts in smart farming and precision agriculture. By combining the many popular AI and IoT technologies, agriculture becomes more sustainable in addition to becoming more intelligent and effective. Modern technologies were discussed in Hashni et al. (2022), with a focus on the implications and implementation issues in the agriculture industry. Moreover, various challenges related to smart farming were discussed and addressed in this study, such as technology selection, connectivity in rural areas, higher operational costs, the quality and durability of sensors, prediction accuracy, and data security and privacy.

The current research trend is to maximize crop productivity by using advanced agricultural technology, such as AI and the Internet of Things. Considering some factors, such as soil temperature, pH, and water availability, can result in higher yields. Therefore, an agriculture cultivation recommender and smart irrigation system (ACRSIS) using an IoT-based precision agriculture system was proposed in Rahman et al. (2023) to offer useful recommendations to farmers interested in utilizing IoT technology to increase crop yields. The recommended approach is effective for keeping track of agricultural produce. The proposed system can also predict future irrigation planning needs. Predictors include soil parameters such as temperature, humidity, and moisture content. The findings of this experiment suggest that intelligent irrigation can boost crop yield while consuming less water. In addition, artificial intelligence (AI) methods like machine learning and deep learning have been used to identify, forecast, and create recommendation systems for plant diseases. The past 20 years have seen great effort in this field, which may affect farmers' livelihoods in the years to come. The methodical multifold examination and analysis of such work, with an emphasis on current AI methods created to fight plant diseases, is presented in Rani et al. (2023). This article addresses the numerous issues that farmers confront and how AI can help. The study examines several AI applications in agriculture and contemporary developments. In addition, this article discusses recent developments in AI for plant disease diagnosis, such as identification model improvement (IMI), few shot learning (FSL), generative adversarial networks (GANs), and self-supervised learning (SSL).

3 Novel Trends in Agriculture

Recent advancements in artificial intelligence (AI) have significantly reshaped the landscape of modern agriculture. According to Kisliuk et al. (2023), AI applications in agriculture have evolved from basic automation to more sophisticated systems capable of real-time decision-making. These innovations cover areas like precision farming, where AI-driven data analytics help farmers optimize crop yields and reduce waste. Kisliuk et al. (2023) highlight the integration of AI technologies with Internet of Things (IoT) devices, enabling continuous monitoring of soil and plant conditions, thus paving the way for adaptive agricultural practices that ensure resource efficiency. Additionally, autonomous machinery, such as drones and robotic harvesters, are transforming traditional farming methods, offering a glimpse into the future of agriculture (Elbasi et al. 2024a, b).

Precision agriculture, driven by AI and data analytics, continues to be a major trend in transforming farming practices. The work of Kisliuk et al. (2023) elaborates how precision farming systems utilize satellite imagery, soil sensors, and predictive analytics to make data-driven decisions. These technologies help optimize the use of water, fertilizers, and pesticides, reducing environmental impacts while maximizing crop yields. By enabling real-time monitoring of farm conditions, precision agriculture also supports adaptive management strategies, allowing farmers to respond quickly to changing environmental factors. This approach not only improves the efficiency of resource use but also contributes to the overall sustainability of agriculture.

In the effort to enhance agricultural outcomes, neutrosophic science has emerged as a novel analytical framework. Paolo (2025) explores the potential of tree soft sets and neutrosophic sets in understanding soil organic transformations in urban agriculture systems. By applying these advanced mathematical tools, researchers can better account for uncertainties and variabilities in soil properties, which are critical for urban farming. This methodology enables more accurate predictions of soil behavior, leading to enhanced management strategies that improve crop health and sustainability. The research in [2] highlights the significance of adopting complex scientific models to address the unique challenges of urban agricultural ecosystems, particularly in densely populated areas. Urban agriculture is becoming increasingly relevant as cities seek to reduce their ecological footprints and provide local food sources. The study by Paolo (2025) emphasizes the significance of adopting innovative soil management practices in urban environments. The application of tree soft sets and neutrosophic models helps researchers analyze the complex interactions between organic matter and urban soil systems. These models provide insights into the nutrient cycles and soil health dynamics that are crucial for successful urban farming. By improving our understanding of soil processes, urban farmers can implement practices that enhance soil fertility and support sustainable food production, even in constrained environments.

Another groundbreaking trend in agriculture involves the integration of neutrosophic sets with deep learning techniques for plant disease recognition. The work by Samah et al. (2025) presents a model using Q neutrosophic soft relations to enhance the precision of disease detection in pepper plants. This work combines the strengths of neutrosophic logic in handling uncertainty with the robust pattern recognition capabilities of deep learning algorithms. The research emphasizes how this hybrid approach can aid farmers in Saudi Arabia by providing real-time, accurate diagnoses, thereby reducing crop losses and promoting sustainable agricultural practices. This technology has the potential to be expanded globally, providing a powerful tool for managing crop health across various environments.

A comprehensive framework for driving agricultural innovation is essential for long-term sustainability. Davoodi et al.'s (2024) research proposes a roadmap for fostering agricultural growth through technology and policy interventions. This roadmap highlights the integration of digital tools, genetic innovations, and climate-smart practices to achieve a resilient agricultural sector. The authors argue that a multifaceted method, which includes research investments and collaboration between governments and private stakeholders, is necessary to unlock the full potential of agricultural technologies. Furthermore, the roadmap stresses the importance of education and training programs to ensure that farmers are equipped with the skills needed to adopt and benefit from these advancements.

Looking forward, the agricultural sector must continue to adopt cutting-edge technologies and frameworks to address global challenges such as climate change and food security. In Davoodi et al.'s (2024) work, they stress on the need for a holistic approach that combines technological advancements with robust policy support. Collaboration between different sectors, including agriculture, technology, and environmental sciences, is essential for creating resilient agricultural systems. The future of agriculture lies in the seamless integration of AI, eutrosophic models, and innovative farming practices, as highlighted across the referenced studies. Together, these novel trends provide a roadmap for a more sustainable and efficient agricultural industry.

The agricultural sector in the Philippines as instance has also witnessed a surge of innovations between 2018 and 2023, as documented by Taer et al. (2024). The review paper identifies key areas of development, including the introduction of drought-resistant crop varieties and the adoption of integrated pest management systems. These advancements have significantly bolstered the country's agricultural productivity and resilience to climate change. The study also outlines successful case studies where community-based initiatives have led to widespread acceptance and implementation of new agricultural technologies. By showcasing these success stories, the authors demonstrate the potential of grassroots efforts to complement national policies, creating a sustainable model for agricultural development.

4 AI Techniques in Agriculture

There are various solutions to improve the productivity of smart farms and solve different disease problems in farms. Farms need to adopt new technologies because of reduced fertilizer lands and lack of rain in vast areas. AI technology plays an impressive role in agriculture compared to the other technical solutions for prosperous agriculture. Many farmers use AI advances in their fields to boost earnings by lowering risks and increasing efficiency. One helpful technology AI can use is the IoT, which feeds created data into machine learning algorithms to resolve the issues effectively.

4.1 Internet of Things

Nowadays, it is critical to have new models in agriculture because of global warming, climate changes, lack of rain needed for irrigation, reducing farmlands for new settlers, and vulnerabilities of food-chain supply during energy and political crises. The regular classical solutions used in the 1990s need to be revised to tackle these issues in farmland. This type of critical problem leads to sophisticated solutions having high efficiency and increased productivity in farmlands called smart farms.

AI solutions in agriculture help this improvement quickly and rapidly by using IoT devices and solutions along the fields.

The main advantage of IoT devices is that they are capable of prolonged usage of devices using built-in batteries, small and cheap components, and reasonable prices. It is also easy to collect data at a high frequency and adjust it accordingly to the data range. The IoT solution market was expected to reach almost US$7 billion by 2025, according to the report by Statista (Vailshery 2023).

They are applying various ML algorithms to collect data through IoT components using sensors placed in smart farms. It is especially critical to have innovative IoT models with starvation and poor irrigation territories such as Africa. The models defined in the work by Effah et al. (2024) indicate that IoT-enabled models using AI provide efficient results to combat the limitations of agriculture. Another study to reduce water consumption in irrigation of the field (Morchid et al. 2024) has a promising result of reducing it by 70% compared to the regular solutions. The sustainability models using IoT-based solutions minimize the risk of irrigation problems.

Drones can be used in smart farms in agricultural environments, using various capabilities such as monitoring crop health, harvesting prediction, and spotting unrevealed diseases. This technology reduces the cost of farming and increases productivity in multiple lands (Islam et al. 2021). Other two works categorize the use of drones in water stress monitoring, management, and detection of pests, weeds, disease, spraying, and nutrient disorders using different types of sensors in drones (He et al. 2021; Rejeb et al. 2022).

Data from IoT devices is transmitted to various servers and processed using AI techniques. Various IoT devices play a critical role in farmers generating different types of data using drones, sensors, and video or image processors on numerous lands. Data quality and diversity are essential to build the suitable and best models using machine learning algorithms. Also, captured images can effectively control disease and improve the quality of new feed seedlings to estimate farm food production.

In many agricultural fields today, the IoT offers the most crucial technological advancement to make conventional farming more intelligent, effective, and sustainable. AI can be used to develop smart farms that use various data sources, including real time, picture, and automation data. This will increase productivity, lower expenses, and, most importantly, benefit the environment.

4.2 Computer Vision

The growing global need for sustainable agriculture has spurred technological advancements aimed at optimizing farming practices, enhancing yields, and reducing environmental impacts. Among these innovations, computer vision—a branch of artificial intelligence (AI)—has emerged as a game-changer in precision agriculture. By enabling machines to analyze and interpret visual data from agricultural settings, computer vision has driven progress in areas such as crop monitoring, pest control, and automation (Elbasi et al. 2023a, b).

One of the most significant applications of computer vision in agriculture is crop health monitoring. Using imagery from drones, satellites, and cameras, farmers can evaluate plant conditions, detect nutrient deficiencies, and identify environmental stresses like drought or heat (Sun et al. 2022). These capabilities enable early intervention, often preventing substantial losses by addressing issues before they escalate (Faragó et al. 2020). Additionally, in pest and disease management, computer vision has demonstrated remarkable effectiveness. For example, convolutional neural networks (CNNs) can classify diseases and detect pests in crops like wheat and cotton with over 90% accuracy (Maharlooei et al. 2021). Furthermore, real-time insect population monitoring through advanced object detection algorithms facilitates targeted pesticide application, minimizing environmental harm. Figure 5.1 demonstrates computer vision application in the agriculture sector in monitoring, prediction, and logistic and remote sensing (Lu and Young 2020).

Harvesting and yield estimation are also critical areas where computer vision plays a vital role. Autonomous systems equipped with vision-based sensors can identify ripe fruits and vegetables, enabling automated harvesting (Zou et al. 2024). These technologies utilize deep learning models to improve yield estimation accuracy, which is essential for effective supply chain management.

Looking ahead, the future of computer vision in agriculture hinges on enhancing the reliability and affordability of these technologies. Integrating multispectral and hyperspectral imaging with AI algorithms can improve the detection of plant traits, pests, and diseases under various environmental conditions (Sun et al. 2022). Moreover, advancements in edge computing offer the potential for real-time data processing, increasing the accessibility of these tools for farmers globally. To tackle specific agricultural challenges, several computer vision techniques have been widely adopted:

- Image classification: Powered by CNNs, these methods are used for plant species identification, disease detection, and health classification (Zou et al. 2024).

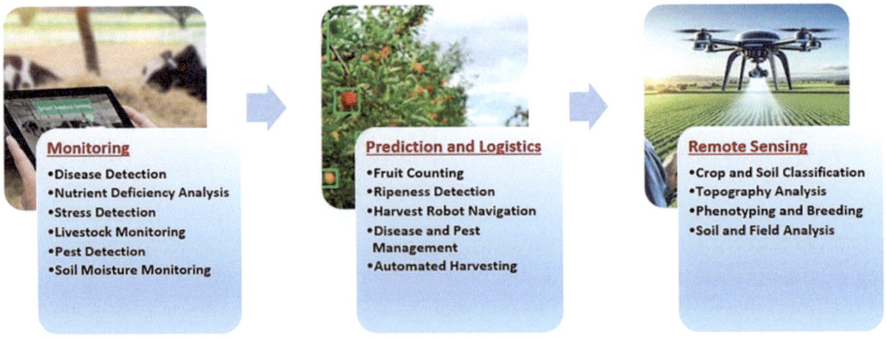

Fig. 5.1 Computer vision applications in agriculture

- Object detection: Algorithms such as Faster R-CNN, YOLO (You Only Look Once), and SSD (single-shot detector) enable the identification of fruits, pests, or weeds within images (Maharlooei et al. 2021).
- Semantic segmentation: This technique divides images into regions based on specific classes like vegetation, soil, or crops, aiding in tasks like weed detection and precision irrigation (Sun et al. 2022).
- Feature extraction: Techniques like edge detection and texture analysis are used to quantify traits such as leaf shape, size, and color, which are essential for plant phenotyping.
- Anomaly detection: Unsupervised learning methods identify deviations from normal patterns, such as signs of disease or pest infestation, allowing for early intervention (Faragó et al. 2020).
- Motion detection: In livestock monitoring, motion detection tracks animal behavior, enabling early identification of health issues and monitoring feeding patterns (Maharlooei et al. 2021).

By combining these techniques, computer vision is revolutionizing agriculture, offering farmers powerful tools to improve productivity, reduce waste, and ensure sustainability.

4.3 Machine Learning

The increasing global demand for sustainable agricultural practices has led to the widespread adoption of advanced technologies aimed at improving productivity, enhancing resource efficiency, and promoting environmental conservation. Among these technologies, machine learning (ML)—a subset of artificial intelligence (AI)—has emerged as a powerful tool for addressing key agricultural challenges, including crop monitoring, disease detection, pest management, and yield prediction. By analyzing vast datasets and identifying intricate patterns, ML models offer actionable insights that go beyond the capabilities of traditional farming methods (Kamilaris and Prenafeta-Boldú 2018). Coupled with innovations in remote sensing, automation, and IoT devices, these techniques are fostering the development of precision agriculture, empowering farmers to make data-driven decisions that optimize resource usage and boost productivity.

Machine learning has proven versatile in addressing a wide range of agricultural needs. Supervised learning techniques excel in tasks like crop classification, while unsupervised methods aid in soil-type clustering. Furthermore, cutting-edge approaches, such as deep learning and reinforcement learning, enable autonomous systems like harvesting robots and drones for real-time crop monitoring. As agriculture confronts pressing challenges like climate change and food security, machine learning offers a pathway to greater efficiency and sustainability (Patel et al. 2021; Liakos et al. 2018). Several machine learning technologies are used in agriculture as follows:

- Supervised learning: Supervised learning models are trained using labeled data to predict specific outcomes. In agriculture, these models—such as support vector machines (SVMs) and random forests—are commonly used for crop classification, disease detection, and yield forecasting. For example, SVMs have achieved high accuracy in identifying wheat diseases from spectral imagery (Pantazi et al. 2016). Similarly, random forests have been applied to predict crop yields based on environmental variables like temperature, soil moisture, and rainfall.
- Unsupervised learning: Unsupervised learning methods, including k-means clustering and hierarchical clustering, group similar data points without requiring labeled inputs. These techniques are valuable in applications such as soil-type classification and agricultural field segmentation (Jin et al. 2020). They also play a critical role in analyzing multispectral and hyperspectral data, enabling precise identification of crop zones and field variability.
- Deep learning: Deep learning, a specialized branch of ML, leverages neural networks with multiple layers to process complex data. Convolutional neural networks (CNNs) are widely employed in image-based applications, such as identifying crop diseases or detecting pests with high accuracy (Zhang et al. 2020). Similarly, recurrent neural networks (RNNs) are used for time-series data analysis, enabling tasks such as weather forecasting and its impact on crop growth.
- Reinforcement learning: Reinforcement learning (RL) is increasingly used to optimize dynamic agricultural processes, such as irrigation scheduling and autonomous machinery operations. These algorithms learn by trial and error, identifying strategies that maximize efficiency. For example, RL has been utilized to develop autonomous tractors capable of performing tasks like sowing and spraying (Zou et al. 2024).
- Hybrid techniques: Combining multiple ML approaches often produces superior results. For instance, integrating deep learning with reinforcement learning has enhanced robotic fruit harvesting systems (Patel et al. 2021). Similarly, hybrid models that merge supervised and unsupervised learning have been employed to improve the accuracy of crop monitoring and pest detection systems.

5 Benefits, Challenges, and Future

This integration of AI into agriculture is fostering a new era of innovation, but it also presents unique challenges and opportunities for future research. In the following section, we explore the benefits, challenges, and future work of AI in agriculture, providing insights into its transformative potential.

5.1 Benefits of AI in Agriculture

The primary benefit of AI in agriculture lies in its ability to optimize farming operations. AI technologies, such as machine learning models and neural networks, are instrumental in analyzing vast amounts of data from satellite images, sensors, and IoT devices deployed in the field. These analyses can predict crop yields, detect plant diseases, and provide insights that help farmers make informed decisions (Mohammed et al. 2024). For example, AI-driven precision agriculture allows for the variable rate application of inputs like water, fertilizer, and pesticides, which tailors the treatment to the specific needs of each plant. This targeted approach not only boosts crop yield but also reduces resource waste, leading to more sustainable farming practices (AlZubi and Galyna 2023).

AI also enhances agricultural productivity through automated and robotic systems. These systems perform a variety of tasks traditionally done by humans, such as harvesting, planting, and weeding, more efficiently and tirelessly around the clock. The use of robots not only speeds up the agricultural processes but also frees up human labor for more complex management tasks that require critical thinking and planning. Moreover, AI models contribute to genetic crop improvement by predicting the traits of plants, consequently, expediting the breeding of crop varieties that are more resistant to pests and climate change (Oliveira and Silva 2023).

5.2 Challenges of AI in Agriculture

Despite its numerous benefits, the application of AI in agriculture faces significant challenges. One of the primary hurdles is the high cost of AI technologies, which can be prohibitively expensive for small-scale farmers and in developing countries. The initial setup for AI-driven systems requires substantial investment in sensors, drones, and other technological infrastructure, not to mention the ongoing costs associated with data processing and analysis (Leong et al. 2023).

Data management in agriculture poses another challenge. Farming generates massive datasets, which are often unstructured and come from heterogeneous sources. Managing and processing this data to extract useful information demands robust computational resources and sophisticated algorithms. Additionally, there is the issue of data accuracy and precision, which are critical for making reliable agricultural predictions and recommendations (Kisliuk et al. 2023).

Another significant challenge is the integration of AI systems into the existing agricultural practices. Farmers must be trained to operate advanced AI tools and understand the insights generated by AI algorithms. This learning curve can be steep, especially in regions with limited access to education and technology. Moreover, the variability in environmental conditions and agricultural practices across different regions makes it difficult to develop universal AI solutions that are applicable everywhere.

6 Future Work in AI in Agriculture

Looking ahead, continuous research and development are crucial to address the current limitations and expand the capabilities of AI in agriculture. One area of future work involves developing low-cost AI solutions that are accessible to smallholder farmers. This could include simplified AI tools that require minimal setup and training or scalable AI applications that can be accessed as services over the Internet.

Advancements in AI algorithms are also essential to improve the accuracy and adaptability of predictions in diverse environmental conditions. Future research could focus on creating more robust models that can handle the variability in data quality and farming practices. Moreover, integrating AI with other emerging technologies like blockchain could enhance data security, traceability, and transparency in agricultural supply chains (Konar 2024).

In addition to technological advancements, interdisciplinary collaborations among agronomists, engineers, data scientists, and policymakers are necessary to ensure that AI solutions are practical and grounded in the real-world needs of agriculture. Policies and frameworks should also be developed to guide the ethical use of AI in agriculture, addressing concerns such as data privacy, labor displacement, and the socioeconomic impacts of automation (Hamed et al. 2024).

In conclusion, AI holds the potential to transform agriculture by making farming more intelligent, efficient, and sustainable. However, realizing this potential requires overcoming significant challenges, particularly in terms of cost, data management, and integration. With sustained investment in research and a collaborative approach to innovation, AI can significantly contribute to the future of agriculture, ensuring food security and environmental sustainability for the coming generations.

7 Conclusion

AI is transforming agriculture by making farming more efficient, sustainable, and productive. Technologies like machine learning, computer vision, and IoT are helping farmers optimize resources, improve crop yields, and reduce environmental impacts. Despite these advancements, challenges such as high costs and limited access to technology remain. Addressing these barriers through affordable solutions and farmer training is essential. With continued innovation and collaboration, AI can play a vital role in ensuring food security and building a sustainable future for agriculture.

References

Abinaya M, Vadivu G, Prasanth A, Dhivyasri G (2024) Role of big data analytics in intelligent agriculture. Stud Comput Intell 1170:19–48

Akter J, Nilima SI, Hasan R, Tiwari A, Ullah MW, Kamruzzaman M (2024) Artificial intelligence on the agro-industry in the United States of America. AIMS Agric Food 9(4):959–979

Alexandratos N, Bruinsma J (2012) World agriculture toward 2030–2050: the 2012 revision 2012. FAO, Rome. pp 1–154. Available online: https://www.fao.org/4/ap106e/ap106e.pdf. (Date-of-access: 13 Oct 2024)
AlZubi AA, Galyna K (2023) Artificial intelligence and internet of things for sustainable farming and smart agriculture. IEEE Access
Anwarul S, Misra T, Srivastava D (2022) An IoT & AI-assisted framework for agriculture automation. In: 2022 10th International Conference on Reliability, Infocom Technologies and Optimization (Trends and Future Directions) (ICRITO), Noida, India. pp 1–6
Bendre MR, Thool RC, Thool VR (2016) Big data in precision agriculture: weather forecasting for future farming. pp 744–750
Benefits of AI in agriculture (2024) Available online: https://intellias.com/artificial-intelligence-in-agriculture/. (Date-of-access: 13 Oct 2024)
Bertoglio R, Corbo C, Renga FM, Matteucci M (2021) The digital agricultural revolution: a bibliometric analysis literature review. IEEE Access 9:134762–134782
Chen J, Begicheva S, Nazarov D (2024) The role of big data in advancing precision agriculture and ensuring food security. BIO Web of Conferences 121
Davoodi Farsani E, Choobchian S, Shirvani Naghani M (2024) Unlocking agricultural innovation: a roadmap for growth and sustainability. J Knowledge Econ 15(1):1–20
Effah E, Thiare O, Wyglinski AM (2024) Hardware evaluation of cluster-based agricultural IoT network. IEEE Access 12:33628–33651
Elbasi E, Mostafa N, AlArnaout Z, Zreikat AI, Cina E, Varghese G, Shdefat A, Topcu AE, Abdelbaki W, Mathew S, Zaki C (2023a) Artificial intelligence technology in the agricultural sector: a systematic literature review. IEEE Access 11:171–202
Elbasi E, Zaki C, Topcu AE, Abdelbaki W, Zreikat AI, Cina E, Shdefat A, Saker L (2023b) Crop prediction model using machine learning algorithms. Appl Sci 13(16):9288
Elbasi E, Mostafa N, Zaki C, AlArnaout Z, Topcu AE, Saker L (2024a) Optimizing agricultural data analysis techniques through AI-powered decision-making processes. Appl Sci 14(17):8018
Elbasi E, Topcu AE, Cina E, Zreikat AI, Shdefat A, Zaki C, Abdelbaki W (2024b) Enhanced plant leaf classification over a large number of classes using machine learning. Appl Sci 14(22)
FAO (2017) The future of food and agriculture: trends and challenges. Food and Agriculture Organization of the United Nations, Rome
Faragó D, Rico-Fernández J, Pérez-Zavala A, Fahmi A (2020) Applications of computer vision in agricultural automation—a review. Inf Process Agric 7(1):1–19
Food and Agriculture Organization of the United Nations (2024) Available online: https://www.fao.org/home/en. Date-of-access: 13 Oct 2024
Hamed MA, El-Habib MF, Sababa RZ, Al-Hanjor MM, Abunasser BS, Abu-Naser SS (2024) Artificial intelligence in agriculture: enhancing productivity and sustainability
Hashni T, Amudha T, Ramakrishnan S (2022) IoT & AI in smart farming: implications and challenges. In: 7th International Conference on Communication and Electronics Systems (ICCES), Coimbatore, India. pp 501–505
He Y, Nie P, Zhang Q, Liu F (2021) Agricultural internet of things. Springer International Publishing
Holzinger A, Fister I, Kaul HP, Asseng S (2024) Human-centered AI in smart farming: toward agriculture 5.0. IEEE Access 12:62199–62214
Islam N, Rashid MM, Pasandideh F, Ray B, Moore S, Kadel R (2021) A review of applications and communication technologies for Internet of Things (IoT) and Unmanned Aerial Vehicle (UAV) based sustainable smart farming. Sustainability 13(4):1821
Jin X, Kumar L, Li Z, Feng H, Xu X, Yang G (2020) Important developments in spectral data analysis for precision agriculture: a review on trends and applications. Sci Total Environ 712:136–149. https://doi.org/10.1016/j.scitotenv.2020.136379
Kamilaris A, Prenafeta-Boldú FX (2018) Deep learning in agriculture: a survey. Comput Electron Agric 147:70–90

KBV, research (2023) Artificial intelligence in agriculture market size by 2028. KBV Research. Retrieved 11 Nov 2024. https://www.kbvresearch.com/artificial-intelligence-in-agriculture-market/

Kisliuk B, Krause JC, Meemken H, Saborío Morales JC, Müller H, Hertzberg J (2023) AI in current and future agriculture: an introductory overview. KI - Künstliche Intelligenz 37(2):117–132

Konar M (2024) Integrating IoT and AI for automated crop with smart urban farming. In: 2024 International Conference on Emerging Innovations and Advanced Computing (INNOCOMP). IEEE, pp 454–462

Kumar R, Farooq M, Qureshi M (2024) Advancing precision agriculture through artificial intelligence. In: A biologist's guide to artificial intelligence: building the foundations of artificial intelligence and machine learning for achieving advancements in life sciences. Academic Press, pp 151–165

Leong YM, Lim EH, Subri NFB, Jalil NBA (2023) Transforming agriculture: navigating the challenges and embracing the opportunities of artificial intelligence of things. In: IEEE International Conference on Agrosystem Engineering, Technology & Applications (AGRETA), pp 142–147

Lu Y, Young S (2020) A survey of public datasets for computer vision tasks in precision agriculture. Comput Electron Agric 178:105760

Liakos KG, Busato P, Moshou D, Pearson S, Bochtis D (2018) Machine learning in agriculture: a review. Sensors 18(8):267

Maharlooei M, Sadeghi-Tehran P, Liu H, Chahl J (2021) Advanced computer vision applications for pest detection and crop health monitoring. J Agric Eng Res 19(3):45–62

Markets and Markets (2023) Agriculture robots market industry analysis | types, advantages, and forecast. Markets and Markets. Retrieved 11 Nov 2024. https://www.marketsandmarkets.com/Market-Reports/agricultural-robot-market-173601759.html

Mohammed MA, Jameel SH, Hussein AJ, Ahmad HK, Ismail LS, Zapryvoda A (2024) The transformation of agriculture by artificial intelligence in smart farming. In: 2024 36th Conference of Open Innovations Association (FRUCT). IEEE, pp 515–526

Morchid A, Alblushi IGM, Khalid HM, El Alami R, Sitaramanan SR, Muyeen SM (2024) High-technology agriculture system to enhance food security: a concept of smart irrigation system using internet of things and cloud computing. J Saudi Soc Agric Sci

Oliveira RC, de Souza e Silva RD (2023) Artificial intelligence in agriculture: benefits, challenges, and trends. Appl Sci 13(13):7405

Pantazi XE, Moshou D, Bravo C (2016) Active learning system for weed species recognition based on hyperspectral sensing. Biosyst Eng 146:193–202

Paolo P (2025) A study using Treesoft set and neutrosophic sets on possible soil organic transformations in urban agriculture systems. Int J Neutrosophic Sci 25:315–328

Patel KK, Kar A, Jha SN (2021) Machine learning and artificial intelligence in food processing and agriculture. IEEE Access 9:11599–11610

Patil SA, Bhosale A, Patil A (2024) Advancements in plant leaf disease detection: a comprehensive review of latest trends and technologies. In: 3rd International Conference for Innovation in Technology (INOCON), Bangalore, India, pp 1–5

Qazi S, Khawaja BA, Farooq QU (2022) IoT-equipped and AI-enabled next generation smart agriculture: a critical review, current challenges and future trends. IEEE Access 10:21219–21235

Rabhi L, Falih N, Afraites L, Bouikhalene B (2021) A functional framework based on big data analytics for smart farming. Indones J Electr Eng Comput Sci 24(3):1772–1779

Rahman M, Chippy T, Dankan Gowda V, Prasad K, Sethi HS, Prasanth PV (2023) Optimizing resource allocation in agriculture through IoT and AI. In: 7th International Conference on Electronics, Communication and Aerospace Technology (ICECA), Coimbatore, India, pp 1573–1578

Rani R, Sahoo J, Bellamkonda S, Kumar S, Pippal SK (2023) Role of artificial intelligence in agriculture: an analysis and advancements with focus on plant diseases. IEEE Access 11:137999–138019

Rattan P, Sharma G, Singh PP (2024) Application of Artificial Intelligence (AI) in the agriculture sector. In: Agriculture and aquaculture applications of biosensors and bioelectronics. IGI Global, pp 45–68

Rejeb A, Abdollahi A, Rejeb K, Treiblmaier H (2022) Drones in agriculture: a review and bibliometric analysis. Comput Electron Agric 198:107017

Samah S, Al Zanin S, Alneil AA, Issaou I (2025) Integrating Q neutrosophic soft relation with deep learning-based pepper leaf disease recognition for sustainable agriculture in KSA. Int J Neutrosophic Sci 25:190–202

Sharma R (2021) Artificial intelligence in agriculture: a review. In: 5th International Conference on Intelligent Computing and Control Systems (ICICCS), pp 937–942

Sharma K, Shivandu SK (2024) Integrating Artificial Intelligence and Internet of Things (IoT) for enhanced crop monitoring and management in precision agriculture. Sens Int 5

Sivakumar SA, Shankar BM, Anuradha B, Karan KA, Karthik A, Karthik R, Kumar JRR (2023) Artificial intelligence based agricultural chatbot and virtual assistant for delivery of harvested crops. Int J Intell Syst Appl Eng 11(8):576–583

Sudha V, Mohan S (2019) Extemporize agriculture yield with predictions based on water and soil properties using multivariate analytics and machine learning algorithm. Int J Eng Adv Technol 8(6):1504–1508

Sugunadevi C (2023) Internet of things for sustainable agriculture, 2023 2nd International Conference on Futuristic Technologies (INCOFT), Belagavi, Karnataka, India. pp 1–6

Sun Y, Hu Z, Tang Z, Li Y, Yang J (2022) Application of computer vision in assessing crop abiotic stress: a systematic review. PLoS One 17(8):e0290383

Taer E, Taer A (2024) Cultivating tomorrow: a comprehensive review of agricultural innovations in The Philippines (2018–2023). Agric Res 13(4):1–15

Tetarwal V, Dashora K, Selvamuthu D (2024) Predictive analysis of market trends in agriculture using ML/AI techniques, 2024 IEEE Applied Sensing Conference (APSCON), Goa, India. pp 1–4

Vailshery LS (2023) Agricultural industrial internet of things (IIoT) market size worldwide from 2020 to 2025, July 27, 2023. https://www.statista.com/statistics/1222813/worldwide-agricultural-industrial-iot-market-value/

Wolfert S, Ge L, Verdow C, Bogaardt M-J (2017) Big data in smart farming—a review. Agric Syst 153:69–80

Zhang X, Li J, Wang M (2020) Deep learning for agricultural image processing: a survey IEEE. Access 8:211085–211107

Zou X, Gan H, Li Z, Tang Y (2024) Recent advances in computer vision technology for agriculture machines, Special Issue. Retrieved from MDPI

Strategies for Strengthening Biosecurity in Indian Agriculture

6

Chetan Chauhan, Shanta Kumari, and Taru Singh

Abstract

This chapter reviews the urgent necessity to strengthen biosecurity measures within the agricultural sector of India, taking into account the increasing threats posed by invasive pests, cross-border disease spread, climate change, and the growth of international trade. India's agricultural sectors, including crops and livestock, are becoming more susceptible to these challenges, which calls for the establishment of thorough, science-driven, and proactive biosecurity systems. The chapter promotes integrated strategies that are vital for ensuring food security, protecting public health, and fostering environmental sustainability. The review is organized around thematic areas that are particularly pertinent to the Indian context. It commences with an analysis of risk assessment and surveillance systems, highlighting the necessity for the adoption of advanced tools such as remote sensing, molecular diagnostics, and geospatial technologies. These innovations are crucial for the early detection of threats and the formulation of effective containment strategies. A significant aspect of the chapter is the advocacy for good agricultural practices (GAP), which encompasses integrated pest management (IPM), organic soil amendments, crop rotation, and enhanced sanitary practices in animal husbandry. These practices contribute to reducing pathogen transmission, boosting biodiversity, and improving both soil and animal health. Furthermore, the review emphasizes ecological and biological control methods, including the utilization of beneficial microbes, predator species, and ecological engineering, as sustainable alternatives to chemical inputs. These approaches are particularly effective within India's varied farming systems,

C. Chauhan (✉)
Department of Chemistry, Sardar Patel University, Mandi, HP, India
e-mail: chetan@spumandi.ac.in

S. Kumari
Department of Economics, Eternal University, Sirmour, HP, India

T. Singh
Amity Institute of Biotechnology, Amity University, Noida, India
e-mail: tsingh17@amity.edu

R. E. Krishnankutty et al. (eds.), *Smart Agriculture*,
https://doi.org/10.1007/978-981-95-8785-8_6

where smallholder farmers depend on maintaining ecological balance for ongoing productivity. The One Health framework serves as a fundamental element of the manuscript, connecting agricultural biosecurity with the management of antimicrobial resistance (AMR). It highlights the necessity of synchronizing veterinary, human, and environmental health practices to tackle AMR issues. This includes promoting responsible antimicrobial usage, advocating for proactive measures, and fostering collaboration across sectors.

Keywords

Biosecurity · Antimicrobial resistance (AMR) · Good agricultural practices (GAP) · Integrated pest management (IPM)

1 Introduction

Biosecurity, as a multidimensional approach, encompasses the policies and protocols used to control, prevent, and mitigate the spread of harmful microorganisms, including pests, diseases, and invasive vectors that threaten human health, agriculture sector, and biodiversity (Rawal et al. 2019). In the Indian scenario, where agriculture forms the backbone of the economy and employs nearly 60% of the population (Srivastava et al. 2016), the execution of robust biosecurity strategies is of immense importance (FAO 2021a, 2021b, 2021c, 2021d, 2021e, 2021f, 2021g, 2021h; MoA and FW 2023). Given the country's agroecological diversity, porous borders, and rising trade connectivity, Indian agriculture is particularly at risk to biological threats that can severely disrupt food production segments and impact rural livelihoods. The importance of biosecurity has grown significantly due to mounting incidences of transboundary plant pests (e.g., Fall Armyworm), zoonotic disease outbreaks (e.g., avian influenza), and the growing challenge of antimicrobial resistance (AMR), which together threatens the national food security, environmental strength, and public health (ICAR-NBAIR 2022a, 2022b; WHO 2020a, 2020b, 2020c, 2020d). Furthermore, climate change has altered the dynamics of pest and pathogen, intensifying their geographical reach and complicating traditional pest control methodologies, thus exacerbating biosecurity risks (IPCC 2022). In the current context, biosecurity in Indian agriculture must not only focus on instantaneous pest or disease threats but also take a proactive, integrated, and systems-based approach. It must encompass early detection, prevention, speedy response, and recovery along with the promoting sustainable practices such as integrated pest management (IPM), ecological farming, and the One Health framework (Sasikumar and Eapen 2025). Such measures are vital to ensure long-lasting agricultural resilience and sustainability, especially in a developing nation like India where smallholder farming dominates the landscape (Chakraborty and Sinha 2023).

1.1 Importance of Biosecurity

Biosecurity plays a vital role in safeguarding the food security, public health, and environmental sustainability in agrarian economies like India as shown in Fig. 6.1 (Anand 2018).

With the mounting globalization of food supply chains, the movement of agricultural commodities has become a major conduit for the spread of transboundary pests and diseases. Effective and potential biosecurity actions are therefore essential to protect the integrity and safety of the food system from farm to fork (Arabska 2021; FAO 2021a, 2021b, 2021c, 2021d, 2021e, 2021f, 2021g, 2021h). In the context of food security, biosecurity helps maintain steady agricultural productivity by preventing and managing outbreaks of crop and livestock diseases, thereby lowering the yield losses and ensuring a steady food supply (Anand and Krishnan 2024). In India, where a significant segment of the population is reliant on subsistence farming, even minor disruptions due to biological threats can result in sizeable economic and nutritional challenges (Richard 2024; Dev and Sengupta 2022). Moreover, biosecurity contributes to reduce the postharvest losses, ensuring safe food and accessibility to consumers. From a public health perception, many zoonotic diseases (like avian influenza, brucellosis, and Nipah virus) have agricultural origins and can further spread to humans through contact with animals or contaminated products (Adebiyi and Oluwayelu 2018). Implementing biosecurity protocols in livestock and poultry segments helps to control the transmission of diseases, thereby protecting both rural and the broader public communities (WHO 2020a, 2020b, 2020c, 2020d). Furthermore, the misuse of antibiotics in agriculture has contributed to the growing antimicrobial resistance (AMR) and global threat that can be addressed through integrated biosecurity approaches rooted in the One Health approach (Campey 2024; OIE 2021a, 2021b). Biosecurity also serves as a pillar of environmental sustainability. The excessive reliance on agrochemical

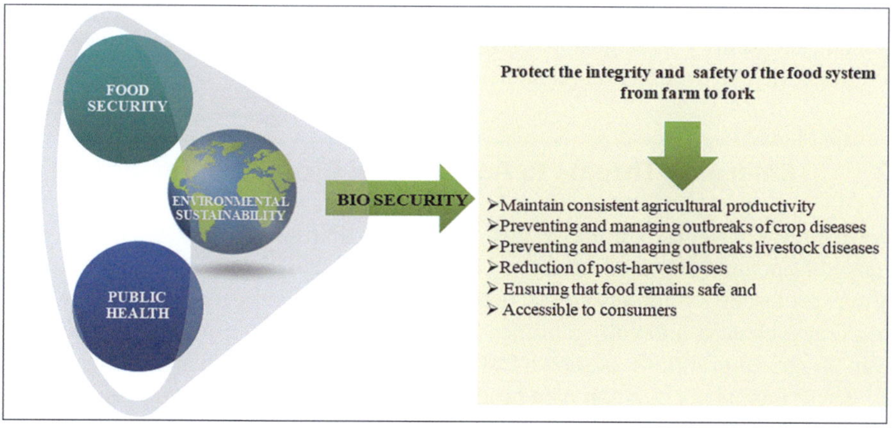

Fig. 6.1 Importance of biosecurity

inputs for pest control, driven by weak preventive measure systems, has led to soil degradation, biodiversity loss, and water bodies of contamination. Emphasizing biological control, habitat conservation, and agroecological resilience through biosecurity strategies can mitigate the environmental impacts (Pretty et al. 2018). Thus, a comprehensive biosecurity framework wires not only the productivity of agriculture but also the long-lasting health of ecosystems and human populations (Babu et al. 2025).

The emergent complexity of biosecurity threats in Indian agriculture ranging from invasive pests and emerging zoonotic diseases to antimicrobial resistance and climate-driven ecological shifts demands toward a systematic and multidimensional measures. Although a number of policies and scientific advancements have contributed to risk mitigation in isolated domains, there remains a critical need for a consolidated, interdisciplinary understanding of biosecurity strategies suited to the Indian context (ICAR 2023a, b; MoEFCC 2022). The primary objective of this chapter is to provide a comprehensive analysis of contemporary challenges and strategic opportunities for enhancing agricultural biosecurity and to bridge that gap between existing knowledge base and identifying practical, science-driven approaches to strengthening biosecurity in the Indian agriculture system. It explores both plant and animal sectors, integrating environmental, human, and veterinary perspectives through the lens of the One Health approach. This chapter evaluates advanced risk detection tools, such as molecular diagnostics and geospatial surveillance, and examines their vital role in early warning systems. It also emphasizes the endorsement of GAP, IPM, and ecological controls as a sustainable means of reducing biological threats and agrochemical dependency (FAO 2021a, 2021b, 2021c, 2021d, 2021e, 2021f, 2021g, 2021h; WHO 2020a, 2020b, 2020c, 2020d). The scope of the chapter spans a wide array of thematic areas, including risk assessment approaches, ecological pest control methodologies, livestock sanitation protocols, antimicrobial stewardship, contemporary policy frameworks, and capacity-building initiatives. By addressing the concerned gaps in institutional coordination, technological deployment, and community-level engagement, this chapter aims to inform stakeholders ranging from policymakers and scientists to farmers and educators on holistic biosecurity strategies essential for achieving public health protection and food security, environmental resilience.

2 Emerging Threats to Agricultural Biosecurity in India

Agricultural biosecurity in the Indian continent is alarmingly challenged by a wide range of upcoming dynamic and overlapping threats that include the rise of invasive alien species, cross-border disease outbreaks, which further accelerated by impacts of climate change, and vulnerabilities introduced all the way through the globalization of agricultural trade as shown in Fig. 6.2 (Quénol and Sturman 2023).

These factors not only threaten productivity of crop and livestock but also further jeopardize public health, food security, and ecological stability.

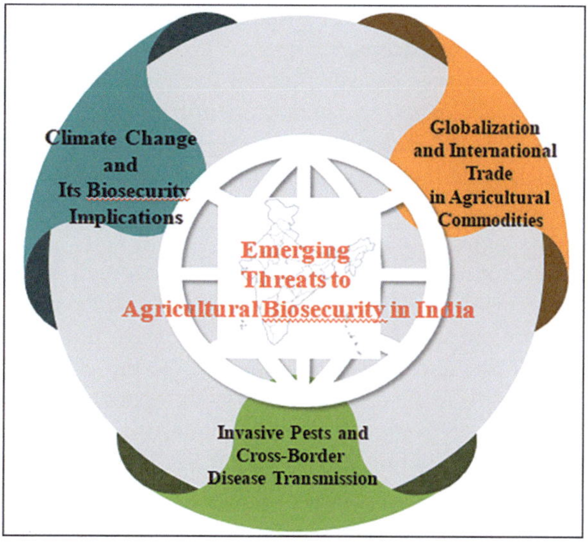

Fig. 6.2 Emerging threats to agricultural biosecurity in India

India has witnessed a sharp rise in the occurrence of invasive pests and trans-boundary diseases in recent years. The introduction of *Spodoptera frugiperda* (Fall Armyworm) in 2018 is a noteworthy example of an invasive species that rapidly extended across multiple agro-climatic zones, causing severe loss to maize and other crops (Maruthadurai and Ramesh 2020; Ganiger et al. 2018). Plant diseases like wheat rust, banana wilt, and citrus greening have emerged as constant threats due to cross-border movement and limited quarantine infrastructure systems. In livestock, diseases such as foot-and-mouth disease (FMD), African swine fever (ASF), and avian influenza highlight the upcoming threat of zoo noses with potential transmission risk (DAHD 2022a, 2022b). Due to porous international borders with neighboring countries, India becomes susceptible and vulnerable to such incursions. Limited coordination and ineffective quarantine protocols among stakeholder and border states aggravate the risks of disease entry and establishment. Furthermore, the increasing frequency of international trade and travel has further accelerated the rate at which pests and pathogens move across borders (FAO 2021a, 2021b, 2021c, 2021d, 2021e, 2021f, 2021g, 2021h).

Climate change plays a vital role as a catalyst in altering pest and disease dynamics. Rising temperatures, shifting precipitation patterns, and extreme weather events influence the life cycle, geographical spread, and virulence of many animal diseases and crop pests (IPCC 2022). Warmer climates can extend the survival and reproduction rates of vectors like mosquitoes, whiteflies, and aphids, raising the incidence of vector-borne animal and plant diseases. In India, shifting agroecological zones due to climate variability have led to the emergence of pest infestations in unaffected regions, challenging traditional management and infrastructure practices. Moreover, warmer winter sessions have lowered the natural mortality of many pathogens and pests, allowing for year-round infestations (Kumar et al. 2021). Climate change also affects livestock health by rising heat stress and varying the epidemiology of viral and parasitic diseases.

Globalization has significantly increased India's agricultural market connectivity and trade. Conversely, such economic integration comes with potential biosecurity risks. Planting materials, including the import and export of seeds, food grains, live animals, and feed, increase the chances of inadvertently introducing harmful pests, pathogens, and invasive weeds (Narayanan and Gulati 2020). Poor enforcement of phytosanitary standards and protocols at entry points like ports and inconsistent inspection measures have often been identified as gaps in the Indian trade-related biosecurity infrastructure system. Moreover, the rising demand for high-value agricultural crops exports encourages intensive production practices that may lead to negotiation and further compromise the biosecurity, like overuse of antibiotics or poor waste management in livestock operations. Such conducive conditions can serve as breeding grounds for resistant pathogens and environmental contamination, thus amplifying biosecurity risks (OIE 2021a, 2021b)

3 Risk Assessment and Surveillance Systems

An effective mechanism and framework for biosecurity relies greatly on robust risk assessment and surveillance approaches (Gershon et al. 2005). These systems serve as the first line of defense against upcoming biological threats, enabling early detection, rapid response, and long-lasting risk mitigation. In India, even though there is the existence of several institutional mechanisms and framework for pest and disease monitoring, the current biosecurity infrastructure faces significant limitations in terms of real-time surveillance, predictive analytics, and integration across sectors (ICAR-NBPGR 2022; FAO 2021a, 2021b, 2021c, 2021d, 2021e, 2021f, 2021g, 2021h) (Fig. 6.3).

In India, the existing surveillance framework is often reactive rather than proactive, responding to outbreaks only after they have escalated. The majority of gaps include limited real-time data collection, fragmented institutional coordination

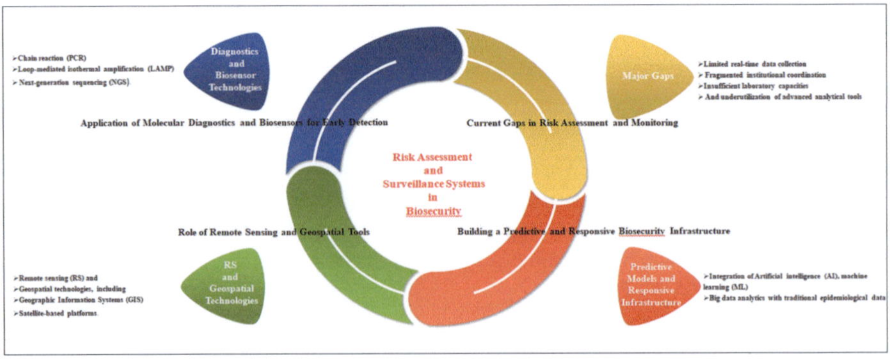

Fig. 6.3 Importance of risk assessment and surveillance systems in biosecurity

mechanism, insufficient laboratory infrastructure, and underutilization of advanced analytical tools and techniques (Singh et al. 2022). Moreover, data generated by different scientific and academic agencies such as agricultural universities, research institutes, and extension departments of agriculture and horticulture are seldom integrated into a unified risk database, weakening the decision-making process. The absence of PRA (standardized pest risk analysis) procedures across states also hampers the effective assessment of emerging threats, especially in regions with high biodiversity and varied agroecological areas. Furthermore, monitoring framework is often underfunded, making it difficult to maintain consistent, nationwide coverage, especially in remote or high-risk areas (MoAFW 2023).

Geospatial technologies, including geographic information systems (GIS), satellite-based platforms, and remote sensing (RS), offer immense transformative potential in the field of biosecurity surveillance in the agriculture sector. Such technological tools facilitate real-time tracking of vegetation health, climatic conditions, pest migration patterns, and habitat changes that may signal biosecurity threats (Bhatia et al. 2020). For example, normalized difference vegetation index (NDVI) data could be used to detect early signs of pest outbreaks in various crops, enabling timely intervention. In India, remote sensing applications are progressively more being adopted for locust forecasting, disease mapping, and crop monitoring under programs like FASAL (Forecasting Agricultural output using Space, Agrometeorology and Land-based observations) (Delince et al. 2017). However, the full potential of RS remains untapped due to a lack of skilled professionals, interdepartmental collaboration, and integration among the local data sources (ISRO 2022).

Biosensor and molecular diagnostics technologies have revolutionized the early detection of pests and pathogens in agriculture. Techniques such as polymerase chain reaction (PCR), loop-mediated isothermal amplification (LAMP), and next-generation sequencing (NGS) enable the speedy and accurate identification of pathogenic organisms even at asymptomatic stages (Liu et al. 2023; Tripathi et al. 2021). These diagnostic tools are especially crucial in identifying quarantinable species and new variants that may not be visually distinguishable.

In recent times, India has made progress in designing portable biosensors for field-level pathogen detection, like lateral flow devices for fungal infections and electrochemical sensors for bacterial contaminants. In spite of their immense potential, biosensor-based tools are still not widely deployed across farming systems, mainly due to cost constraints and lack of awareness among extension workers and farmers.

A futuristic biosecurity strategy necessitates the design of predictive models and responsive infrastructure framework capable of anticipating and counteracting threats before they become uncontrollable. Integrating artificial intelligence (AI), machine learning (ML), and big data analytics with traditional epidemiological data can enable dynamic risk forecasting module tailored to India's agroecological diversity (Sharma et al. 2022; Rajaram et al. 2024). In addition, establishing a centralized national biosecurity information system linked to state-level surveillance framework can boost coordination among streamline response protocols and stakeholders. Capacity-building initiatives, continuous training of frontier personnel, and community-level awareness campaigns are also crucial to ensure that technological innovations translate into real-world resilience.

4 Integrated Approaches to Pest and Disease Management

Sustainable agricultural biosecurity depends on integrated approaches that not only control pests and diseases but also build long-lasting ecological resilience. In India, where diverse agroecological framework support smallholder farmers and mixed farming system, integrated approaches like crop diversification, integrated pest management (IPM), organic amendments, and biosecure livestock practices are mainly relevant, as shown in Fig. 6.4.

Such practices lower the dependency on agrochemicals, improve ecosystem services, and contribute significantly to overall national biosecurity frameworks (FAO 2021a, 2021b, 2021c, 2021d, 2021e, 2021f, 2021g, 2021h; ICAR 2023a, 2023b).

IPM is a foundation of sustainable pest control, combining biological, cultural, mechanical, and chemical practices to manage pests in an economically and eco-friendly manner (Babu et al. 2025). IPM practices contribute directly to biosecurity by minimizing the development of pest resistance, lowering pesticide residues in food, and further reducing the risk of secondary pest outbreaks (Sasikumar and Eapen 2025; Kumar et al. 2022). In India, IPM has been successfully implemented in a variety of crops like cotton, rice, and vegetables through initiatives such as Farmer Field Schools (FFS) and national IPM programs. These techniques include biopesticides, resistant crop varieties, pheromone traps, and mechanical weed removal. Even though successful, wider adoption is limited due to poor awareness, lack of training, and minimal extension support from authorities in many regions (Yadav and Singh 2021).

Healthy soil is fundamental to plant resistance against pests and pathogens. Organic amendments like compost, farmyard manure, and green manures improve soil microbial diversity, nutrient cycling, and suppress soil-borne diseases (Gupta

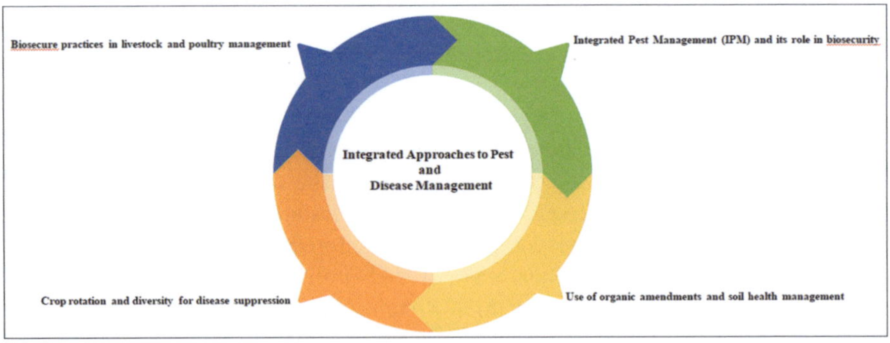

Fig. 6.4 Integrated approaches to pest and disease management

et al. 2020). Such input efforts also promote beneficial soil organisms that outcompete pathogens, thereby sustainable contributing to natural pest control. The Indian government initiatives like "Paramparagat Krishi Vikas Yojana" and "National Mission on Soil Health" have been emphasized on organic farming and soil testing, yet adoption remains fragmented (Tripathi et al. 2023). Scaling and upscaling such practices through incentives and integrated soil management strategies can significantly improve agricultural biosecurity at the grassroots level (MoAFW 2022). Intercropping and crop rotation are time-tested ecological strategies that fracture the life cycle of pests and pathogens. These methods lower the buildup of host-specific pathogens and insect populations while improving soil fertility along with ecosystem resilience (Altieri et al. 2015). Incorporating deep-rooted crops and legumes in crop rotations improves nutrient balance and lowers disease incidence. Indian farming systems, especially in hill and dryland regions, are naturally well suited for poly cropping and seasonal rotations. An effective buffer against pest invasions and emerging plant diseases can be provided, especially in the context of climate change, by promoting diversified cropping systems that are adapted to the local conditions (Ramesh and Ramana 2021). Livestock and poultry production systems are some of the most vulnerable to transboundary animal diseases and zoonotic infections that can compromise public health and national food security. Biosecurity practices like quarantine protocols, regular health monitoring, controlled farm access, and proper waste disposal are essential in preventing disease outbreaks in animals (OIE 2021a, 2021b). In India, backyard poultry and smallholder livestock holdings frequently lack formal biosecurity infrastructure, making them hotspots for disease transmission and spreading. Improving animal health services, vaccinating against common pathogens, and training farmers in hygiene practices are crucial steps toward safeguarding animal health and integrating livestock management into the broader spectrum of biosecurity agenda (DAHD 2022a, 2022b).

5 Adoption of Good Agricultural Practices (GAPs)

Good agricultural practices (GAPs) are a set of principles aimed at ensuring food safety, environmental sustainability, and farm productivity through science-based, practical guidelines (Gereffi and Lee 2009). From the perspective of agricultural biosecurity, GAPs play a crucial role in minimizing pest and disease risks, maintaining soil and water quality, improving animal health, and ensuring safe handling of produce across the value chain (Chauhan et al. 2023). As agriculture in India modernizes amidst mounting biosecurity threats, the systematic adoption of GAP is fundamental practice for building resilience in both crop and livestock sectors (FAO 2021a, 2021b, 2021c, 2021d, 2021e, 2021f, 2021g, 2021h; ICAR 2022) (Fig. 6.5). Implementing GAP in crop production includes the application of certified seeds, soil fertility management, timely pest and disease monitoring, and scientific and judicious use of agrochemicals. Animal health involves proper housing, nutrition, vaccination, and biosecure handling of livestock to prevent the spread and transmission of zoonotic and transboundary diseases (DAHD 2022a, 2022b). These

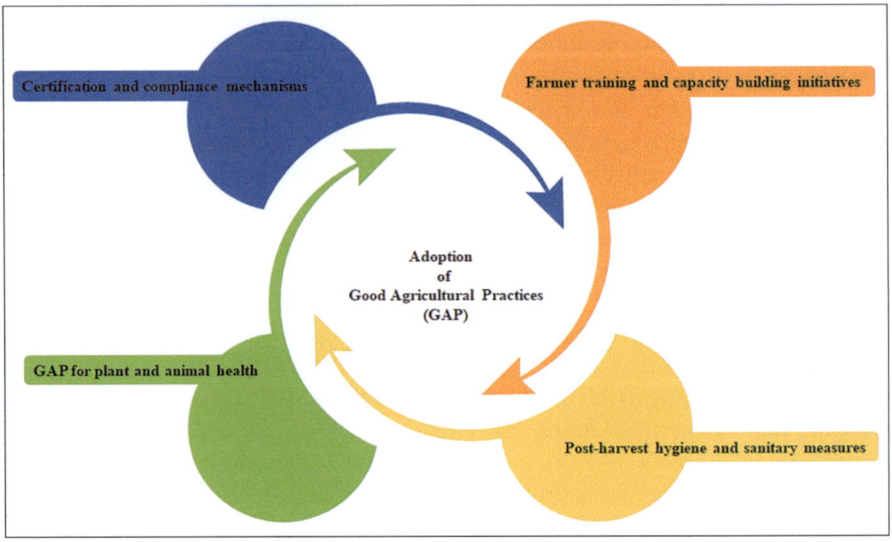

Fig. 6.5 Adoption techniques for good agriculture practices

practices lower biological contamination and promote an effective secure food production system. Indian government initiatives like the National Mission on Sustainable Agriculture (NMSA) and guidelines issued by the National Centre for Organic Farming have promoted GAP adoption, mainly among organic and high-value crop growers (Meena et al. 2022). Conversely, wide-spectrum application and implementation remain inconsistent, especially among smallholder farmers due to wide knowledge gaps and resource limitations (Singh et al. 2021; Yadav and Singh 2021).

Postharvest handling is another critical stage where lapses in hygiene can lead to biosecurity breaches through pathogen proliferation. Practices such as proper cleaning of equipment, use of clean water for washing produce, sanitary packaging, and temperature-controlled storage significantly lower the microbial contamination and spoilage (Codex Alimentarius 2020). These intervention steps are of utmost importance for perishables like fruits, vegetables, milk, and meat, which are mostly vulnerable to cross-contamination during transport and retail. Though there is increasing awareness, infrastructure for postharvest sanitation like cold chains and grading facilities is still lacking in many regions of India, especially in rural and periurban areas. The biosecurity profile of Indian agriculture can be improved by strengthening logistics and providing incentives for hygienic handling (APEDA 2023).

Training and knowledge dissemination are key inputs to successful GAP application and implementation. Capacity-building programs empower farmers with skills to monitor plant and animal health, identify early signs of pests and diseases, and adopt environmentally responsible practices. Government-supported initiatives like Krishi Vigyan Kendras (KVKs), Farmer Field Schools (FFS), and Agri Clinics have been instrumental in biosecurity practices (ICAR-ATARI 2022). However, the reach and quality of these programs vary widely, and many smallholders remain outside the domain of formal agricultural training. Tailored, community-level awareness programs using native languages or dialects and ICT-based tools (e.g., mobile apps) can enhance farmer participation and compliance (Patel and Desai 2021).

Certification schemes like India GAP, ISO 22000, and GLOBALG.A.P ensure that producers strictly adhere to food safety level and biosecurity protocols across production, processing, and distribution stages. These certification schemes build consumer trust, facilitate market access, especially exports, and provide traceability in the event of a biosecurity breach (GLOBALG.A.P. 2022). While certification uptake is mounting among large farms and agribusinesses, smallholders often find the process costly and complex. Cluster-based certification models, subsidies, and digital monitoring platforms are being explored to bridge the gap under schemes like Pradhan Mantri Formalisation of Micro Food Processing Enterprises (PM-FME) (MoFPI 2022; Patil and Naik 2021). Strengthening the enforcement and incentivizing compliance through market linkages can further drive broader adoption of GAP across India's agricultural landscape.

6 Ecological and Biological Control Strategies

In response to upcoming threats from synthetic agrochemicals and emerging pest dynamics, ecological and biological control strategies have gained prominence as environmentally sustainable alternatives that align with the goals of biosecurity (Fig. 6.6).

The objectives of such strategies are to harness natural biological processes to suppress pest populations, improve soil, plant health, and improve the resilience of farming systems against biotic stress. With diverse agroecological zones in India and strong tradition of agroecology, such methods offer scalable and context-specific solutions (FAO 2021a, 2021b, 2021c, 2021d, 2021e, 2021f, 2021g, 2021h; ICAR-NBAIR 2022a, 2022b). Bioinoculants comprising beneficial actinomycetes, fungi, and bacteria are predominantly used in Indian agriculture for plant growth promotion and disease suppression. Organisms like *Trichoderma harzianum*, *Bacillus subtilis*, and *Pseudomonas fluorescens* help to control soil-borne pathogens and enhance the plant immunity (Kumar et al. 2020). These microbes compete with pathogens for space and nutrients, produce antifungal active compounds, and induce systemic resistance in crops. India has been witnessing significant advancements in the commercialization of biofertilizers and biopesticides under schemes like the National Mission on Sustainable Agriculture (NMSA) and Paramparagat Krishi Vikas Yojana (PKVY) (Mohapatra and Pohit 2024). Despite such advances, quality control, storage, and farmer awareness remain critical bottlenecks in wide-spectrum application and adoption (Choudhary and Sharma 2021). Natural enemies such as ladybird beetles (*Coccinella septempunctata*), green lacewings (*Chrysoperla carnea*), and parasitoid wasps (*Trichogramma* spp.) serve as effective biological

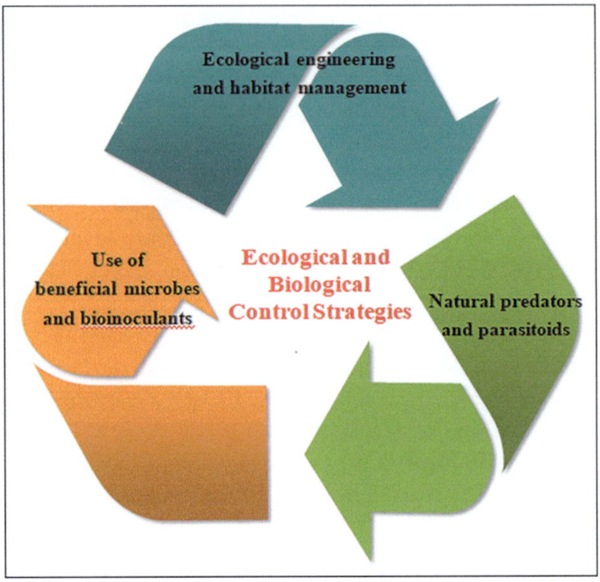

Fig. 6.6 Ecological and biological control strategies

control agents in cropping systems across India. These predators and parasitoids regulate insect pest populations without disrupting ecological balance, thereby lowering the demand for synthetic insecticides (Singh and Yadav 2019). Field releases and mass rearing of such bioagents have been explored by the Indian Council of Agricultural Research National Bureau of Agricultural Insect Resources (ICAR-NBAIR), especially in cotton, sugarcane, and rice ecosystems, for maintaining ecological niches and avoiding indiscriminate pesticide use, which can harm beneficial populations (ICAR-NBAIR 2022a, 2022b; Dhobi et al. 2020). Ecological engineering refers to the strategic design of agroecosystems to favor natural enemies and discourage pest proliferation. Techniques include planting nectar-rich flowering borders, maintaining alternate host plants for parasitoids, and practicing crop diversification (Gurr et al. 2017). These approaches improve biological control efficacy by supporting beneficial organisms throughout their life cycles. In India, ecological engineering has shown promising results in rice cultivation through initiatives led by institutions like the Directorate of Rice Research (DRR), where bund crops like marigold and sunflower support predator populations. Such practices are cost-effective and suitable for smallholder conditions (Rajendran et al. 2020; Kumar et al. 2020).

6.1 Case Studies from the Indian Farming Systems

Several successful case studies demonstrate the efficacy of ecological and biological control strategies in India:

(i) Sikkim's Organic Farming Model: As the first fully organic state, Sikkim has institutionalized the application of biopesticides and vermin compost, significantly reducing pest incidence and soil degradation (Mehta et al. 2019).
(ii) Tamil Nadu IPM Programs: In vegetables and cotton, state-sponsored IPM and ecological control programs have led to a 40–60% reduction in pesticide usage while maintaining yields (Ramaswamy and Jayaraj 2021).
(iii) Punjab's Ecological Approaches in Basmati Rice: Introduction of trap crops and habitat manipulation has helped in controlling stem borers and leaf folders, thus enhancing export-quality production with minimal residues (Gill et al. 2020).

These examples highlight the potential of integrating traditional knowledge with scientific innovations to build ecologically sound biosecurity strategies.

7 One Health Approach and Antimicrobial Resistance (AMR)

The growing interconnection among agricultural systems, public health, and environmental integrity necessitates a One Health approach, an integrative framework that recognizes the links between the health of people, animals, and ecosystems. In the context of agricultural biosecurity, One Health is particularly relevant to addressing the alarming threat of antimicrobial resistance (Stephen 2020; Soares 2024). AMR arises from the excessive and improper application of antimicrobials in livestock, aquaculture, and crop production that ultimately poses a global threat to health and food security (WHO 2020a, 2020b, 2020c, 2020d; FAO 2021a, 2021b, 2021c, 2021d, 2021e, 2021f, 2021g, 2021h).

7.1 Linkages Between Agricultural Biosecurity, Human, and Animal Health

Agriculture has been playing a central role in the emergence and transmission of resistant pathogens in current evolving agriculture practices (Pearce-Duvet 2006). The use of antimicrobials in animal husbandry and crop protection can result in residues that contaminate soil, water, and food chains, leading to the spread of resistant genes to human and animal pathogens (EFSA Panel on Biological Hazards) (BIOHAZ 2021). So, the role played by the environment in the emergence and spread of antimicrobial resistance (AMR) through the food chain is significant (*Efsa Journal*, *19*(6), e06651.). Zoonotic diseases such as avian influenza, leptospirosis, and brucellosis can also emerge from poor biosecurity in animal farming, underlining the need for integrated surveillance systems (OIE 2021a, 2021b; NCDC 2022). In India, the increasing incidence of antibiotic residues in food products, water sources, and farm environments highlights the urgent need for stringent cross-sectoral interventions (Sharma et al. 2021).

7.2 Responsible Antimicrobial Usage in Livestock and Aquaculture

India is among the largest consumers of veterinary antimicrobials, and nontherapeutic use such as growth promotion in poultry and aquaculture but their widespread use lacks proper regulation and awareness (Van Boeckel et al. 2019). Responsible and judicious antimicrobial application involves administering drugs only when necessary, ensuring proper dosages, and observing withdrawal periods before products or residues enter the food chain. National guidelines, such as the National Action Plan on AMR (2017–2021), advocate for the exclusion of critical antibiotics for growth promotion and further the drive should be carried out to strengthen the veterinary diagnostic capacity and training programs for farmers to spread awareness regarding the scientific and judicious drug use (MoHFW 2017).

7.3 Surveillance and Mitigation Strategies for AMR

Effective AMR control requires robust surveillance framework that monitors and tracks antimicrobial usage and resistance patterns across the human–animal–environment interface. India has established initiatives like the Indian Council of Medical Research's Antimicrobial Resistance Surveillance Network (ICMR-AMRSN) and National AMR Surveillance Network for Animals (NARSNet) to boost and advance national monitoring (ICMR 2020; Kaur et al. 2021). ICMR's Antimicrobial Resistance Surveillance system (i-AMRSS) is a potential system for global antimicrobial resistance surveillance. Mitigation strategies include promoting alternatives measures like probiotics, herbal antimicrobials, and vaccines in animal rearing, along with hygiene and sanitation measures that prevent disease outbreaks and lower dependency on antimicrobials (FAO 2021a, 2021b, 2021c, 2021d, 2021e, 2021f, 2021g, 2021h).

7.4 Intersectoral Collaboration and Policy Alignment

The complexity of AMR requires urgent efforts for multisectoral coordination among ministries of agriculture, health, environment, and fisheries. The One Health model encourages harmonization of surveillance protocols, antimicrobial stewardship programs, research funding, and regulatory frameworks (Tripartite Collaboration 2022). India's engagement on global platforms like the Global Leaders Group on AMR and national collaborations such as the One Health Consortium led by DBT represents vital steps toward integrated policy action (DBT 2021). However, scaling up such efforts to grassroots levels through state policies, veterinary public health programs is of immense impact but farmer education remains a critical gap.

8 Policy Framework and Institutional Coordination

India's biosecurity landscape is reshaped by a complex web of national policies, regulatory authorities, and institutional entities. The country has made substantial progress in establishing a policy architecture to address biosecurity threats through frameworks such as the National Action Plan for Biosecurity by the Ministry of Agriculture and Farmers' Welfare (MoAFW), the Vision 2050 document of the Indian Council of Agricultural Research (ICAR), and guidelines formulated by Food Safety and Standards Authority of India (FSSAI) (Mukherjee et al. 2024; MoAFW 2023; ICAR 2023a, 2023b). These policies aim to manage risks associated with transboundary pests, zoonoses, invasive species, and antimicrobial resistance (AMR), aligning with the One Health Approach (FAO 2021a). In spite of contemporary efforts, India's biosecurity enforcement remains fragmented. Overlapping mandates among the ministries, including the Ministry of Environment, Forest and Climate Change (MoEFCC), Ministry of Fisheries, Animal Husbandry and Dairying (DAHD), Ministry of Health and Family Welfare (MoHFW), and Ministry of Agriculture and Farmers' Welfare (v), often lead to poor coordination and implementation (Chakraborty and Sinha 2023). For Example, AMR surveillance falls under ICMR, DAHD, FSSAI, and MoHFW, creating jurisdictional overlaps that hinder unified response strategies (ICMR 2020; WHO 2020a, 2020b, 2020c, 2020d). State governments can play a pivotal role in implementation but also face challenges due to resource constraints, lack of training, and inconsistent adoption of central guidelines. Institutions like ICAR, FSSAI, and the National Centre for Disease Control (NCDC) have been instrumental in designing diagnostics, surveillance networks, and regulatory frameworks (NCDC 2022; ICAR-NBAIR 2022a, 2022b). However, their work often operates in silos, limiting synergy and knowledge sharing (Sharma et al. 2022). To overcome surging challenges, there is a critical need for a Unified National Biosecurity Framework (UNBF) that integrates existing mechanisms under a single coordinating authority. Such a framework would harmonize the roles of stakeholders across sectors like agriculture, health, environment, and trade and further promoting inter-agency collaboration, resource optimization, and rapid response mechanisms. It should be aligned with international standards like the OIE Terrestrial Code and Codex Alimentarius to facilitate safe trade and cross-border cooperation (OIE 2021a, 2021b; Codex Alimentarius 2020). Moreover, policy review and reforms must mandate the regular biosecurity audits and improve digital surveillance system at local levels. Training programs, risk communication strategies, and community-based surveillance can further reinforce the enforcement and preparedness (Patel and Desai 2021; ICAR-ATARI 2022).

9 Capacity-Building, Awareness, and Community Engagement

Strengthening biosecurity in Indian agriculture requires a public-centric approach that empowers all stakeholders, especially those at the grassroots level with the prerequisite knowledge, tools, and networks. Capacity-building through structured training, stakeholder awareness, and participatory engagement is crucial for early detection, rapid response, and adoption of good biosecurity practices. Training programs play a pivotal role in building competencies among frontline personnel such as farmers, veterinarians, and agricultural extension workers. Organizations like the Indian Council of Agricultural Research (ICAR), National Bureau of Agricultural Insect Resources (NBAIR), and Agricultural Technology Application Research Institutes (ATARIs) conduct regular field-based training sessions, demonstration programs, and workshops on pest identification, disease surveillance, and integrated pest management (ICAR-NBAIR 2022a, 2022b; ICAR-ATARI 2022). For example, the Krishi Vigyan Kendras (KVKs) network provides hands-on training on biosecurity protocols and agroecological management practices tailored to local conditions (DARE 2021). Public awareness campaigns through radio, mobile apps, social media, and local events are increasingly effective in disseminating information about invasive species, antimicrobial resistance, and zoonotic risks. Collaborative efforts under schemes like the National Animal Disease Control Programme (NADCP) and One Health India demonstrate the potential of community-based biosecurity education (DBT 2021; DAHD 2022a, 2022b). Participatory approaches, where communities codevelop surveillance and response strategies and measures, lead to greater ownership and effective compliance with biosecurity norms across the nation (Sharma et al. 2022). Nongovernmental organizations (NGOs), Farmer Producer Organizations (FPOs), cooperatives, and agri-tech startups are instrumental in bridging policy and strategies. NGOs like BAIF Development Research Foundation and PRADAN support rural livelihoods through pest management literacy and disease prevention education, while startups such as AgNext and CropIn offer digital platforms for early warning systems, remote diagnostics, and data-driven interventions (Pandey et al. 2025; Patel and Desai 2021; Gill et al. 2020). These players help democratize biosecurity information and technologies across socioeconomic divides. There is also a growing consensus on the need to integrate biosecurity into academic curricula. Agricultural universities, veterinary colleges, and extension institutions should mainstream biosecurity as a cross-cutting discipline within plant protection, animal health, food safety, and environmental science programs (Choudhary and Sharma 2021). Biosecurity courses at the undergraduate and vocational levels can ensure the next generation of professionals is biosecurity literate, thereby strengthening preparedness at systemic levels.

10 Future Perspectives and Recommendations

As the agricultural sector becomes increasingly vulnerable to global trade dynamics, climate change, and transboundary pest outbreaks, strengthening biosecurity is the need of the hour that must be envisioned through long-lasting, futuristic strategies. A future-ready biosecurity system in India should emphasize research-driven innovation, digital transformation, institutional synergy, and inclusive governance (Fig. 6.7).

10.1 Research and Innovation Priorities

Cutting-edge research and innovation in the field of pest and pathogen genomics, microbial inoculants, early-warning biosensors, and climate-resilient cropping systems are vital to strengthen India's biosecurity architecture. Emerging and advanced technologies such as remote sensing, AI, machine learning, and CRISPR-based diagnostics offer immense potential and possibilities for predictive biosecurity surveillance (Siddiqui 2024; Tripathi et al. 2021; Bhatia et al. 2020). Investments must be pointed toward multidisciplinary R&D programs, especially in the interface among plant health, animal health, food safety, and environmental biosecurity (Hulme 2020; ICAR 2023a, 2023b).

10.2 Strengthening Digital Infrastructure and Data Sharing

The integration of digital platforms for surveillance, real-time reporting, geospatial tracking, and risk assessment can catalyze and revolutionize India's biosecurity response framework system. Various programs such as FASAL by ISRO and National Animal Disease Reporting System (NADRS) demonstrate early steps toward digitization in biosecurity framework (ISRO 2022; DAHD 2022a, 2022b).

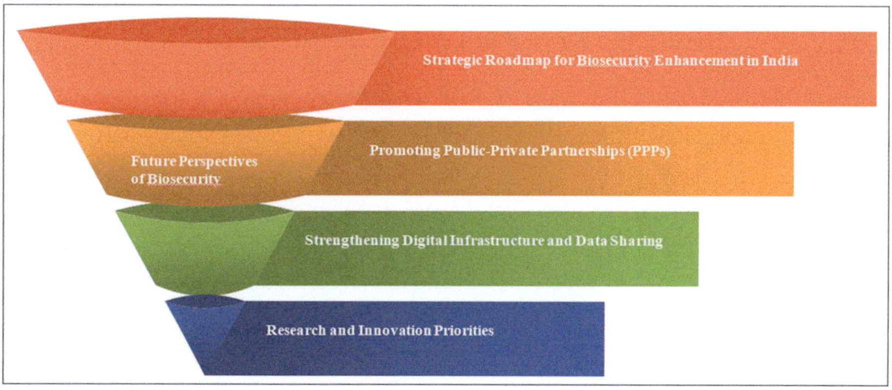

Fig. 6.7 Future perspectives of biosecurity

On the other hand, wide-spectrum application and implementation require a standardized, interoperable data sharing and managing framework across agencies, including ICAR, DAHD, FSSAI, and state departments (Sharma et al. 2022).

10.3 Promoting Public–Private Partnerships (PPPs)

PPPs are pivotal in bridging resource and knowledge gaps within permissible framework. Collaboration among research institutions, agri-tech startups, industrial players, and farmer cooperatives can further accelerate the development and deployment of biosecurity technologies across the domains. For example, startups like Stellapps, CropIn, and AgNext are working on precision livestock management and crop health diagnostics (Patel and Desai 2021). Incentivizing PPPs through financial instruments and policy support will foster innovation ecosystems focused on biorisk mitigation.

10.4 Strategic Roadmap for Biosecurity Enhancement in India

India must formulate a comprehensive National Biosecurity Strategy to institutionalize the efforts that integrates legal, scientific, and operational frameworks. The roadmap should involve the following:

(i) Establishment of a Central Biosecurity Coordination Authority for interagency coherence
(ii) Mandate periodic risk assessments and national inventories of critical bioresources
(iii) Development of standard operating procedures (SOPs) for outbreak containment and border inspection
(iv) Alignment of national efforts with global standards like the Codex Alimentarius, OIE guidelines, and the International Plant Protection Convention (IPPC) (Henao 2012; Codex Alimentarius 2020; OIE 2021a, 2021b; FAO 2021a, 2021b, 2021c, 2021d, 2021e, 2021f, 2021g, 2021h)

Ultimately, the successful implementation of national biosecurity strategies will depend on political will, sustained funding, and active stakeholder participation from grassroots communities to the central level. The future of agricultural resilience in India lies in embracing biosecurity as a core element of food system governance and sustainability.

11 Conclusion

Indian agriculture is at a critical juncture, facing multifaceted emerging threats from mounting incidences of invasive pests, zoonotic diseases, and environmental disruptions exacerbated through globalization and climate change. This chapter has comprehensively examined the strategic imperatives required to strengthen biosecurity systems in India, emphasizing that the society health and food security cannot be ensured without a robust and integrated biosecurity framework. Key insights from the chapter highlight the urgent need to modernize risk assessment and surveillance infrastructure through the adoption of digital tools like remote sensing, geospatial mapping, and molecular diagnostics. The inclusion of GAP, notably IPM, organic inputs, and improved hygiene measures in livestock systems, offers cost-effective, sustainable practices to lower the pathogen load and transmission. Ecological approaches, including the promotion of biological control agents and habitat diversification, have been recognized as viable alternatives to chemical-dependent methods, particularly in India's heterogeneous and smallholder-dominated agricultural landscapes. The One Health approach linking human, animal, and environmental health emerges as a unifying framework to combat antimicrobial resistance (AMR) and enhance cross-sector coordination. The way forward requires a coordinated national biosecurity strategy that aligns policy instruments, institutional mechanisms, and stakeholder responsibilities. State governments, ICAR, FSSAI, and emerging agri-tech innovators must co-create implementation pathways that integrate scientific research with community engagement. Investments in capacity-building for farmers, extension workers, and local institutions are equally critical. To build a resilient and futuristic biosecurity system, India must prioritize and put focus on interdisciplinary research and innovation, strengthen digital infrastructure for real-time response monitoring, encourage public–private partnerships for scalable solutions, and finally foster a culture of awareness and shared responsibility from grassroots to governance levels. Ultimately, safeguarding Indian agriculture from biosecurity threats is not just a matter of policy framework but of national resilience. Sustainable and scientific-based biosecurity measures are crucial to secure livelihoods, ensure better health, safe food supplies, and uphold ecological balance for generations to come.

References

Adebiyi AI, Oluwayelu DO (2018) Zoonotic fungal diseases and animal ownership in Nigeria. Alexandria J Med 54(4):397–402

Altieri MA, Nicholls CI, Henao A, Lana MA (2015) Agroecology and the design of climate change-resilient farming systems. Agron Sustain Dev 35:869–890

Anand M (2018) A systems approach to agricultural biosecurity. Health Secur 16(1):58–68

Anand AJ, Krishnan S (eds) (2024) Smart technologies for sustainable development goals: no poverty. CRC Press

APEDA (2023) Post-harvest infrastructure and quality standards in Indian horticulture. Agricultural and Processed Food Products Export Development Authority, Government of India, New Delhi

Arabska E (2021) From farm to fork: human health and well-being through sustainable agri-food systems. J Life Econ 8(1):11–27
Babu S, Das A, Rathore SS, Singh R, Singh R (eds) (2025) Ecological solutions to agricultural land degradation. Springer
Bhatia VS, Singh AK, Das DK (2020) Remote sensing and GIS-based pest and disease monitoring in Indian agriculture. Curr Sci 119(3):400–407
Campey AN (2024) Understanding the impact of environmental structure and sub-inhibitory antibiotic concentrations on the emergence of resistance. Doctoral dissertation, University of Exeter, UK
Chakraborty A, Sinha R (2023) Biosecurity threats in Indian agriculture and the need for integrated policy frameworks. J Agric Policy Res 15(2):95–107
Chauhan APS, Singh D, Sharma OP, Kushwah N, Kumhare A (2023) Agronomic practices for enhancing resilience in crop plants. Plant Sci Arch 8:1–3
Choudhary DK, Sharma K (2021) Role of microbial inoculants in sustainable agriculture. Indian J Microbiol 61(2):125–133
Codex Alimentarius (2020) General principles of food hygiene (CXC 1-1969). FAO/WHO, Rome/Geneva
DAHD (2022a) Annual report 2021–22. Department of Animal Husbandry and Dairying, Government of India, New Delhi
DAHD (2022b) Manual on biosecurity for livestock and poultry. Department of Animal Husbandry and Dairying, Government of India, New Delhi
DBT (2021) One Health India program: coordinated surveillance of zoonotic diseases. Department of Biotechnology, Government of India, New Delhi
Delince J, Lemoine G, Defourny P, Gallego J, Davidson A, Ray S, Achard F (2017) In: Delincé J (ed) Handbook on remote sensing for agricultural statistics
Dev SM, Sengupta R (2022) Food security and agriculture in India: post-pandemic priorities. Indian J Agric Econ 77(1):15–29
Dhobi CB, Zala MB, Verma HS, Sisodiya DB, Thumar RK, Patel MB, Borad PK (2020) Evaluation of bio-pesticides against fall armyworm, Spodoptera frugiperda (JE Smith) in maize. Int J Curr Microbiol Appl Sci 9(8):1150–1160
EFSA Panel on Biological Hazards (BIOHAZ), Koutsoumanis K, Allende A, Álvarez-Ordóñez A, Bolton D, Bover-Cid S, Peixe L (2021) Role played by the environment in the emergence and spread of antimicrobial resistance (AMR) through the food chain. EFSA J 19(6):e06651
FAO (2021a) Biosecurity: a strategic and integrated approach. Food and Agriculture Organization of the United Nations, Rome
FAO (2021b) The state of food and agriculture 2021: making agrifood systems more resilient to shocks and stresses. Food and Agriculture Organization of the United Nations, Rome
FAO (2021c) Biosecurity toolkit. Food and Agriculture Organization of the United Nations, Rome
FAO (2021d) The global framework for the progressive control of transboundary animal diseases (GF-TADs). Food and Agriculture Organization of the United Nations, Rome
FAO (2021e) Integrated pest management: guidelines and best practices. Food and Agriculture Organization of the United Nations, Rome
FAO (2021f) Good agricultural practices for food safety and environmental protection. Food and Agriculture Organization of the United Nations, Rome
FAO (2021g) Biological control: a global strategy for sustainable agriculture. Food and Agriculture Organization of the United Nations, Rome
FAO (2021h) Addressing antimicrobial resistance through a one health approach. Food and Agriculture Organization of the United Nations, Rome
Ganiger PC, Yeshwanth HM, Muralimohan K, Vinay N, Kumar ARV, Chandrashekara K (2018) Occurrence of the invasive fall armyworm Spodoptera frugiperda in maize in Karnataka, India. Pest Manag Hortic Ecosyst 24(1):23–29
Gereffi G, Lee J (2009) A global value chain approach to food safety and quality standards. Global Health Diplomacy for Chronic Disease Prevention Working Paper Series

Gershon RR, Qureshi KA, Barrera MA, Erwin MJ, Goldsmith F (2005) Health and safety hazards associated with subways: a review. J Urban Health 82(1):10–20

Gill SS, Kaur H, Singh G (2020) Agroecological management for sustainable basmati rice production in Punjab. J Eco-Friendly Agric 15(2):50–58

GLOBALG.A.P (2022) Integrated farm assurance: crops and livestock standards. GLOBALG.A.P, Cologne

Gupta RK, Sharma S, Thakur P (2020) Organic amendments and their effect on soil microbial diversity and disease suppression. J Environ Biol 41(3):532–538

Gurr GM, Wratten SD, Snyder WE (2017) Biodiversity and insect pest management in agroecosystems. Springer, Cham

Henao LP (2012) 16th Inter-American meeting at ministerial level on health and agriculture (RIMSA 16)

Hulme PE (2020) One biosecurity: a unified concept to integrate human, animal, plant, and environmental health. Emerg Top Life Sci 4(5):539–549

ICAR (2022) Good agricultural practices: Indian context and global opportunities. Indian Council of Agricultural Research, New Delhi

ICAR (2023a) Vision 2050. Indian Council of Agricultural Research, New Delhi

ICAR (2023b) National biosecurity policy framework. Indian Council of Agricultural Research, New Delhi

ICAR-ATARI (2022) Farmer capacity building through KVKs and field schools. Agricultural Technology Application Research Institute, Ludhiana

ICAR-NBAIR (2022a) Annual report 2021–22. Indian Council of Agricultural Research—National Bureau of Agricultural Insect Resources, Bengaluru

ICAR-NBAIR (2022b) Annual report on biological control programmes. Indian Council of Agricultural Research—National Bureau of Agricultural Insect Resources, Bengaluru

ICAR-NBPGR (2022) Annual report 2021–22. National Bureau of Plant Genetic Resources, ICAR, New Delhi

ICMR (2020) Annual report on antimicrobial resistance surveillance network. Indian Council of Medical Research, New Delhi

IPCC (2022) Climate change 2022: impacts, adaptation and vulnerability. Sixth assessment report. Intergovernmental Panel on Climate Change, Geneva

ISRO (2022) FASAL programme overview. Indian Space Research Organisation, Department of Space, Government of India, Bengaluru

Kaur J, Dhama AS, Buttolia H, Kaur J, Walia K, Ohri V, Singh H (2021) ICMR's antimicrobial resistance surveillance system (i-AMRSS): a promising tool for global antimicrobial resistance surveillance. JAC-Antimicrob Resist 3(1):dlab023

Kumar M, Verma A, Sharma S (2020) Bioinoculants for plant health: mechanisms and field applications. Indian J Agric Sci 90(8):1350–1356

Kumar S, Singh R, Sharma A (2021) Climate change and insect pest dynamics: implications for Indian agriculture. J Agrometeorol 23(2):195–202

Kumar S, Yadav B, Kumar R (2022) Advances in IPM technologies and their relevance in sustainable agriculture. Indian J Agric Sci 92(7):987–993

Liu Q, Jin X, Cheng J, Zhou H, Zhang Y, Dai Y (2023) Advances in the application of molecular diagnostic techniques for the detection of infectious disease pathogens. Mol Med Rep 27(5):104

Maruthadurai R, Ramesh R (2020) Occurrence, damage pattern and biology of fall armyworm, Spodoptera frugiperda (JE Smith) (Lepidoptera: Noctuidae) on fodder crops and green amaranth in Goa, India. Phytoparasitica 48(1):15–23

Meena RS, Rao CS, Kumar A (eds) (2022) Plans and policies for soil organic carbon management in agriculture. Springer

Mehta P, Bhattacharya M, Khanna R (2019) The organic transformation of Sikkim: a policy-driven success story. Agric Econ Res Rev 32(1):67–75

MoAFW (2022) Mission on soil health card and organic farming initiatives. Ministry of Agriculture and Farmers' Welfare, Government of India, New Delhi

MoAFW (2023) National biosecurity framework: draft policy report. Ministry of Agriculture and Farmers' Welfare, Government of India, New Delhi
MoEFCC (2022) India's state of environment report. Ministry of Environment, Forest and Climate Change, Government of India, New Delhi
MoFPI (2022) PM-FME scheme: guidelines for micro food processing enterprises. Ministry of Food Processing Industries, Government of India, New Delhi
Mohapatra S, Pohit S (2024) Charting the path to a developed India: Viksit Bharat 2047 (No. 167). National Council of Applied Economic Research, New Delhi
MoHFW (2017) National action plan on antimicrobial resistance (2017–2021). Ministry of Health and Family Welfare, Government of India, New Delhi
Mukherjee A, Gaikwad K, Pal A (2024) Sustainable food systems and India's trade agreements (No. 25). Indian Council for Research on International Economic Relations (ICRIER), New Delhi
Narayanan S, Gulati A (2020) Globalization and Indian agriculture: reforms, challenges, and the way forward. Indian Council for Research on International Economic Relations (ICRIER), New Delhi
NCDC (2022) Zoonotic disease surveillance and the One Health approach in India. National Centre for Disease Control, New Delhi
OIE (2021a) Biosecurity guidelines for animal production. World Organisation for Animal Health, Paris
OIE (2021b) OIE standards on antimicrobial resistance and use in animals. World Organisation for Animal Health, Paris
Pandey MK, Singh M, Sulekha, Rana P, Tripathi S, Vishwakarma SK, Pandey A, Shastri S (2025) Public-private partnerships in agricultural extension: opportunities and challenges. J Exp Agric Int 47(6):597–612. https://doi.org/10.9734/jeai/2025/v47i62373
Patel D, Desai K (2021) Enhancing farmer engagement in sustainable agriculture through ICT tools. Indian J Ext Educ 57(2):25–30
Patil RT, Naik R (2021) Food processing in India—Present status and way forward
Pearce-Duvet JM (2006) The origin of human pathogens: evaluating the role of agriculture and domestic animals in the evolution of human disease. Biol Rev 81(3):369–382
Pretty J, Benton TG, Bharucha ZP, Dicks LV, Flora CB, Godfray HCJ et al (2018) Global assessment of agricultural system redesign for sustainable intensification. Nat Sustain 1:441–446
Quénol H, Sturman A (2023) Climate change: impacts and adaptation at regional and local scales
Rajaram S, Gupta S, Medhi B (2024) Digital medicine, intelligent medicine, and smart medication system. Indian J Pharmacol 56(3):159–161
Rajendran R, Ramesh R, Sundaram M (2020) Ecological engineering for pest management in rice ecosystems: an Indian perspective. Indian J Entomol 82(1):41–47
Ramaswamy M, Jayaraj S (2021) Evaluation of ecological pest control practices in Tamil Nadu vegetable cultivation. Indian J Plant Prot 49(3):225–230
Ramesh P, Ramana DBV (2021) Agro-biodiversity and crop rotation for disease suppression: an Indian perspective. Indian J Agron 66(1):14–20
Rawal V, Bansal V, Thokchom D (2019) Commission on genetic resources for food and agriculture. In: Biodiversity for food and agriculture and food security
Richard IE (2024) The impact of sustainable agriculture practices on food security and environmental conservation. Int J Dev Res 14(12):67180–67202. https://doi.org/10.37118/ijdr.29039.12.2024
Sasikumar B, Eapen SJ (2025) Historical perspectives on adulteration and contamination of spices. In: Spices production to products, pp 3–10
Sharma M, Gupta N, Joshi A (2021) Emerging concerns of AMR in Indian agriculture. J Environ Biol 42(5):1234–1240
Sharma N, Raj A, Kapoor P (2022) Predictive analytics in agricultural biosecurity: an Indian perspective. J Agric Inform 13(2):45–58
Siddiqui SM (2024) Insights on serology, CRISPR diagnostics, and machine learning architectures for biological sequences. Doctoral dissertation, Massachusetts Institute of Technology

Singh R, Yadav M (2019) Use of natural enemies in sustainable pest management. Indian Farming 69(5):34–38

Singh M, Rathi P, Kumar A (2021) Challenges in the implementation of good agricultural practices in Indian farming. J Agri Policy Dev 9(4):110–117

Singh R, Kumar S, Bhardwaj A (2022) Challenges in biosecurity surveillance in Indian agriculture. Indian J Plant Prot 50(1):10–17

Soares BC (2024) More-than-one health: human, animals, and the environment post-COVID. Nord J Sci Technol Stud 12(1)

Stephen C (ed) (2020) Animals, health, and society: health promotion, harm reduction, and health equity in a one health world. CRC Press, Boca Raton

Tripartite Collaboration (FAO, WHO, OIE) (2022) Global action plan on antimicrobial resistance: progress report. FAO, Rome

Tripathi RK, Singh M, Rathore R (2021) Molecular diagnostics and biosensor technologies for detection of plant pathogens: advances and prospects. Plant Pathol J 37(1):1–11

Tripathi G, Dhodia A, Giri A, Rathore V, Verma A, Shukla A, Verma LK (2023) Government agriculture schemes in India: a review. Asian J Agric Ext Econ Sociol 41(11):58–67

Van Boeckel TP et al (2019) Global trends in antimicrobial use in food animals. Science 347(6224):564–567

WHO (2020a) Antimicrobial resistance: global report on surveillance. World Health Organization, Geneva

WHO (2020b) Managing zoonotic public health risks at the human-animal-ecosystem interface. World Health Organization, Geneva

WHO (2020c) One Health approach: protecting health of people, animals and ecosystems. World Health Organization, Geneva

WHO (2020d) Antimicrobial resistance: a global health threat. World Health Organization, Geneva

Yadav SK, Singh R (2021) Barriers to the adoption of integrated pest management in Indian agriculture. Agric Rev 42(3):250–256

Plant Disease Management Through Artificial Intelligence Tools

7

Govindh A. Nair, Arsha Kamal, and N. S. Radhika

Abstract

Plant diseases represent a severe and persistent threat to global agriculture, impacting crop yields and food security. Traditional methods for disease detection and management are frequently inefficient, labor-intensive, and prone to error, making them inadequate for the scale of modern farming. Artificial intelligence (AI) and machine learning (ML) are emerging as transformative solutions, offering the potential for accurate, scalable, and real-time diagnostics. AI leverages advanced image recognition, Internet of Things (IoT) sensor data, and sophisticated imaging modals to enable early disease detection, often with capabilities that surpass human visual assessment. Within this domain, deep learning (DL), particularly using convolutional neural networks (CNNs), has become a milestone for precise, image-based disease identification. The application of AI extends beyond diagnostics; predictive analytics models forecast disease outbreaks, while integration with precision agriculture hardware optimizes treatment strategies and minimizes resource consumption. Furthermore, AI-driven computational methods like molecular docking are accelerating the discovery of novel compounds for disease management. Despite these significant benefits, widespread adoption is hindered by challenges such as the availability of high-quality field data, significant computational demands, data privacy concerns, and the need for interpretable models. Future research is focused on integrating diverse AI techniques, such as combining CNNs with vision transformers (ViTs), and deploying privacy-preserving frameworks. These advancements aim to develop sophisticated, autonomous systems that leverage emerging technologies to enhance plant health and foster a more sustainable agricultural future.

G. A. Nair · A. Kamal · N. S. Radhika (✉)
Department of Plant Pathology, College of Agriculture, Kerala Agricultural University, Thiruvananthapuram, Kerala, India
e-mail: radhika.ns@kau.in

R. E. Krishnankutty et al. (eds.), *Smart Agriculture*,
https://doi.org/10.1007/978-981-95-8785-8_7

Keywords

Artificial Intelligence · Machine Learning · Plant Diseases · Precision Agriculture · Deep Learning · Convolutional Neural Networks · Vision Transformers · Disease Detection · Molecular Docking · Crop Management

1 Introduction: The Growing Threat of Plant Diseases and the Promise of Artificial Intelligence

Plant diseases represent a persistent and escalating threat to global agriculture, significantly impacting crop yields, food security, and economic stability. Estimates suggest that annual crop losses due to these diseases can amount to as much as 40% of overall production, a figure that underscores the critical need for effective and innovative management strategies. The Food and Agriculture Organization of the United Nations has confirmed that plant pests and diseases are responsible for a substantial portion of global agricultural production loss, highlighting the magnitude of the challenge (IPPC Secretariat 2021). In some cases, without proactive intervention, diseases can devastate up to 90% of a crop (Yang et al. 2025). The economic and social consequences of these losses can be catastrophic, potentially leading to food shortages, weakening national economies, and severely damaging the agricultural sector.

Traditional methods of plant disease detection and management often fall short in addressing the complexities and scale of modern agriculture. These approaches frequently rely on manual scouting by human experts and subsequent laboratory-based assays, which are time-intensive, prone to human error, and often fail to detect infections in their early, most treatable stages. The reliance on a limited pool of skilled plant pathologists makes manual detection inherently unscalable and inadequate for the vast acreages of farming operations. This operational model, where expertise is physically bound to an individual, creates a fundamental bottleneck. Modern agriculture demands a system where expertise can be decoupled from physical presence and scaled across immense geographical areas, a requirement that traditional methods cannot meet. This mismatch between the linear scalability of human expertise and the exponential needs of modern agriculture creates a powerful economic and operational momentum for technological innovation.

Artificial intelligence and machine learning have emerged as an innovative solution, offering highly accurate, scalable, and real-time diagnostic capabilities that overcome the limitations of conventional approaches (Almasoudi 2024; Pun et al. 2025). AI-powered systems can analyze vast quantities of data from sources like smartphone cameras, drones, and satellites to automatically detect disease symptoms, often with an accuracy that surpasses human visual assessment (Pandey et al. 2024; Bharwal and Choudhary 2025). Studies have demonstrated that AI models can achieve diagnostic accuracies exceeding 98% in various applications (Mahabub et al. 2024). This capability enables a shift toward precision agriculture, where interventions are targeted and resource use is optimized, thereby reducing the

Table 7.1 A comparison of traditional and advanced methods for plant disease detection

Feature	Traditional methods	Advanced methods
Primary method	Manual field scouting by human experts and subsequent laboratory-based assays	Automated analysis of data from images, Internet of Things (IoT) sensors, and advanced imaging technologies like hyperspectral imaging
Speed	Slow and time-intensive	Real-time or near real-time, enabling rapid diagnostics
Scalability	Poor. Inherently unscalable due to reliance on a limited pool of human experts	Highly scalable. Expertise can be decoupled from physical presence and scaled across vast geographical areas
Accuracy and timing	Prone to human error and often fails to detect infections in their early stages	Often surpasses human visual assessment and enables pre-symptomatic detection before symptoms are visible to the naked eye
Data source	Visual inspection by the naked eye and laboratory analysis	Diverse sources including smartphone cameras, drones, satellites, in-field IoT sensors, and hyperspectral/multispectral cameras
Key limitation	Labor-intensive, inefficient, and fundamentally constrained by the physical presence and availability of human experts	Requires significant computational resources, high-quality data for training, and can be hindered by data privacy concerns

reliance on broad-spectrum pesticides and improving overall crop management. Recognizing the promising potential of these technologies, this chapter provides a comprehensive review of the application of AI tools in managing plant diseases. It covers the full spectrum from detection and diagnosis to management, including foundational ML and DL concepts, the innovative use of molecular docking, and a look toward future trends (Table 7.1).

2 Foundational Concepts: AI, Machine Learning, and Deep Learning in Plant Pathology

To understand the application of artificial intelligence in plant pathology, it is essential to first define the core technologies that form the bas for these advancements. AI, machine learning (ML), and deep learning (DL) represent a hierarchy of interconnected fields, each contributing unique capabilities to the challenge of managing plant diseases.

2.1 Artificial Intelligence, Machine Learning, and Deep Learning

Artificial intelligence (AI) is the broadest concept, referring to the ability of machines to perform tasks that typically require human intelligence, such as

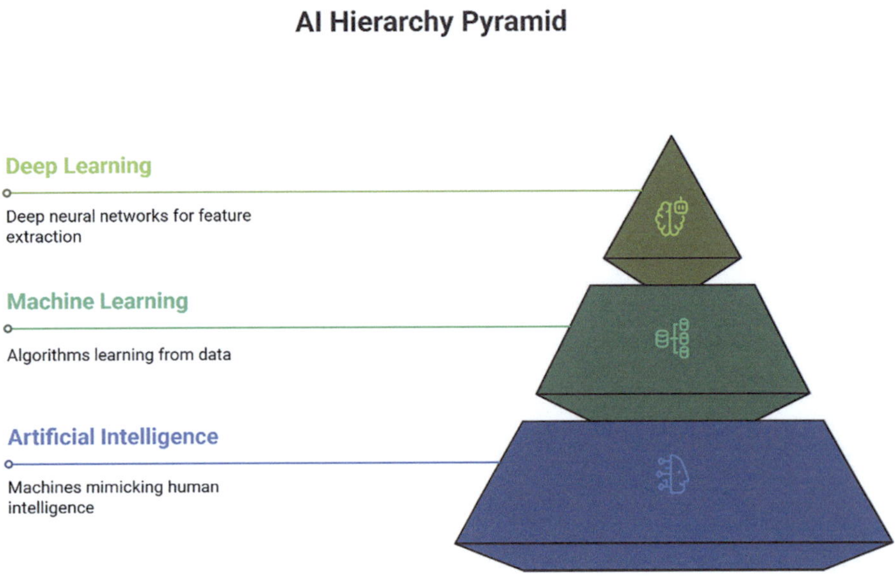

Fig. 7.1 The AI hierarchy, where the broad field of artificial intelligence (AI) encompasses machine learning (ML), a subfield in which algorithms learn from data, which in turn includes deep learning (DL), a specialization using deep neural networks to automatically learn features from data

learning, problem-solving, and decision-making (Almasoudi 2024). Machine learning (ML) is a pivotal subfield of AI that involves the development of algorithms that allow computers to learn from and make predictions or decisions based on data, without being explicitly programmed for the task. In plant pathology, ML algorithms are trained on agricultural data that includes images of leaves, environmental sensor readings, and weather patterns to identify patterns indicative of disease.

Deep learning (DL) is a further specialization within ML characterized by the use of artificial neural networks with multiple layers and hence the term "deep." These deep neural networks are particularly proficient at automatically learning intricate, hierarchical features from raw data. This capability is a significant departure from traditional ML methods, which often require a human expert to manually define and extract relevant features from the data, a process known as feature engineering (Fig. 7.1 and Table 7.2).

2.2 Key Algorithms and Architectures

The evolution of algorithms used in plant pathology reflects a clear trajectory toward greater model autonomy and the capacity to understand increasingly complex spatial relationships within data. This progression can be understood by examining the different classes of algorithms employed.

Table 7.2 A comparison of key differences between machine learning (ML) and deep learning (DL)

Feature	Machine learning (ML)	Deep learning (DL)
Core concept	A subfield of AI where algorithms learn from data to make predictions without being explicitly programmed	A specialization within ML that uses artificial neural networks with multiple layers ("deep" networks)
Feature engineering	Relies heavily on manual feature engineering, where a human expert defines and extracts relevant features (e.g., color, texture, shape) from the data	Performs automatic feature extraction, learning intricate, hierarchical features directly from raw data like pixels, which eliminates the human bottleneck in feature design
Architecture	Uses algorithms like support vector machines (SVMs), decision trees, and random forests	Employs deep neural network architectures like convolutional neural networks (CNNs), recurrent neural networks (RNNs), and vision transformers (ViTs)
Primary limitation	Performance is fundamentally constrained by the quality of the human-defined features; the process requires significant domain expertise and is time-consuming	Often requires large, diverse datasets to achieve high performance and can be computationally demanding to train. Models can also be a "black box," making them difficult to interpret

2.2.1 Traditional Machine Learning

Before the dominance of deep learning, traditional ML algorithms were widely used for disease classification. These methods, which fall under the category of supervised learning, are trained on datasets where each data point (e.g., an image) is labeled with a known outcome (e.g., "healthy" or "leaf spot infection"). Key algorithms include

- *Support vector machines (SVMs)*: These algorithms find an optimal hyperplane that separates data points into different classes. In plant pathology, SVMs analyze manually extracted features like color, texture, and shape to differentiate between healthy and diseased plants (Cervantes et al. 2020; Koshariya et al. 2025).
- *Decision trees and random forests*: A decision tree is a model that uses a tree-like structure of decisions. A random forest is an ensemble method that builds multiple decision trees and merges their outputs to improve predictive accuracy and control over-fitting, making it a robust choice for classification tasks (Fretello and Tagliaferri 2019; Koshariya et al. 2025).
- *Other supervised methods*: Algorithms such as K-nearest neighbors (KNN), which classifies a data point based on the labels of its nearest neighbors, have also been used (Cervantes et al. 2020).

The primary limitation of these methods is their reliance on manual feature engineering. This process is not only time-consuming and requires significant domain expertise but also acts as a knowledge bottleneck, as the model's performance is fundamentally constrained by the quality of the human-defined features.

2.2.2 Deep Learning Architectures

Deep learning has largely overcome the feature engineering bottleneck by learning relevant features automatically. Several key architectures have become central to modern plant pathology.

- *Convolutional neural networks (CNNs)*: CNNs are the cornerstone of modern computer vision and the most widely used DL architecture for image-based plant disease detection (Al-Amin et al. 2019; Jafar et al. 2024). They are specifically designed to process grid-like data, such as images, using specialized layers called convolutional and pooling layers to automatically and hierarchically learn features like edges, textures, and shapes directly from pixel data. This automated feature extraction removes the human bottleneck in feature design.
- *Recurrent neural networks (RNNs) and LSTMs*: Recurrent neural networks (RNNs) are a class of deep learning models designed to effectively process sequential data by maintaining an internal state (memory), a key distinction from feedforward neural networks (Mienye et al. 2024). This characteristic makes them suitable for applications where the order and context of data points are crucial, such as natural language processing or time series forecasting. While early RNNs faced challenges like the vanishing gradient problem, a significant advancement came with long short-term memory (LSTM) networks, introduced by Hochreiter and Schmidhuber in 1997. LSTMs overcome this by employing sophisticated gating mechanisms (input, forget, and output gates) that regulate information flow, enabling them to capture and maintain long-term dependencies over extended periods. This makes them ideal for analyzing time-series data from weather stations or IoT sensors to model and predict disease progression over time.
- *Vision transformers (ViTs)*: A more recent and powerful alternative to CNNs, ViTs have shown remarkable success in computer vision (Dosovitskiy et al. 2021; Thakur et al. 2022). Instead of using the localized filters of CNNs, ViTs treat an image as a sequence of patches and employ a mechanism called self-attention to weigh the importance of all patches relative to each other (Salman et al. 2025). This allows ViTs to capture long-range, global dependencies within an image, a capability that is crucial for identifying diseases with subtle or widely distributed symptoms that a CNN's local view might miss.
- *Hybrid models*: An emerging trend is the development of hybrid architectures that combine the strengths of CNNs and ViTs. Models such as MaxViT and PlantXViT leverage CNNs for their robust ability to extract local features while using ViT components to integrate global context (Thakur et al. 2022). This synergistic approach aims to create models that are superior at capturing both local and global spatial relationships, pushing the boundaries of diagnostic accuracy.

This architectural evolution represents a conceptual shift in what is being asked of the model. Traditional ML required humans to tell the model what features to look for. CNNs allowed the model to discover local features on its own. ViTs and hybrid models now permit the model to decide that parts of the entire image are most important in a global context. This progression is vital for tackling the most difficult diagnostic cases, such as fine-grained identification of diseases with very similar or subtle symptoms, where long-range dependencies are key to a correct classification. Table 7.3 provides a comparative overview of these foundational technologies.

2.2.3 The Architectural Workflow of a CNN for Disease Identification

The capacity of a convolutional neural network (CNN) to accurately identify plant diseases from images is not a single process but a structured workflow executed through a sequence of specialized architectural layers. Each layer performs a specific transformation on the input data, progressively abstracting information from raw pixels to high-level diagnostic features. The technical workflow can be broken down into the following key stages.

A. *Input Layer and Data Representation*

The process initiates at the input layer, where an image is provided to the network. This image is converted into a numerical format that the network can process: a multidimensional array, or tensor. For a standard color image, this is

Table 7.3 Examples of machine learning and deep learning architectures in plant disease management

Architecture type	Specific model/ algorithm	Learning paradigm	Primary application	Key strengths/ weaknesses
Traditional ML	Support vector machine (SVM)	Supervised	Feature-based classification	Interpretable; requires manual feature engineering
Traditional ML	Random forest	Supervised (ensemble)	Robust feature-based classification	High accuracy for tabular data; requires manual feature engineering
Deep learning	ResNet-50	Supervised, transfer learning	Image classification	High accuracy; residual connections prevent vanishing gradients
Deep learning	MobileNet	Supervised, transfer learning	Mobile/edge image classification	Lightweight and fast; slightly lower accuracy than larger models
Deep learning	LSTM	Supervised	Time-series analysis, forecasting	Models temporal dependencies; not ideal for static images
Deep learning	Vision transformer (ViT)	Supervised, transfer learning	Image classification with global context	Captures long-range dependencies; requires large datasets to train effectively
Deep learning	MaxViT (hybrid)	Supervised, transfer learning	Image classification	Combines local (CNN) and global (ViT) strengths; complex architecture

typically a 3D tensor of size (height × width × 3), where the three channels correspond to the red, green, and blue (RGB) pixel intensity values.

B. *Convolutional Layers: Hierarchical Feature Extraction*

The core of a CNN's analytical capability resides in its convolutional layers. These layers are responsible for automatically detecting and learning features directly from the image data.

- *Filters and the convolution operation*: Each convolutional layer contains a set of learnable filters (also known as kernels), which are small matrices of weights. These filters are designed to detect specific features. The filter systematically slides, or convolves, across the input tensor, performing an element-wise multiplication with the patch of the image it covers and summing the results. This operation produces a feature map, a 2D representation indicating where in the image the filter's specific feature was detected.
- *Hierarchical learning*: The power of the CNN architecture comes from stacking these layers. The initial layers learn to detect very low-level features, such as edges, color gradients, or simple textures. The feature maps from these layers are then passed as input to subsequent layers, which learn to combine the simpler features into more complex representations. For example, a deeper layer might learn to identify the circular pattern of a "leaf spot" lesion or the fine texture of powdery mildew by combining the edge and color information from earlier layers.
- *Activation function*: Following the convolution, an activation function, most commonly the rectified linear unit (ReLU), is applied to the feature map. This introduces nonlinearity into the model, which is critical for learning the complex, nonlinear relationships that define visual symptoms.

C. *Pooling Layers: Dimensionality Reduction and Invariance*

After a convolutional layer, a pooling layer is often introduced to reduce the spatial dimensions of the feature maps. This serves two primary purposes:

1. *Computational efficiency*: It decreases the number of parameters in the network, reducing computational load and helping to prevent overfitting.
2. *Translational invariance*: It makes the feature detection more robust to the exact location of the feature in the image. For instance, Max Pooling, the most common technique, takes the maximum value from a small window of a feature map. This ensures that the network recognizes a feature as present within a region, regardless of its precise pixel coordinates.

Fully Connected Layers and Classification

After passing through the sequence of convolutional and pooling layers, the final high-level feature maps are flattened into a one-dimensional vector. This vector represents a compressed, feature-rich summary of the original image.

- *Classifier*: This vector is fed into a fully connected layer, which is a standard feedforward neural network. This part of the architecture acts as a classifier,

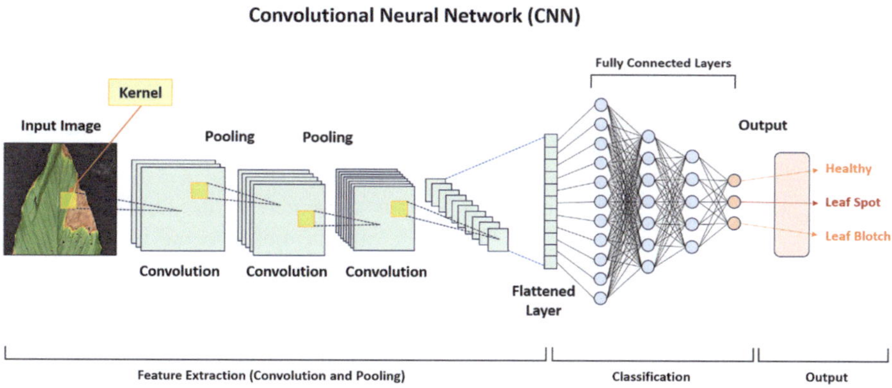

Fig. 7.2 The architectural workflow of a convolutional neural network (CNN) for plant disease identification

learning to weigh the importance of the different abstract features and map their combinations to a specific disease.

- *Output layer*: The final layer, typically using a Softmax activation function, produces the network's output. It calculates a probability distribution across all predefined classes (e.g., "Healthy," "Leaf Spot," "Leaf Blotch"). The class assigned the highest probability is the model's final diagnosis for the input image (Fig. 7.2).

3 AI-Powered Plant Disease Detection: Early Warning Systems

Early and accurate detection of plant diseases is essential for minimizing crop losses and enabling timely, effective interventions. Artificial intelligence facilitates a multimodal approach to early detection, creating sophisticated warning systems by integrating data from a variety of sources, including conventional images, advanced remote sensing platforms, and in-field IoT devices.

One of the primary applications of AI is in the analysis of images captured from diverse platforms such as smartphones, unmanned aerial vehicles (UAVs or drones), and satellites (Pun et al. 2025). AI models trained on extensive datasets of healthy and diseased plants can analyze high-resolution images to identify subtle visual patterns such as changes in color, texture, or the appearance of spots that are indicative of specific diseases, often before they become apparent to the naked eye (Jafar et al. 2024). Drone-based imaging systems are particularly effective, allowing for efficient and automated surveillance of large agricultural areas, providing farmers with wide-ranging aerial views that facilitate targeted management strategies (Abbas et al. 2023).

AI also enables early detection through the analysis of nonvisual data streams from the Internet of Things (IoT). Sensors deployed in fields can continuously monitor a host of environmental conditions like temperature, humidity, and soil moisture, as well as plant physiological parameters (Khan et al. 2022). AI algorithms can process this real-time data to detect anomalies or patterns that are known to precede the visual manifestation of symptoms, thereby providing early alerts to farmers about potential outbreaks (González-Rodríguez et al. 2024; Pandey et al. 2024).

The capability for early detection is further enhanced by combining AI with advanced imaging technologies like hyperspectral and multispectral imaging (Pun et al. 2025). These techniques capture information from across the electromagnetic spectrum, far beyond the range of human vision or standard cameras. Hyperspectral data, in particular, can reveal subtle changes in a plant's chemical composition and physiological state that are early indicators of stress from a pathogen infection. Deep learning models trained on this spectral data can identify unique signatures associated with different diseases, enabling pre-symptomatic and highly precise diagnosis.

The true power of these systems, however, lies in data fusion for predictive analytics. AI-powered forecasting models serve as the cornerstone of modern early warning systems by integrating these different data sources (Pandey et al. 2024). A predictive model might learn, for instance, that a specific combination of rising humidity detected by an IoT sensor, a 10-day weather forecast predicting rain, and a history of blight in the region creates a high probability of an outbreak within the coming week. This multimodal approach enables a fundamental shift in disease management philosophy, moving farmers from a reactive stance ("I see spots on my leaves, what should I do?") to a proactive one ("The system predicts a high risk of blight next week, so I will apply a preventative treatment now"). This transition from detection to prediction has profound economic and environmental consequences. Proactive, targeted prevention is far more efficient and consumes fewer resources, such as water and pesticides, than large-scale reactive treatment of an established outbreak, directly supporting the goals of sustainable and precision agriculture. Several AI-powered platforms are already making this a reality. For example, Plantix is a mobile application that uses deep learning to analyze images of symptomatic crops and provides diagnoses and management options, while also using anonymized metadata to track disease outbreaks. Similarly, Farmonaut integrates satellite technology with AI to offer comprehensive crop health monitoring, and PlantVillage Nuru provides phone-based diagnostics to support growers in making informed decisions (Bolakhe 2023).

4 Machine Learning for Accurate Plant Disease Diagnosis

While deep learning has become the dominant factor in image-based diagnostics, traditional machine learning algorithms remain highly relevant and effective. These methods are particularly valuable in scenarios where computational resources are limited or when working with well-defined, engineered features derived from domain expertise.

The performance of traditional ML algorithms like support vector machines (SVMs), decision trees, and random forests is critically dependent on the quality of the features extracted from the images (Koshariya et al. 2025). This process, known as feature extraction, involves a human expert defining and calculating specific metrics from an image that are believed to be discriminative for disease identification. These features typically fall into three categories:

- *Spectral features*: These relate to the color information in the image, such as the mean and standard deviation of pixel values in the red, green, and blue channels, or color histograms.
- *Texture features*: These describe the surface patterns of the leaf or lesion. A common method for extracting texture features is the gray-level co-occurrence matrix (GLCM), which captures spatial relationships between pixels and can be used to calculate metrics like contrast, correlation, and energy.
- *Morphological features*: These capture the shape and size of lesions or affected areas, including metrics like area, perimeter, circularity, and aspect ratio.

Once these features are extracted, they are fed into an ML classifier. SVMs have been effectively used for feature-based classification, while ensemble methods like random forests are known for their robustness and high accuracy (González-Rodríguez et al. 2024; Koshariya et al. 2025). Studies have shown that these systems can achieve high diagnostic accuracy, often exceeding 90%, and can outperform the subjectivity of human visual inspection (Otero et al. 2022). For example, one system for cotton disease detection combined GLCM texture features with a CNN classifier to achieve 87% accuracy (Meena et al. 2024).

The distinction between traditional ML and deep learning is fundamentally about the division of labor between the human and the machine. In traditional ML, the intellectual effort is "front-loaded" into the human-led design of features. This requires deep domain expertise but can be advantageous when labeled training data is scarce. In deep learning, the effort is "front-loaded" into acquiring massive labeled datasets and the computational power needed to train the model to learn features automatically. This also directly impacts model interpretability. Because the features in a traditional ML model are human defined (e.g., "lesion redness," "texture contrast"), the model's decision-making process is more transparent. We can inspect the model to see which features it deemed most important. The "black box" nature of many deep learning models arises precisely because the features they learn are complex, abstract, and not always intuitive to humans (Jafar et al. 2024). Therefore, for applications where high levels of trust and explainability are required, traditional or hybrid ML approaches may remain a preferred choice.

5 Deep Learning Architectures in Image-Based Plant Disease Identification

Deep learning architectures, particularly convolutional neural networks (CNNs), have set the performance benchmark for image-based plant disease identification and form the technological core of most modern agricultural diagnostic systems. Their success stems from their exceptional ability to automatically learn complex, hierarchical features directly from raw image data, thereby eliminating the need for the laborious and expertise-dependent process of manual feature engineering (Al-Amin et al. 2019; Agarwal et al. 2020; Jafar et al. 2024).

CNNs have become the standard and dominant architecture for this task. A wide array of specific CNN architectures have been successfully applied, including early models like AlexNet and VGGNet (VGG16, VGG19), as well as more advanced networks such as GoogLeNet, ResNet (e.g., ResNet-50), InceptionV3, MobileNet, DenseNet, and Xception (Barbedo 2018; Al-Amin et al. 2019; Jafar et al. 2024). On benchmark datasets like PlantVillage, these models consistently achieve classification accuracies exceeding 95%, with some studies reporting success rates over 99% (Koshariya et al. 2025). For instance, an early landmark study by Mohanty et al. (2016) achieved 99.35% accuracy using a CNN on the PlantVillage dataset.

A critical factor in the success and rapid proliferation of these models is the use of transfer learning. This technique leverages models that have been pretrained on massive, general-purpose image datasets like ImageNet. The underlying premise is that the initial layers of these networks have already learned to recognize universal, low-level visual features such as edges, colors, textures, and simple shapes. This foundational visual understanding is largely domain-agnostic and thus transferable to other image recognition tasks. By using a pretrained model, researchers can simply "fine-tune" the final layers on a smaller, domain-specific dataset of plant disease images. This approach dramatically reduces the amount of data and computational resources required to train an effective model from scratch, as the network does not need to learn how to "see" all over again.

This insight that the basic building blocks of vision are universal has democratized the development of specialized AI tools. It allows research groups to achieve state-of-the-art results for specific regional crops or diseases without needing millions of their own labeled images, thereby accelerating innovation across the agricultural sector. Furthermore, the development of lightweight architectures like MobileNet and SqueezeNet, which are designed with fewer parameters and lower computational demands, has been crucial for deploying these powerful diagnostic tools on resource-constrained edge devices such as smartphones and drones for real-time, in-field analysis.

6 AI in Plant Disease Management: Predictive Analytics and Precision Agriculture

The role of artificial intelligence in agriculture extends far beyond simple diagnosis. It is a cornerstone of proactive crop management, where predictive analytics and integration with precision agriculture hardware enable optimized, targeted, and sustainable interventions that are impossible to achieve with traditional methods.

AI-powered predictive models analyze a confluence of data, including historical weather patterns, real-time environmental factors from IoT sensors, and data on pathogen life cycles, to forecast the likelihood of disease outbreaks with remarkable accuracy (Pandey et al. 2024). These forecasts, which can predict risks weeks in advance, are a core function of modern agricultural decision support systems (DSS). By providing early warnings, these systems allow farmers to shift from a reactive to a proactive management strategy, applying preventative measures before a disease can establish itself and cause significant damage.

This predictive capability is tightly integrated with the principles of precision agriculture. AI systems process high-resolution data from drones, satellites, and ground-based sensors to generate detailed "disease maps" of a field, identifying specific areas where infection is present or imminent. This enables highly targeted interventions, such as the variable-rate application of pesticides or fungicides. Instead of spraying an entire field uniformly, automated equipment can apply treatments only to the affected zones. This approach drastically reduces the overall amount of chemicals used, which not only lowers costs for the farmer but also minimizes environmental contamination and slows the development of pesticide resistance.

This entire process represents a fundamental shift in the scale of agricultural management. Traditional farming has historically managed populations, treating the entire field as a single, uniform unit. AI-driven precision agriculture enables the management of subpopulations, treating specific zones within a field differently based on their needs. The future trajectory of this technology points toward the management of individuals. The development of autonomous agricultural robots, equipped with sophisticated AI vision systems and multispectral sensors, is paving the way for real-time, plant-by-plant monitoring and treatment. These robots can autonomously navigate fields, identify individual diseased plants with high precision, and apply a micro-dose of treatment directly to the affected plant, further reducing resource use and labor costs. This "personalization of agriculture," analogous to personalized medicine in healthcare, is driven by AI's ability to process data at a previously unattainable level of granularity. It creates a virtuous cycle: more granular data leads to better models, which enable more precise actions, which in turn generate more detailed data on treatment efficacy, further refining the models. This self-improving loop is the engine driving the future of smart, efficient, and sustainable agriculture.

7 Molecular Docking: An AI-Driven Approach to Discovering Novel Disease Management Compounds

In addition to field-level management, artificial intelligence is contributing to plant health at the molecular level by accelerating the discovery of novel disease control agents. Molecular docking, a computational technique heavily augmented by AI, is at the forefront of this effort. It offers a rational, targeted approach to identifying new fungicides and bactericides, moving beyond the traditional trial-and-error screening of chemicals.

Molecular docking is an in silico method that predicts the preferred orientation and binding affinity of one molecule (a ligand, such as a potential fungicide) to the active site of another (a target protein, typically an enzyme crucial for a pathogen's survival. By simulating these molecular interactions, researchers can rapidly screen vast virtual libraries containing thousands of natural or synthetic compounds to identify candidates with the highest potential to inhibit the pathogen's function (Ma et al. 2024). AI algorithms are increasingly integrated into these pipelines to enhance the speed of screening, improve the accuracy of binding affinity predictions, and even aid in the de novo design of new compounds with optimized properties (Fig. 7.3).

A common and effective strategy is to target enzymes that are essential for a pathogen's metabolism or virulence. For example, succinate dehydrogenase (SDH)

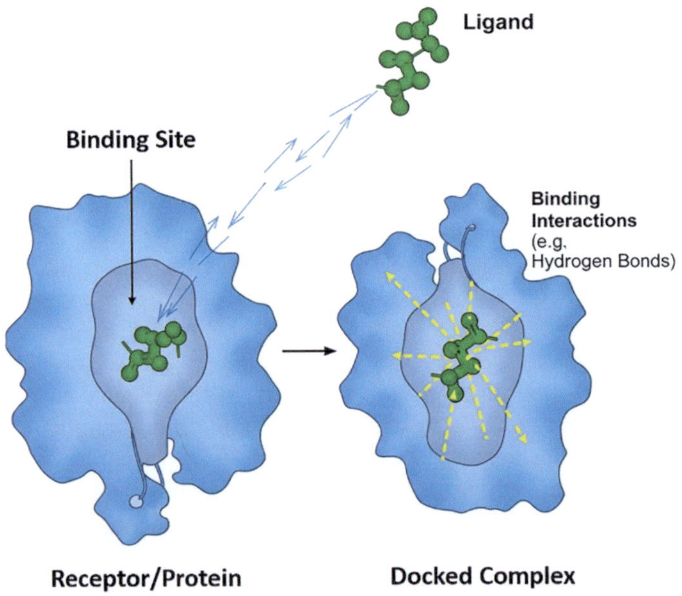

Fig. 7.3 An illustration of the molecular docking workflow. This computational technique predicts the preferred orientation of a ligand as it binds to a receptor to form a stable complex. The process evaluates numerous potential poses to find the most favorable one, which is stabilized by key binding interactions

is a critical enzyme in the fungal respiratory cycle, and SDH inhibitors (SDHIs) represent a major and rapidly growing class of modern fungicides (Ma et al. 2024). Several studies have successfully used molecular docking to identify potential SDHIs and other antifungal compounds:

- In a study targeting wheat tan spot disease, molecular docking was used to screen plant-derived bioactive compounds against the pathogen's Tox-A protein. The analysis identified lycopene as having the highest binding affinity, suggesting its potential as a novel bio-fungicide (Ullah et al. 2024).
- Another research effort designed and synthesized a new class of spirotryprostatin A derivatives. Molecular docking simulations confirmed that these compounds bind effectively within the active site of the SDH enzyme, validating them as promising candidates for a new family of chiral fungicides (Ma et al. 2024).
- Similarly, researchers investigating flavones as potential fungicides against the soil-borne pathogen *Rhizoctonia solani* used molecular docking to reveal that a specific chemical structure, the bromophenyl moiety, was a key feature for effective binding to the SDH enzyme, providing critical guidance for future chemical synthesis (Onnom et al. 2024).

This planned design method is similar to how drug discovery has changed in human medicine, moving from lucky finds to focused and purposeful development. This is particularly important for addressing the growing problem of fungicide resistance. As pathogens evolve and mutate their target proteins, rendering existing chemicals ineffective, molecular docking can be used to rapidly screen for new compounds that can bind to the mutated protein or to a different essential target altogether. This provides a more agile and responsive pipeline for developing the next generation of crop protection agents, helping to ensure the long-term viability of chemical disease control.

8 Successful Applications of AI in Managing Specific Plant Diseases: Case Studies

The practical value and versatility of AI in agriculture are best demonstrated through its successful application in detecting and managing economically significant diseases across a wide range of major crops. These case studies highlight the tangible benefits of AI, from improved accuracy to enabling new management strategies.

Case Study 1: Black Pepper Disease Identification Using a Customized CNN

A study conducted by Sreethu et al. (2025) presented a deep learning-based approach for classifying black pepper diseases based on foliar symptoms. Black pepper is a crucial global spice, but its production is severely threatened by diseases such as foot rot and anthracnose. The research underscored the urgency of early and precise disease diagnosis, as traditional methods relying on visual inspection by experts are often time-consuming, labor-intensive, and inaccessible to many farmers in remote areas.

The researchers built a comprehensive dataset of 2786 original leaf images from various black pepper farms in Kerala, India, covering healthy leaves and both initial and advanced stages of anthracnose, as well as foot rot. To ensure the model's robustness and handle data imbalance, image augmentation techniques (e.g., rotation, flipping, brightness adjustment) were applied, generating a total of 18,234 images. A customized convolutional neural network (CNN) model was developed and rigorously optimized by tuning hyperparameters like batch size and learning rate. This developed CNN model achieved an outstanding classification accuracy of 98.72% on the unseen test dataset, along with high precision (99.28%), recall (97.65%), and F1-score (98.66%). Critically, this customized model demonstrated superior performance compared to eight other pretrained state-of-the-art models, including VGG16, VGG19, ResNet50, MobileNet V2, DenseNet121, InceptionV3, and Xception. Notably, ResNet50 exhibited the lowest test accuracy among the pre-trained models in this specific comparative analysis, at 65.13%. To facilitate practical application, the study also involved the development of a web application titled "Black pepper Disease Identification App," enabling farmers and extension personnel to upload leaf images for instant diagnosis and classification.

Case Study 2: AI-Powered Banana Disease and Pest Detection Across Plant Parts

A study conducted by Selvaraj et al. (2019) introduced an AI-based system for detecting banana diseases and pests by analyzing visual symptoms on different parts of the banana plant. Given that bananas are a critical staple food and commercial crop, and are highly susceptible to numerous diseases and pests that cause significant yield losses, especially for smallholder farmers, the research aimed to provide novel and rapid detection methods. The study leveraged deep convolutional neural networks (DCNNs) and transfer learning as promising avenues for timely crop disease detection.

A distinguishing feature of this research was the collection of a large dataset of approximately 18,000 expert prescreened real-field images of banana disease and pest symptoms from various hotspots in Africa and Southern India. This approach countered the limitations of existing datasets, which often feature detached leaves on plain backgrounds, by incorporating images captured under diverse environmental conditions and with heterogeneous backgrounds, thereby enhancing the model's robustness for real-world application. The system was designed to detect five major banana diseases (Xanthomonas wilt, Fusarium wilt, black sigatoka, yellow sigatoka, banana bunchy top disease) and the banana corm weevil pest, with symptoms analyzed across six different plant parts: the entire plant, leaves, pseudo stem, fruit bunch, cut fruits, and corm. The researchers retrained three different CNN architectures—ResNet50, InceptionV2, and MobileNetV1—using a transfer learning approach. Their findings indicated that ResNet50 and InceptionV2-based models performed better than MobileNetV1, achieving accuracies of more than 90% in most of the tested models. Models focusing on pseudostem, fruit bunch, and corm showed particularly high accuracy (over 95%) due to more focused images with less background variation. This robust detection system represents a significant step toward developing a fully automated mobile application that can provide real-time assistance to millions of banana farmers in developing countries.

Case Study 3: DeepCrop—A Web Application for Multicrop Disease Prediction

A study conducted by Islam et al. (2023) focused on developing a deep learning-based system for crop disease prediction, culminating in a smart web application. The motivation behind this research was to combat the significant economic losses experienced by farmers due to various plant diseases, emphasizing the critical need for timely and efficient detection to prevent waste and conserve resources. The researchers investigated the performance of several convolutional neural network (CNN) models, including CNN, VGG-16, VGG-19, and ResNet-50, using a large dataset of 10,000 plant-village images to identify crop infections. Their extensive experiments revealed that the ResNet-50 model significantly outperformed the others, achieving an accuracy rate of 98.98%. This surpassed the accuracy of CNN (98.60%), VGG-16 (92.39%), and VGG-19 (96.15%).

Based on ResNet-50's superior performance, it was selected as the foundational model for the development of a smart web application. This application is designed to aid farmers in diagnosing plant diseases by allowing them to upload photos of affected plant leaves for analysis. Beyond merely identifying whether a leaf is healthy or infected, the application aims to classify the specific disease type and provide preliminary treatment recommendations. The target crops for this system include potatoes, tomatoes, and peppers, which are vital vegetables frequently affected by diseases like early and late blight. This deep learning-guided system is intended to mitigate crop loss, conserve resources, and ultimately enhance agricultural productivity, particularly beneficial in regions like Bangladesh and India where access to agricultural specialists may be limited and traditional methods are time-consuming.

Some of the other major studies are as follows:

- *Rice*: As a staple food for a large portion of the world's population, rice health is critical. AI technologies, including CNNs and deep neural networks (DNNs), have been successfully employed for disease detection. A convolutional neural network (CNN) approach for automated rice leaf disease detection, proposed by Kumar et al. (2025), achieved an overall accuracy of 96.5% on its test set. This model specifically focused on classifying healthy rice leaves and those affected by blast, brown spot, and bacterial leaf blight. For individual disease categories, the model demonstrated strong performance with F1-scores (a balance between precision, which measures the accuracy of positive predictions, and recall, which measures the ability to identify all actual positive cases) of 95.6% for blast, 96.7% for brown spot, 97.0% for bacterial leaf blight, and 97.1% for healthy leaves. In a broader context of artificial intelligence and machine learning applications in rice disease detection, Aggarwal et al. (2022), in their survey of various techniques, highlighted diverse results from other CNN-based studies. They noted one CNN model that achieved 94% accuracy in classifying healthy versus ailing rice leaves and 78.44% accuracy in categorizing different types of sick leaves. Furthermore, their review indicated that accuracies for CNNs in rice disease detection ranged from 91.23% (using AlexNet for three diseases) to a high of 99.53% (with Stacked-CNN and VGG-16 for nine diseases). These results collectively underscore the high efficacy of CNNs in accurately identifying and classifying rice leaf diseases for improved agricultural management.

- *Wheat*: AI techniques are being used to detect devastating wheat diseases like rust and powdery mildew. Deep learning models like VGG19, ResNet152, DenseNet169, InceptionNetV3, and MobileNetV2, have achieved classification accuracies for wheat rusts ranging from 91% to over 98% (Nigam et al. 2023). Beyond detection, molecular docking has been explored to identify potential fungicides against pathogens like the one causing tan spot disease (Ullah et al. 2024).
- *Maize*: AI has been successfully applied for the early detection of diseases like maize leaf blight, enabling timely interventions to reduce crop losses. In a novel application addressing data privacy, federated learning has been used to train various CNNs (including AlexNet and VGG-11) for maize leaf disease detection, achieving accuracies over 97% without centralizing sensitive farm data (Antico et al. 2022).
- *Tomato*: Tomato plants have been a major focus of AI research. Numerous deep learning models have been developed, with reported accuracies often exceeding 98% on benchmark datasets. For example, novel architectures like the T-Net model and optimized compact convolutional transformers (CCT) have achieved accuracies of 98.97% and 98.87%, respectively, demonstrating the power of tailored models (Batool et al. 2024; Mahabub et al. 2024).
- *Cotton*: To address the need for in-field deployment, researchers have developed lightweight, parameter-efficient models for cotton disease classification. One such model, based on MobileNet, can classify seven different cotton diseases with 98.42% accuracy, making it suitable for mobile applications (Patra and Gajurel, 2025). Other approaches have successfully used ResNet architectures and hybrid methods combining GLCM features with CNNs (Meena et al. 2024).
- *Grapes*: The grape industry has also benefited from AI. Deep learning frameworks using transfer learning (e.g., with AlexNet) and ensemble models have detected diseases like powdery mildew with accuracies up to 98% (Patil and Jadhav, 2024).

9 Challenges and Limitations in the Implementation of AI for Plant Disease Management

Despite the significant advancements and promising applications of AI in plant disease management, several formidable challenges and limitations hinder its widespread, real-world adoption. Addressing these issues is crucial for realizing the full potential of AI to revolutionize agricultural practices and ensure global food security.

One of the most significant barriers is the issue of *data availability and quality* (Almasoudi 2024; Pun et al. 2025). Deep learning models, in particular, are data-hungry, requiring large, diverse, and accurately labeled datasets to achieve high performance. While public datasets like PlantVillage have been instrumental for research, they predominantly consist of images taken in controlled laboratory settings with uniform backgrounds. This creates a "lab-to-field" gap; models trained on this clean data often perform poorly when deployed in real-world agricultural environments, which are characterized by complex backgrounds, variable lighting, weather conditions, and occlusions from other leaves or branches.

Another significant challenge in automated disease detection is that *plants often suffer from multiple issues at once*. It is a mistaken assumption that an image will only show one ailment, as it is common for a plant to be affected by a combination of different diseases, pest infestations, and nutritional deficiencies simultaneously (Jafar et al. 2024). This complexity of overlapping symptoms complicates the accurate identification of any single condition. Furthermore, the way symptoms present themselves can differ based on the plant's geographical location, which must be factored into any analysis. This challenge is compounded by the visual similarity between the symptoms of different diseases. Many diagnostic systems rely on standard images in the visible light spectrum, which can make it difficult to distinguish between ailments that present with similar patterns or colors. While expanding analysis to include infrared spectral bands could offer a solution for better differentiation, this approach introduces significant technical and financial obstacles (Jafar et al. 2024). Consequently, current diagnostic models can remain prone to misclassification when faced with diseases that have overlapping visual characteristics.

Computational resource constraints also pose a major challenge (Almasoudi 2024). Training complex deep learning models demands powerful hardware, such as high-end GPUs, and extensive cloud computing capabilities. These resources can be prohibitively expensive and inaccessible for many farmers, researchers, and institutions, particularly in developing countries. This digital divide drives the ongoing research into more computationally efficient, lightweight models.

The inherent "black box" nature of many deep learning models raises critical concerns about *model interpretability*. It can be exceedingly difficult to understand the internal reasoning behind a model's prediction. This lack of transparency can affect the trust among end-users and makes it challenging to diagnose and correct potential biases or errors in the model's decision-making process (Almasoudi 2024).

These technical hurdles are compounded by a deeply intertwined socio-technical challenge: *data privacy*. The best AI models require massive amounts of diverse, real-world data, which can only be sourced from a multitude of farms. However, farmers are often reluctant to share their operational data due to legitimate concerns about privacy, security, and the potential for competitive disadvantage (Janga and Dave 2022; Koshariya et al. 2025). This creates a difficult impasse: without data sharing, robust models cannot be built, and without robust models, there is little incentive for farmers to adopt the technology and share data.

To break this cycle, the field is increasingly turning to privacy-preserving technologies, with *federated learning (FL)* emerging as a leading solution. FL is a distributed machine learning paradigm that enables collaborative model training without centralizing the raw data (Truex et al. 2019; Liu et al. 2023; Janga and Dave 2022). In an FL system, a global model is sent to individual clients (e.g., farms or research institutions). Each client trains the model locally on its own private data. Then, instead of sending the sensitive data back, the client sends only the updated model parameters (weights) to a central server. The server aggregates these updates from multiple clients to improve the global model, which is then sent back to the clients for another round of training. This process preserves data sovereignty, as the raw data never leaves the owner's premises. FL has been successfully demonstrated for maize disease detection and is being actively explored for cotton, among other crops, proving its potential to build robust, collaborative intelligence while respecting data privacy (Antico et al. 2022). The rise of FL signals a maturation of the field, moving beyond a singular focus on accuracy to engineer systems that are technically sound, socially responsible, and built on a foundation of trust.

10 Future Directions and Emerging Trends in AI-Driven Plant Health

The field of AI-driven plant disease management is in a state of rapid evolution, with several emerging trends and future directions assured to further revolutionize plant health and agricultural sustainability. The overarching trajectory is toward the integration of multiple advanced technologies to create systems that are more robust, intelligent, private, and autonomous.

A key technical trend is the development of hybrid deep learning architectures. The fusion of convolutional neural networks (CNNs) for their powerful local feature extraction with vision transformers (ViTs) for their ability to capture global context is a promising frontier. Models like MaxViT, PlantXViT, and compact convolutional transformers (CCTs) are designed to leverage the complementary strengths of both architectures, aiming for superior accuracy and robustness, especially in complex visual scenes (Ashqar and Abu-Naser 2019; Thakur et al. 2022; Mahabub et al. 2024).

The principles of federated learning (FL) and Edge AI are set to become foundational. FL will enable the creation of more powerful and generalizable models by facilitating privacy-preserving collaborative training on diverse, real-world datasets from multiple sources. This will be increasingly combined with Edge AI, which involves deploying lightweight, optimized models directly onto on-farm devices like drones, sensors, or smartphones. This combination allows for real-time, on-site data processing and decision-making, reducing latency and dependency on constant Internet connectivity, which is critical for many rural areas.

The integration of AI with robotics and full automation represents the next leap in precision agriculture. Autonomous systems, including unmanned ground vehicles (UGVs) and UAVs, equipped with AI-powered vision systems and treatment applicators, will enable fully autonomous crop monitoring, diagnosis, and targeted interventions. This will dramatically reduce labor costs and further improve the efficiency of resource use.

Beyond immediate diagnostics, AI is making significant inroads into genomic selection for breeding disease-resistant crops. By analyzing vast genomic datasets, AI algorithms can identify genetic markers associated with disease resistance much faster than traditional methods, accelerating the development of more resilient and productive crop varieties (Koshariya et al. 2025).

The trajectory of these trends points toward the eventual creation of a "digital twin" for the farm. This concept involves building a dynamic, virtual representation of the physical farm, continuously updated with real-time, multi-modal data from every available source: visual and hyperspectral data from drones, environmental readings from IoT sensors, soil health data, genomic information about the crop varieties, and long-range weather forecasts. This integrated digital model would serve as a comprehensive platform for simulation, prediction, and optimized decision-making. A farmer could use the digital twin to run "what-if" scenarios. For example, "What is the projected impact on yield and cost if I delay fungicide application by three days?" and test management strategies in a virtual environment before deploying them in the real world. This represents the ultimate realization of data-driven, precision agriculture, where risk is minimized and efficiency is maximized through a holistic, intelligent, and predictive understanding of the entire farming ecosystem.

11 Conclusion: The Transformative Potential of AI in Plant Disease Management

Artificial intelligence has potentially emerged as a powerful and versatile force in the global effort to manage plant diseases, offering profound advancements in detection, diagnosis, and treatment. By moving beyond the constraints of traditional methods, AI-powered systems provide the speed, accuracy, and scalability necessary to address the challenges of modern agriculture. Early and precise detection is now possible through the intelligent analysis of images, sensor data, and advanced imaging techniques. Machine learning and deep learning algorithms, especially convolutional neural networks and emerging vision transformer architectures, have demonstrated remarkable accuracy in diagnosing a wide array of plant diseases from visual symptoms, often outperforming human experts.

Furthermore, the role of AI extends into the realm of proactive management. Predictive analytics models are enabling farmers to anticipate disease outbreaks, while the integration of AI with precision agriculture hardware facilitates optimized, targeted treatments that reduce pesticide use and enhance agricultural sustainability. At a more fundamental level, AI-driven computational methods like molecular docking are accelerating the discovery of novel, safer compounds for disease control.

While the potential of AI is immense, its journey from the laboratory to widespread field deployment is not without significant challenges. Issues related to the availability of diverse, high-quality data, the need for immense computational resources, the "black box" nature of some models, and critical concerns over data privacy must be systematically addressed. Promising solutions like federated learning are paving the way for collaborative and trustworthy AI ecosystems.

Looking ahead, the future of plant disease management will be increasingly shaped by the synergistic integration of AI with other emerging technologies. The development of more sophisticated hybrid models, the deployment of privacy-preserving federated and edge systems, the fusion of AI with autonomous robotics, and the application of AI in genomic selection all point toward a transformative future for plant health. By continuing to innovate and overcome the existing hurdles, artificial intelligence holds the potential to not only revolutionize how we manage plant diseases but also to make a substantial and lasting contribution to global food security and the long-term sustainability of our agricultural systems.

References

Abbas A, Zhang Z, Zheng H, Alami MM, Alrefaei AF, Abbas Q, Naqvi SAH, Rao MJ, Mosa WFA, Abbas Q, Hussain A, Hassan MZ, Zhou L (2023) Drones in plant disease assessment, efficient monitoring, and detection: a way forward to smart agriculture. Agronomy 13(6):1524. https://doi.org/10.3390/agronomy13061524

Agarwal M, Singh A, Arjaria S, Sinha A, Gupta S (2020) ToLeD: tomato leaf disease detection using convolution neural network. Procedia Comput Sci 167:293–301. https://doi.org/10.1016/j.procs.2020.03.225

Aggarwal S, Suchithra M, Chandramouli N, Sarada M, Verma A, Vetrithangam D, Pant B, Ambachew Adugna B (2022) Rice disease detection using artificial intelligence and machine learning techniques to improvise agro-business. Sci Program 2022(1):1757888. https://doi.org/10.1155/2022/1757888

Al-Amin M, Bushra TA, Nazmul Hoq M (2019) Prediction of potato disease from leaves using deep convolution neural network towards a digital agricultural system. In: 2019 1st International Conference on Advances in Science, Engineering and Robotics Technology (ICASERT), pp 1–5

Almasoudi NM (2024) Advancing crop health: the role of artificial intelligence in disease management. Int J Phytopathol 13(3):279–291. https://doi.org/10.33687/10.33687/phytopath.013.03.5526

Antico TM, Moreira LFR, Moreira R (2022) Evaluating the potential of federated learning for maize leaf disease prediction. In: Encontro Nacional de Inteligência artificial e Computacional (ENIAC). SBC, pp 282–293. https://doi.org/10.5753/eniac.2022.227293

Ashqar B, Abu-Naser S (2019) Image-based tomato leaves diseases detection using deep learning. Int J Eng Res 2:10–16

Barbedo JGA (2018) Factors influencing the use of deep learning for plant disease recognition. Biosyst Eng 172:84–91. https://doi.org/10.1016/j.biosystemseng.2018.05.013

Batool A, Kim J, Lee S-J, Yang J-H, Byun Y-C (2024) An enhanced lightweight T-Net architecture based on convolutional neural network (CNN) for tomato plant leaf disease classification. PeerJ Comput Sci 10:e2495. https://doi.org/10.7717/peerj-cs.2495

Bharwal H, Choudhary A (2025) A comprehensive literature review on real time wheat leaf disease detection using Convolutional Neural Network (CNN). IJEECE 14(1&2):48–50

Bolakhe S (2023) AI for better crops. Knowable Magazine. https://doi.org/10.1146/knowable-011723-1

Cervantes J, Garcia-Lamont F, Rodríguez-Mazahua L, Lopez A (2020) A comprehensive survey on support vector machine classification: applications, challenges and trends. Neurocomputing 408:189–215. https://doi.org/10.1016/j.neucom.2019.10.118

Dosovitskiy A, Beyer L, Kolesnikov A, Weissenborn D, Zhai X, Unterthiner T, Dehghani M, Minderer M, Heigold G, Gelly S, Uszkoreit J, Houlsby N (2021) An image is worth 16x16 words: transformers for image recognition at scale (No. arXiv:2010.11929). arXiv. https://doi.org/10.48550/arXiv.2010.11929

Fratello M, Tagliaferri R (2019) Decision trees and random forests. In: Ranganathan S, Gribskov M, Nakai K, Schönbach C (eds) Encyclopedia of bioinformatics and computational biology. Academic Press, Oxford, pp 374–383

González-Rodríguez VE, Izquierdo-Bueno I, Cantoral JM, Carbú M, Garrido C (2024) Artificial intelligence: a promising tool for application in phytopathology. Horticulturae 10(3):197. https://doi.org/10.3390/horticulturae10030197

IPPC Secretariat (2021) Scientific review of the impact of climate change on plant pests—a global challenge to prevent and mitigate plant pest risks in agriculture, forestry and ecosystems. FAO on behalf of the IPPC Secretariat. https://doi.org/10.4060/cb4769en

Islam MM, Adil MAA, Talukder MA, Ahamed MKU, Uddin MA, Hasan MK, Sharmin S, Rahman MM, Debnath SK (2023) DeepCrop: deep learning-based crop disease prediction with web application. J Agric Food Res 14:100764. https://doi.org/10.1016/j.jafr.2023.100764

Jafar A, Bibi N, Naqvi RA, Sadeghi-Niaraki A, Jeong D (2024) Revolutionizing agriculture with artificial intelligence: plant disease detection methods, applications, and their limitations. Front Plant Sci 15. https://doi.org/10.3389/fpls.2024.1356260

Janga R, Dave R (2022) Enhancing smart farming through federated learning: a secure, scalable, and efficient approach for AI-driven agriculture. Artif Intell Appl. https://doi.org/10.47852/bonviewAIA52025089

Khan AA, Faheem M, Bashir R, Wechtaisong C, Abbas M (2022) Internet of things (IoT) assisted context aware fertilizer recommendation. IEEE Access:1–1. https://doi.org/10.1109/ACCESS.2022.3228160

Koshariya AK, Alok, Hiremath SS, Nigam R, Shankar U, Kansara SS, Sinha SS, Pratibha (2025) Detection and management of plant disease using artificial intelligence and machine learning applications: a review. J Sci Res Rep 31(3):281–298. https://doi.org/10.9734/jsrr/2025/v31i32902

Kumar KK, Sasikala S, Maheswari G, Suthar MKB, Sridevi P, Kavitha D, Vengatesh T (2025) A convolutional neural network approach for rice leaf disease detection in India using deep learning. J Neonatal Surg 14(4S):1024–1028. https://doi.org/10.52783/jns.v14.1909

Liu W, Zhang Y-H, Li Y-F, Zheng D (2023) A fine-grained medical data sharing scheme based on federated learning. Concurr Comput: Pract Exp 35(20):e6847. https://doi.org/10.1002/cpe.6847

Ma Y-M, Miao X, Jia B, Sun Z-Y, Ma S-Y, Yan C (2024) Design, synthesis, antifungal evaluation, structure–activity relationship (SAR) study, and molecular docking of novel Spirotryprostatin a derivatives. Molecules 29(4):864. https://doi.org/10.3390/molecules29040864

Mahabub S, Jahan I, Hasan MN, Islam MS, Akter L, Foysal MMR, Onik MKR (2024) Efficient detection of tomato leaf diseases using optimized Compact Convolutional Transformers (CCT) model. Magna Scientia Adv Res Rev 12(2):039–053. https://doi.org/10.30574/msarr.2024.12.2.0183

Meena P, Gupta P, Agarwal S (2024) Integrated GLCM texture features and CNN for automated cotton disease identification. J Theor Appl Inf Technol 102(21)

Mienye ID, Swart TG, Obaido G (2024) Recurrent neural networks: a comprehensive review of architectures, variants, and applications. Information 15(9):517. https://doi.org/10.3390/info15090517

Mohanty SP, Hughes DP, Salathé M (2016) Using deep learning for image-based plant disease detection. Front Plant Sci 7. https://doi.org/10.3389/fpls.2016.01419

Nigam S, Jain R, Marwaha S, Arora A, Haque MA, Dheeraj A, Singh VK (2023) Deep transfer learning model for disease identification in wheat crop. Eco Inform 75:102068. https://doi.org/10.1016/j.ecoinf.2023.102068

Onnom P, Chotpatiwetchkul W, Meesin J, Thapanapongworakul P, Kittiwongwattana C, Chotsaeng N (2024) Synthesis and anti-plant pathogenic fungal activity of Flavokawain-derived flavones and related flavones against Rhizoctonia solani. Curr Appl Sci Technol:e0258374. https://doi.org/10.55003/cast.2023.258374

Otero M, Velasquez LF, Basile B, Onrubia JR, Pujol AJ, Pijuan J (2022) Data driven predictive models based on artificial intelligence to anticipate the presence of Plasmopara Viticola and Uncinula Necator in Southern European winegrowing regions. In: Artificial intelligence research and development. IOS Press, pp 164–167

Pandey K, Mishra S, Sree VLS, Kushwaha R, Pathak R, Yadav S (2024) Disease prediction models in plant pathology: leveraging ai and machine learning for early intervention. Int J Adv Biochem Res 8(12):362–368. https://doi.org/10.33545/26174693.2024.v8.i12e.3121

Patil P, Jadhav P (2024) A novel approach for detection of grape leaf disease using CNN and Alexnet. ICTACT J Data Sci Machine Learn 5(2):598–603. https://doi.org/10.21917/ijdsml.2024.0128

Patra AK, Gajurel T (2025) Improved cotton leaf disease classification using parameter-efficient deep learning framework:737–742. https://doi.org/10.1109/ECTIDAMTNCON64748.2025.10961983

Pun KB, Katiyar H, Midde S, Paul M, Vigneshwaran K, Kannan S, Desouza VA, Nughal J, Das S, Bhattacharyya UK, Yadav A (2025) Artificial intelligence and drone-assisted plant disease diagnosis: a review of advancements and challenges. Int J Adv Biochem Res 9(3):358–366. https://doi.org/10.33545/26174693.2025.v9.i3e.3959

Salman Z, Muhammad A, Han D (2025) Plant disease classification in the wild using vision transformers and mixture of experts. Front Plant Sci 16. https://doi.org/10.3389/fpls.2025.1522985

Selvaraj MG, Vergara A, Ruiz H, Safari N, Elayabalan S, Ocimati W, Blomme G (2019) AI-powered banana diseases and pest detection. Plant Methods 15(1):92. https://doi.org/10.1186/s13007-019-0475-z

Sreethu PT, Paul MM, Gopinath PP, Shahana IL, Radhika NS (2025) Foliar symptom-based disease detection in black pepper using convolutional neural network. Phytopathol Res 7(1):21. https://doi.org/10.1186/s42483-024-00305-1

Thakur PS, Khanna P, Sheorey T, Ojha A (2022) Explainable vision transformer enabled convolutional neural network for plant disease identification: PlantXViT (No. arXiv:2207.07919). arXiv. https://doi.org/10.48550/arXiv.2207.07919

Truex S, Baracaldo N, Anwar A, Steinke T, Ludwig H, Zhang R, Zhou Y (2019) A hybrid approach to privacy-preserving federated learning (No. arXiv:1812.03224). arXiv. https://doi.org/10.48550/arXiv.1812.03224

Ullah F, Khan S, Naeem I, Fahad S, Nawaz N, Rahman KU (2024) Using molecular docking develops plant-bioactive compounds as fungicides against Tox-a protein in wheat tan spot disease. The Sciencetech 5(2):50–62

Yang Y, Mali P, Arthur L, Molaei F, Atsyo S, Geng J, He L, Ghatrehsamani S (2025) Advanced technologies for precision tree fruit disease management: a review. Comput Electron Agric 229:109704. https://doi.org/10.1016/j.compag.2024.109704

Microbiomes in Weed and Postharvest Management

8

P. Sukrith, Gholamreza Salehi Jouzani, and Radhakrishnan Edayileveetil Krishnankutty

Abstract

Agricultural productivity has already been known to be affected by a range of factors and the success in yield enhancement and quality will depend on effective management of the various challenges. At the same time, methods used for the same should be aligned with sustainable development goals.. Here comes the relevance of exploration of microorganisms associated with plant in the form of plant microbiome for diverse agricultural applications. The current chapter specifically acquires information on the effectiveness of utilization of plant microbiome for weed control and post harvest protection. Identifying microbial candidates with the potential to control weeds and also to increase the shelf life of fresh produce will have significant impact on effective utilization of space in cropland cultivation and also to ensure the availability of fresh produce without any damage to consumers.

Moreover, a systematic examination of microbial flora in the agroecosystems is crucial to providing substantial agricultural advantages. Microbiome studies have created a framework for sustainable postharvest biology and handling, avoiding the loss of crops and improving food security. The study of the plant microbiome (primarily highlighting postharvest and weedy species) adds a unique, functional element to agricultural sustainable development.

Keywords

Microbiome · Post-harvest management · Green technology · Eco-friendly

---

P. Sukrith · R. E. Krishnankutty (✉)
School of Biosciences, Mahatma Gandhi University, Kottayam, Kerala, India
e-mail: radhakrishnanek@mgu.ac.in

G. S. Jouzani
Microbial Biotechnology Department, Agricultural Biotechnology Research Institute of Iran (ABRII), Agricultural Research, Education and Extension Organization (AREEO), Karaj, Iran

R. E. Krishnankutty et al. (eds.), *Smart Agriculture*,
https://doi.org/10.1007/978-981-95-8785-8_8

1 Introduction

The field of microbiology was virtually nonexistent prior to Anton van Leeuwenhoek's pioneering work with microscopes in the 1600s. Then, Robert Koch's innovative groundbreaking first efforts in quality research elevated the field of microbiology. Microbes are cosmopolitan and are present everywhere. A microbiome is a collection of all the different microbes including groups of bacteria, archaea, fungi, viruses, protozoa, and their genes that can be present in a particular environment (Yadav 2020). Microbes exist in nearly every habitat that can sustain life—water, soil, plants, and animals. In any given habitat, microorganisms perform important roles in ecosystem like nutrient cycling functions, disease control, insect, pest and weed regulation or suppression of harmful organisms (Pascale et al. 2020), and represent a sumptuous resource for improving crop farming and postharvest production. Microbiomes provide a multitude of benefits, one of the most important being maintenance of soil health. Fungal and bacterial interactions are involved in the formation of compounds such as the glomalin and polysaccharides like pullulan that glue soil particles together to create aggregates. Consequently, aggregation alters soil structure, increases porosity and water infiltration ability, allowing for root growth, as well as, reducing soil erosion, and increasing the movement of air and water through soil. Nutrient cycling is also important for the soil because it adds minerals to soil, by favoring the decomposition of organic materials. Over 100 different microorganisms are now recognized to have potential in biocontrol of weeds and hence the interest in biological weed control mechanisms are expanding. The three major strategies for biological control include classical, augmentative, and integrated management for weed suppression. Research on microbial management of weeds has focused on mycoherbicides derived from fungal pathogens, as they have broad spectrum action in leaf weed control. The success of commercial mycoherbicides illustrates the possibilities of microorganisms to develop effective weed control methods. The weedicidal nature of several other fungal groups are still to be investigated for their ability to be used for biocontrol. In addition, while the possible roles of bacteria in assisting with weed management have been investigated in previous studies, the possibilities are often overlooked (Kennedy and Kremer 1996). Regarding postharvest disease control, mixed microbial cultures or consortia appear to have more promise than unique cultures or single strains of microorganisms. The effectiveness of microbial agents might also be enhanced through the application of low fungicide doses, or salt, or physical treatments like hot water dips or UV treatment. Microbial agents have been cultivated in large volumes as an alternative biocontrol strategy and products such as Aspire, BiobSave, or Shemer have been successfully developed and licensed (Sharma et al. 2009). Microbes associated with plants are collectively called plant microbiomes—which comprises of unique alliances of various microorganisms associated with the plant. The term *plant microbiome* (Goyal et al. 2021) may refer to microbes that are either closely associated with plant's immediate surroundings, inside the plant parts, on the surface of the plant. Plant tissues (roots, stem, leaves, fruits, and flowers) harbor hundreds of different microbes that could be critical to plant growth and health, as well as their ability to cope with biotic and abiotic stress factors (Trivedi et al. 2020). Microbiome can provide key services to endure biotic and abiotic challenges: nutrient cycling, biocontrol of pests, promoting plant growth, developing stress tolerance, and also disease resistance. Organic matter in the soil is converted by microorganisms to macro and micronutrients. Beneficial soil bacteria support-plant growth by making available the organic matter nutrients in a form that can be taken up by plants and ultimately helps in promoting growth (Goyal et al. 2021). The genera Rhizobium and Bradyrhizobium are known to contribute to the nitrogen fixation by having symbiotic relationships with the roots of plants. Mycorrhizal fungi also enable plants to take up phosphorus through their association with roots of the plant. Azospirillum assists cereals and grasses with the uptake of nitrogen, and the genus Pseudomonas produces antibiotics that help to suppress some plant pathogens; through starving the pathogen of nutrients and also by induced resistance in the plant. Trichoderma and actinomycetes can also stimulate root development through enhanced nutrient availability and also through enhanced soil structure.

2 Factors Affecting the Plant Microbiome

There are multiple factors that affect plant microbial association, where inter-microbial interactions. are *interactions within a microbial community*. These interactions within microbial communities can either be neutral, cooperative, or competitive interactions (Aguirre de Cárcer 2019).

1. 2.1 Competition: These type of interactions serve to shape the microbial community, build community structure, and change the community composition. One example is the fact that plants produce organic compounds that serve as carbon and energy sources for the surrounding microorganisms. There are many different soil microorganisms that are competing for the same plant exudates, and typically those with competitive advantages like faster growth rates or will dominate, those versatile microorganisms that are capable of efficiently using plant exudates dominate.
2. 2.2 Predation: Here, one microorganism may kill another microorganism for their food and nutrition. A recent study demonstrated the impact of a bacterial predator *Candidatus sp.* that destroyed cyanobacterial colonies leading to changes in the soil texture making the soil unsutiable for weed growth. Additionally, root parasitic weeds can be controlled by altering soil microbiome.
3. 2.3 Syntrophy: This is a type of biological interaction among two or more dissimilar microorganisms. A great example would be anaerobic bacteria that has a syntrophic interaction with fermentative species which produce hydrogen and maintain the low oxygen levels in the surroundings.
4. 2.4 Parasitism: These are negative interactions between organisms which can lead to decay of other, for example, a few fungi can parasitize a plant weed by merely extracting nutrient from the plant which can lead to weed suppression.

5. 2.5 Commensalism: This is a longer-term biological interaction system in which one organism benefits and the other organism remains unaffected. An example would include the Pseudomonas and Bacillus that would grow extracellularly on the plant leaf of unwanted herbs like cheatgrass.
6. 2.6 Mutualism: In this, both the microbes are benefited; for example, mutualistic microbial coloinies of lactic acid bacteria for biofilms over surface of fruits that can protect them from spoilage by fungal species (Gamalero et al., 2022).

3 Key Microbial Groups in the Plant Microbiome

Previous reports have demonstrated that healthy plants harbour diverse groups of microbial communities within their internal and external tisuues. This section lists some of the key microbial groups (Trivedi et al., 2020). Archaea: They contribute to the plants by fixing nitrogen and the studies on these bacteria are not so common like other bacteria due to its extreme habitats.

Bacteria and fungi are well studied candidates of plant microbiome. Protists are also extremely important in nutrient cycling; they are single celled eukaryotes that occur in both the plant rhizosphere and soil.

Viruses: They potentially influence the regulation of bacteria and fungi (e.g., bacteriophages, mycoviruses).

The main internal factors affecting the plant microbiome are the plant genetics, its associations with other species of plants, and adaptations (Clouse & Wagner, 2021). These adaptations involve production of different proteins or attracting the microflora, as a source of food and shelter. As the plant undergoes developmental changes over time periods, the microbiome will also change accordingly. Aspects of these changes are implied with changing root exudation patterns, nutrient uptake, and growth factors which affect the growth of the plant at different life stages, from germination through vegetative and reproductive, then into senescence. The factors affecting plant microbiome also include climate (e.g., weather), aspects of sunlight intensity, amount of rainfall, and heat (Fig 8.1). The sustainability of the plant microbiome associations is heavily considered dependent on the conditioning of the abiotic factors in the soil including salinity, pH, moisture, and biomass.

In an agricultural operation, the fertilizer regime varies according to the demands of the crop phytobiome. Pesticide, tillage, rotation cropping, cover cropping, and any other agronomic practice can have impacts on microbiome populations. Indeed, the combination of on-farm practices may provide a useful pathway of intervention to modify the classical farming practices used, and ultimately enrich the benfeicial plant microbiome for quality crops.

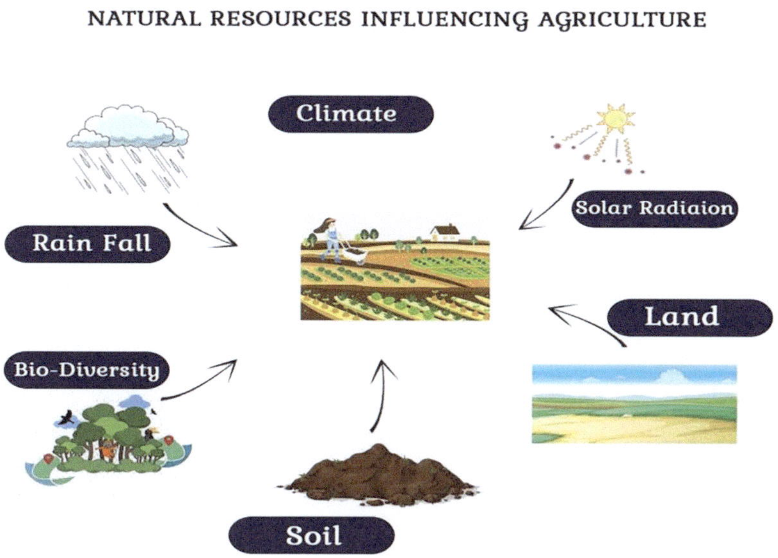

Fig. 8.1 Natural resources influencing agriculture

4 The Significance of the Plant Microbiome for Productivity

Researches confirm that plant health and agricultural productivity are driven by the composition and function of the plant microbiome (Trivedi et al. 2022). As per a previous report (Goyal et al. 2021), several biological processes supported by microbes provide access to soil nutrients, support plant growth, increase resistance to disease and stress. Highlighting an emphasis on changing the advanced capabilities of phytobiome knowledge into actionable microbial tactics (Mukhtar et al. 2019) is significant to support food security. Plant microbiome has a key role in

- Nutrient Acquisition/Solubilization: From solubilising soil nutrients such as phosphate, nitrates and increasing the uptake of iron, manganese, zinc, and silicon to enhanced mineral nutrition plant microbiome plays a key role in mobilising nutrients to the host plant and to increase the crop growth.
- Phytohormone Modulation: Microbial production of plant phytohormones like auxins, cytokinins, and other hormones is highly significant as it can affect plant architecture, growth, and development. Beneficial microbes can oppose plant pathogens by competing for defensive mechanisms of organic matter and nutrient resources in the soil niche.

Phytohormones serve as essential signals that govern the plant growth and development and can affect plant vigor and viability under many stress conditions. Plant growth-promoting bacteria (PGPB) produce phytohormone which significantly impact plant growth. PGPB also produce numerous antagonistic compounds that could modulate plant physiology (Orozco-Mosqueda et al. 2023).

- Tolerance to abiotic stress: Abiotic stress is the most significant cause limiting crop development and yield at the global level. Reports from physiology, genetics, and molecular biology offers considerable evidence on microbial role in plant responses to abiotic stress. For example, the susceptibility of rice plants to abiotic stress has already been known to be supported by the rice microbiome (Gao et al. 2007).

Mechanistically, plant growth enhancement, disease suppression, stress tolerance, or increased access to the soil resources are influenced by plant microbiomes which play roles in the survival and productivity of plants. Repeated increases in yield potential have been achieved over time through plant cultivation which involves enrichment of plant microbiome. It is apparent that through innovating within the realms of plant-microbe systems, we can increase yield of crops in a sustainable manner (Hitch et al. 2022). Some examples of prevalent microorganisms in the plant microbiome and their effects on plant growth have been listed in Table 8.1.

4.1 Weed Management Using Microbiomes

Weed management depends on associating natural enemies of weeds along with chemical and control methods along with cultivation of crops. There are vast diversity of microbes that have the potential to control weeds. The new sequencing data and protocols in soil microbiome research have generated extensive insight on how beneficial microorganisms support plant growth. This research is valuable for the crop plant and weed exploration of how microorganisms could affect crop yield and health and prevent weed dispersal. This could lead to a greater knowledge base, to create new biocontrol methods which will be able to target weeds (Trognitz et al. 2016).

4.2 Microbes that Inhibit Weed Growth

A variety of microorganisms associated with agricultural land can inhibit weed growth. These microorganisms include mycorrhizal fungi and beneficial bacteria, fungi and oomycetes, which can limit the germination, establishment, or fecundity of weeds through complex mechanisms. Investigating model bioherbicidal microbes has created an avenue for microbiome engineering and bioformulation development (Hitch et al. 2022). A few of the most well-studied bioherbicidal microbes are foliar fungal pathogens, such as *Sclerotinia sclerotiorum* which can infect susceptible broadleaf weeds and *Fusarium oxysporum*, which is a soil-borne fungus that releases phytotoxins that inhibit the establishment of new weed seedlings. Many bacteria produce herbicidal natural products, such as *Xanthomonas campestris* pv. Poae, and *Streptomyces spp.*, that can be toxic to plants outside of their normal host range, in addition to noninfectious rhizosphere organisms that do not behave as a

Table 8.1 Some of the most important microorganisms associated with the plant microbiome

Microbial group	Category	Examples of roles and impacts	References
Rhizobium	Bacteria	Nitrogen-fixing bacteria help in plant growth and enrichment of particular nutrient	Trivedi et al. (2022)
		Root nodule formation and nutrient uptake are stimulated by microbes.	Albright et al. (2022)
Pseudomonas	Bacteria	Contributes to plant growth by producing plant growth regulators.	
		Thwart the soil-borne diseases and infections.	Rai et al. (2023)
		Helps in a systematic resistance	
Bacillus	Bacteria	Phosphorus intake becomes easier due to the soluble form.	Saleem (2021)
		Secretion of antimicrobial metabolites	
		Promoting the shoot and root growth	Nerva et al. (2022)
Trichoderma	Fungi	Parasitises the pathogenic fungi and destroys them	Tan et al. (2021)
		Reduces the severity of fungal diseases and damages caused by it	Shayanthan et al. (2022)
		Enhances the active nutrient uptake	
Arbuscular mycorrhizal fungi	Fungi	Forms symbioses with a majority of crops	Trivedi et al. (2022)
		Facilitates phosphorus and nitrogen acquisition	Xiong et al. (2021)
		Improves soil structure stability	
Phytophthora	Oomycetes		
		Alters soil microbiome composition	Tan et al. (2021)
		Reduces yields of susceptible crops	
Spiroplasma	Bacteria	Colonizes phloem and xylem tissues	Xiong et al. (2021)
			Trivedi et al. (2022)
Gaeumannomyces	Fungi		Ali et al. (2023)
		Persists in soil for decades	Tan et al. (2021)

pathogen e.g., the oomycete Phytopythium. These organisms exist in rhizospheres and support weed management by depleting nutrients required by the weeds for their germination and growth Jha et al. (2015) This competition for nutrients can sometimes be deleterious to growth and development of cultivated plants. Hence precise management of nutrient supply is crucial for favoring crop growth and suppressing weed growth to improve productivity. The management mechanisms of weed involve many modalities and all are scientifically based approaches. Just as all chemical management options have optimal doses, so can all microbial management options have an optimal dose at which it is most effective (Hitch et al. 2022).

The broadcast of microbial interaction would provide a safer alternative to chemicals that are typically used in agricultural operations.

4.3 Mode of Action of Anti-weed Microorganisms

Microbes have adapted to develop contrasting strategies that have weed-inhibitory properties (Trivedi et al. 2022). Understanding the mechanisms that underly microbial inhibition of weeds will give insight into methods for developing biocontrol and microbiome engineering in order to make the cultivated crops more competitive.

Toxins, enzymes, and inhibitory volatiles like phenazines, cyanide, and gloesporone of microbial origin are known to disrupt weed physiological processes directly. Cell wall-degrading enzymes may disrupt some root tissue (Trivedi et al. 2022).

- Resource Competition—Microbes may deprive germinating weed seeds of soil resources (e.g., iron) due to rapid colonization of roots and high-affinity uptake systems (Xiong et al. 2021).
- Antibiosis—Bioherbicidal microbes produce antimicrobials, volatile organics, and other inhibitory compounds that inhibit microbial symbionts (e.g., mycorrhizae) growth of weeds.
- Host Mimicry—Mimesthesia by certain microbes help in inhibiting seed germination of weeds like Striga plants. The microbes mimic certain plant exudates like strigolactone which results in triggering of plant-seed germination. The weed seeds germinate without attachment to the original plant hosts and this results in premature death of the weeds. Triggers by microbes could facilitate weed seed germination and recruitment to the soil seedbank (Trivedi et al. 2022).
- Hyperparasitism—Microbial natural enemies of plant pathogens may ameliorate the attack of root flora pathogens (soil-borne fungi) that could diminish weed fitness in the presence of the crop plant.

4.4 Possibility of Bioengineering Microbiome for Biocontrol

The plant microbiome includes a primary group of plant growth-promoting bacteria (PGPB), found typically in the root zone (the rhizosphere), on above-ground plant tissues (the phyllosphere), or as endophytic bacteria residing inside plant tissues. The last few years have seen new opportunities to manipulate the plant microbiome. There are mainly five different approaches for microbiome introduction:

1. Host-mediated multi-generational selection of the microbiome.
2. Inoculation—in soil or in the rhizosphere.
3. Inoculation—tissue (the seed or seedling).
4. Aerosolization of tissues.
5. Injection to tissues or wounds.

Overall, microbiome engineering presents a viable option to solve a range of agricultural threats to an agricultural enterprise, including weed management and to minimize the negative environmental effects on crops (Shiny et al. 2024). The containment of recombinant microbes is a critical step in validating its performance and efficacy. Microbiome engineering for weed management not only involves methods to increase biocontrol efficacy, but also to competitively promote and allow the crop-associated communities to establish dominance over weeds. Thus, microbiome in plants could improve efficiency of resource capture and confer stress tolerance which may significantly alter weed-management strategies by tipping the equilibrium towards the crops.

4.5 Challenges with Microbial Biocontrol

Laboratory and greenhouse studies have shown the potential of experimental bioherbicide products, but when piled together in datasets of many trials, field application of the bioherbicide products did not provide consistent weed suppression. There are numerous variable factors within sites including the soil characteristics, moisture, temperature, microbial activity, etc., and variation among sites, that can affect the activity and interactions with the microbial agents, or effects on the intended weed species and other species as well. Cultural practices and issues can also influence the intended weed communities within locations and seasons. At molecular level, this is certainly a more difficult proposition, but how a strain of microorganism interacts with a specific weed species may differ if there are variations in the target weed among locations. In some cases, the activity spectrum (number of weed species affected) of the same strain has been seen to differ for different soil types. Without taking into account unintended ecological perturbations caused by adding microorganisms to the rhizosphere via microbiome engineering and avoiding negative effects on taxa already present, there may be another challenge for implementation. The temporal aspect is also uncertain, including how long the desired compositional change (microbiome) will last in the rhizosphere after intervention;. However, the increasing sophistication of tools to monitor or model the trajectory of an assembled microbiome in situ or to relate this to crop health in reality is improving the capability of an ecosystem-based management approach that considers plants, soils, and microbial agents simultaneously to implement biological weed control on a large scale. This is achieved through careful management and monitoring of both the nonliving microbial matter along with the living microbial cells (Table 8.2).

5 Postharvest Microbiome

It has been estimated that about 30–50% of fruit and vegetable production can be lost due to many diseases along with preharvest and postharvest processes. Improved control systems in postharvest storage facilities could greatly reduce the threat from

Table 8.2 Microbial weed management strategies

Types of microbial formulation	Specific applications and limitations	References
Fungal mycelium granules	Specific humidity and temperatures are used to act on the targeted foliage application for susceptible broadleaf weeds.	Goyal et al. (2021)
Bacterial suspensions	Seed treatment for soil-borne weed infection Variable persistence in the rhizosphere	Darabi et al. (2021)
Microbial consortia	Induces broad-spectrum weed suppression Complex registration requirements	Goyal et al. (2021)
Microencapsulated strains	Improved survival of seeds in soils High cost compared to chemicals	Greenspan et al. (2020)
Transgenic rhizobacteria	Engineered pathway expression gives sustained weed control Stringent containment protocols are mandatory	Hitch et al. (2022)
Metabolite extracts	Defined bioactive compounds enable reproducibility Secondary metabolites are prone to biodegradation	Mu et al. (2021)
Mycoherbicidal baits	Stimulate germination before crop emergence Phytotoxic effects differ between soil types	Nerva et al. (2022)
Designer prebiotics	Stimulate the growth of native beneficial Soil moisture influences selectivity	Tan et al. (2021)

pathogens and insect pests. However, spoilage and storage rot continue to be a global issue, largely due to weak technology, poor harvesting and handling, and unavoidable changes during storage such as sprouting or loss of sugars and moisture (Wassermann et al. 2022). Understanding postharvest microbiome is still relatively new for most and there remain large areas of uncertainty regarding how composition and function of microbiome impact fresh product postharvest. Not much is understood regarding core microbiome and their function in the host–microbiome interaction between different commodity types. Viewing harvested produce as more of a metaorganism, employing the holobiont concept, and a fresh approach is likely to shed some light on the postharvest microbiome considering the harvested commodities are complex systems with host–microbiome interactions (Wisniewski and Droby 2019). Endophytes as part of the plant microbiome exhibit many inhibitive mechanisms against pathogens, including, but not limited to, competition for space and nutrients, siderophore (iron-chelation) production, bioactive antimicrobial and antibiosis production, mycoparasitism and lytic enzymes, endotoxins and lipopolysaccharides, redox homeostasis in harvested fruit. They also have quorum sensing, and biofilm formation and dissolution, phytohormone mitigation such auxin, gibberellins, cytokines, ethylene, nitric oxide, and ACC (1-aminocyclopropane-1-carboxylic acid) deaminase, induction of disease resistance in plants, and regulation of mycotoxins produced by fungi (Kumari et al. 2022). The study on the postharvest microbiome is a relatively new area of research and there is much to learn regarding how the fresh produce microbiome and its composition might influence the postharvest microbiome. Secondly, although some work has been done on the composition of the fresh produce microbiome, it is not yet known if some (or all) produce types have a core microbiome that might connect to some relationship to host–microbiome. The postharvest microbiome perspective is slow to develop, and there is much to explore in terms of the composition and functioning of the fresh produce microbiome postharvest. Considering the harvested produce as a meta-organism, or holobiont, it may help us navigate towards the postharvest microbiome aspect—as harvested commodities are both complex systems involving the diversity of associations related to microbial communities as living subjects of interest, as well as the parts of the host–host, and host–microbiome structures (Kusstatscher et al. 2020).

5.1 Role of Microbiome in Postharvest Losses

Microorganisms are responsible for the majority of postharvest losses of produce that occur during storage and distribution. This results in loss of millions of tons and serious consequences on economics and nutrition. During the weaving stage of the grape harvest, severe damage is caused by opportunistic microorganisms The primary microbial colonization can happen with various species of fruits, vegetables, and grains resulting in loss of nutrients and rotting tissues if the plants primary defense relies on the resinous materials within the plant. Some fungi also infect the postharvest produce via entry of mycelia through punctures, abrasions, the stomata, and straight penetrations, for example, in genera like *Botrytis sp., Cladosporium sp.,* and *Alternaria sp.*. At the same time, some bacterial microorganisms, for example, *Pseudomonas sp., Ralstonia sp.*, and *Erwinia sp.*, that cause soft rots or necrosis are able to establish infections postharvest produce. Once the soft tissue has decayed secondary microbes invade which mainly consist of fungi, and significantly reduce quality and health of the produce. Each fresh produce has a native microbiome, consisting of billions of microorganisms, influenced by the cultivar, maturity stage of harvest, production inputs, and environmental conditions at the time of harvest. A wide variety of bacteria not only live in soil but also colonize fresh produce grown in that soil. These microbes occupy infection sites and form stable communities that can suppress pathogen, thereby reducing decay during storage (Deng et al. 2022). Resident bacteria can prevent colonization from pathogens by competitive exclusion and production of antimicrobials. Such functional microbes include *Pseudomonas sp.*, *Bacillus sp.*, *Stenotrophomonas* sp., *Enterobacter* sp., and *Cryptococcus sp.* that are able to bestow protective abilities to fresh fruits and vegetables against pathogen colonization, disease, and by extension of tissue decay (Greenspan et al. 2020). Promoting crop-associated microbiomes and these aforementioned protective taxa will provide an unmet biological alternative to synthetic postharvest controlling fungicides.

Microorganisms have the potential to colonize various host plant tissue types and the opportunity to interact with each other and host plants to modify the physiological and metabolic pathways, which may be integrated, or re-purposed for effective and sustainable management of postharvest diseases. Endophytic strains of bacteria, actinomycetes, and fungi have been investigated extensively as biocontrol agents against many plant pathogens in both preharvest and postharvest contexts (Kumari et al.

2022). Omics technology, including but not limited to genomics, transcriptomics, proteomics, metagenomics, metatranscriptomics, etc., can be harnessed to identify and characterize the microbes which may be utilized for management of postharvest diseases. Continued exploration of combining microbial agents into consortia may be a promising and productive path in disease management. Reports have indicated that endophytes have multiple mechanisms for limiting the proliferation of pathogenic hosts and providing physiological support for fruit health. With numerous documented examples of using successful bioformulations of microbial endophytic strains inhibiting the pathogenicity of the pest or pathogen during the preharvest stages, the application of microbial endophytes for controlling pathogens postharvest is still in its infancy. The continued use of endophytic microbiomes will facilitate an ongoing decrease in, or eventual eradication of, our reliance on chemical pesticides and fungicides for management of postharvest diseases (Kumari et al. 2022).

5.2 Manipulation of Microbiome to Improve Storage

There were several options that seem to be advantageous for controlling postharvest losses through the microbiome and resistant varieties of crops are one among these. These may have candidates that are resistant to pathogenic activity or can reduce the negative effects associated with storage duration.

Microbiome transplantations are also an option, but the ecological aspect of the community integration needs to be considered in terms of how each community was impacted by the donor, receptor communities, and the environment (Jiang et al. 2022).

Prebiotics and probiotics can help in providing specific nutritional substrates as part of the composition in packaging materials, or wash solutions, to selectively enrich the beneficial, native, and protective microbes on produce surfaces after the postharvest phase.

Microbial inoculants can be sprayed on, dry dipped, or wash treated to prevent pathogen development. Individualized (in encapsulated or lyophilized for) or prepared as microbial consortia, microbial inoculants would also improve stability of the harvested products during storage. Resistance and therefore efficacy in storage.

The integrated approach of using resistant varieties, along with prebiotics, probiotics and specific inoculants targeted at the native microbiota, can be a promising strategy for microbiome-based postharvest disease-management practices with improved sustainability.

6 Future Outlook

6.1 Integration with Precision Agriculture Techniques

The rapidly advancing technology of precision agriculture, including GPS, remote sensing, robotics, artificial intelligence and big data analytic tools, is also being extended to potential microbiome management. The integration of such tools can allow the fine tuning of microbial applications (bioinoculants, biostimulants, or community metabolites) that match the specific soil properties and crop requirements, at different locations and times similar to the methods used in precision agriculture. Pest could be identified using pest-management systems currently available that include infrared, thermal, audio sensors, and image recognitions of insects. The potential efficiencies with using sensors to develop management systems, along with interpretation through software will likely make the production more efficient while improving sustainability. Digital images of crops obtained through satellites or aerial (manned or unmanned) platforms, could be processed through high-resolution images from the same date of acquisition for a larger and more complex crop situation to maximize information extraction. The analyzed information will help to add value for the decision processes involved in pest control. As technology evolves, augmented and virtual reality visualization will ultimately help to clarify plant–microbiome relationships and build on our knowledge for management purposes (Tan et al. 2021). More targeted microbial consortium are also likely to be better biocontrol agents than single strains, because it is more likely to augment biocontrol. At the same time, accelerated breeding approaches can also be beneficial and are being utilized to develop precision crop varieties representing target hosts of these microbial communities.

6.2 New Microbiome Biocontrol Strategies

In addition to enhancing the potential biocontrol efficacy of microbially mediated pest-management function, engineered microbial communities could offer more consistently favorable outcomes. Researchers have suggested that detailed studies on the complex microbial interactions between the gut microbiota of insects and pests with the plant microbiome of the host plants can help reduce the uncertainity of microbial treatments (Mason 2020). Over time, adaptive evolution species will hence be selected to function as beneficial root endophytes, leaf-dwelling epiphytes resulting in symbiotic complexes. By developing these new rapid-breeding strategies, crop varieties can be developed to specifically host microbial communities (Rändler-Kleine et al. 2020), and adaptive laboratory evolution can also select beneficial root endophyte for managing both rooting-associated pests and leaf-associated pests. New rapid-breeding methods can be utilised for commercial production of crop varieties and breeding populations identified as hosts for beneficial groups of plant-associated microbes (Trivedi et al. 2022). These exciting and innovative new developments in microbially mediated plant management and breeding should provide more plant-associated microbial communities that can perform at their full biocontrol capabilities they have developed from field-based performance.

6.3 For Additional Issues Associated with Ecological Effects

Modern-day agriculture prefers to minimise agrochemical inputs like inorganic fertilisers and pesticides. Besides the financial and environmental costs, pesticides are often applied without clear guidelines or monitoring. Farmers frequently use them in excess or in inappropriate ways, leading to health risks, environmental pollution, and pesticide resistance (Sharma et al. 2019; Pretty and Pervez Bharucha 2015). The true terrifying part is with soil ecology; inappropriate and indefinite applications of pesticide products can alter, or erase and thereby destroy beneficial or plant probiotic soil microflora which functions in the top-soil layer as described in Pineda et al. (2020). Once top-soil or weathered soil are degenerated (or eroded), harvests and yields will decrease even with adequate inputs needed for crop and these inputs will render the soil incapable of supporting the production of crops (Kalia and Gosal 2011). Hence, synthetic pesticides can be replaced by novel safer formulations like biochemical pesticides (which include natural product pesticides, as well as products based on naturally occurring minerals or plant extracts), semiochemicals (e.g., pheromones), macroorganism pesticides (e.g., nematodes, insects), biotechnology products (plant-incorporated protectants), and microbial pesticides (bacteria, fungi) (Borges et al. 2021). The increase in microbial pesticides and their growing importance in the plant protection industry means it is important to develop general regulatory requirements that include and always address the unique nature of these products. A typical ecological risk-assessment framework involves determining potential toxicity to relevant nontarget organisms of concern in a variety of environmental compartments. This assessment of toxicity might be combined with the assessment of likely exposure to determine if a potential risk exists for microbial pesticides (Borges et al. 2021).

7 Summary and Conclusion

This chapter has highlighted the often overlooked role of the soil and plant microbiome, which is not just a passive resident in soil, but a dynamic participant in the ecological functioning of agroecosystems. Terrestrial life is largely dependent upon a thin layer of soil, which indicates the importance of soil microbiome in maintaining ecological balance. (Tkacz and Poole 2020). The structure of endophyte communities in the plant is dynamic and is also in constant flux through a number of processes, and stochastic events which include biogeography, plant species, soil conditions, microbe–microbe interactions and plant–microbe interactions. Both the soil microbiome and plant microbiome play crucial roles in enhancing nutrient availability to crops, suppressing weed growth, increasing stress resilience in crops as well as extending the post harvest shelf life which are essential to increasing agricultural productivity in a sustainable way (Yadav, 2020). As per the emerging research on the microbial communities of plants, the impact of plant microbiome is not just limited to providing nutrients to plants, but extends to influence weed management and even protection of the post harvest plant materials. Hence the optimal use of plant microbiome to control weeds and to minimise post harvest loss is a critical issue that needs to be addressed.

References

Aguirre de Cárcer D (2019) A conceptual framework for the phylogenetically constrained assembly of microbial communities. Microbiome 7(1):142. https://doi.org/10.1186/s40168-019-0754-y

Albright MB, Louca S, Winkler DE, Feeser KL, Haig SJ, Whiteson KL et al (2022) Solutions in microbiome engineering: prioritizing barriers to organism establishment. ISME J 16(2):331–338. https://doi.org/10.1038/s41396-021-01088-5

Ali S, Tyagi A, Bae H (2023) Plant microbiome: an ocean of possibilities for improving disease resistance in plants. MDPI. https://www.mdpi.com/2076-2607/11/2/392

Borges S, Alkassab AT, Collison E, Hinarejos S, Jones B, McVey E, Wassenberg J (2021) Overview of the testing and assessment of effects of microbial pesticides on bees: strengths, challenges and perspectives. Apidologie 52(6):1256–1277

Gamalero E, Bona E, Glick BR (2022) Current techniques to study beneficial plantmicrobeinteractions. Microorganisms *10*(7):1380. https://pubmed.ncbi.nlm.nih.gov/35889099/

Mason CJ (2020) Complex relationships at the intersection of insect gut microbiomes and plant defenses. J Chem Ecol 46(8):793–807. https://pubmed.ncbi.nlm.nih.gov/32537721/

Clouse KM, Wagner MR (2021) Plant genetics as a tool for manipulating crop microbiomes: opportunities and challenges. Front Bioeng Biotechnol 9:567548. https://www.frontiersin.org/articles/10.3389/fbioe.2021.567548/full

Darabi M, Majeed H, Diehl A, Norton J, Zhang Y (2021) A review of microplastics in aquatic sediments: occurrence, fate, transport, and ecological impact. Curr Pollut Rep 7(1):40–53. https://doi.org/10.1007/s40726-020-00171-3

Deng X, Zhang N, Li Y, Zhu C, Qu B, Liu H et al (2022) Bio-organic soil amendment promotes the suppression of Ralstonia solanacearum by inducing changes in the functionality and composition of rhizosphere bacterial communities. New Phytol 235(4):1558–1574

Gaiero JR et al (2013) Inside the root microbiome: bacterial root endophytes and Plant Growth Promotion. Am J Bot 100(9):1738–1750

Gao J, Chao D, Lin H (2007) Understanding abiotic stress tolerance mechanisms: recent studies on stress response in Rice. J Integr Plant Biol 49(6):742–750. https://doi.org/10.1111/j.1744-7909.2007.00495.x

Goyal A, Wang T, Dubinkina V, Maslov S (2021) Ecology-guided prediction of cross-feeding interactions in the human gut microbiome. Nat Comm 12(1). https://doi.org/10.1038/s41467-021-21586-6

Greenspan SE, Migliorini GH, Lyra ML, Pontes MR, Carvalho T, Ribeiro LP, Moura-Campos D, Haddad CF, Toledo LF, Romero GQ, Becker CG (2020) Warming drives ecological community changes linked to host-associated microbiome dysbiosis. Nat Clim Chang 10(11):1057–1061. https://doi.org/10.1038/s41558-020-0899-5

Gressel J (2018) Microbiome facilitated pest resistance: potential problems and uses. Pest Manag Sci 74(3):511–515. https://doi.org/10.1002/ps.4777

Hitch TCA, Hall LJ, Walsh SK, Leventhal GE, Slack E, de Wouters T, Walter J, Clavel T (2022) Microbiome-based interventions to modulate gut ecology and the immune system. Mucosal Immunol 15(6):1095–1113. https://doi.org/10.1038/s41385-022-00564-1

Jiang G, Zhang Y, Gan G, Li W, Wan W, Jiang Y, Dini-Andreote F (2022) Exploring rhizo-microbiome transplants as a tool for protective plant-microbiome manipulation. ISME Commun 2(1):10

Jha C. K, Saraf M (2015) Plant growth promoting rhizobacteria (PGPR). J Agric Res Dev 5(2): 108–119

Kalia A, Gosal SK (2011) Effect of pesticide application on soil microorganisms. Arch Agron Soil Sci 57(6):569–596. https://doi.org/10.1080/03650341003787582

Kennedy AC, Kremer RJ (1996) Microorganisms in weed control strategies. J Prod Agric 9(4):480–485. https://doi.org/10.2134/jpa1996.0480

Kumari M, Qureshi KA, Jaremko M, White J, Singh SK, Sharma VK et al (2022) Deciphering the role of endophytic microbiome in postharvest diseases management of fruits: opportunity areas in commercial up-scale production. Front Plant Sci 13:1026575

Kusstatscher P, Cernava T, Abdelfattah A, Gokul J, Korsten L, Berg G (2020) Microbiome approaches provide the key to biologically control postharvest pathogens and storability of fruits and vegetables. FEMS Microbiol Ecol 96(7):fiaa119. https://doi.org/10.1093/femsec/fiaa119

Massart S, Martinez-Medina M, Jijakli MH (2015) Biological control in the microbiome era: challenges and opportunities. Biol Control 89:98–108. https://doi.org/10.1016/j.biocontrol.2015.06.003

Mu A, McDonald D, Jarmusch AK, Martino C, Brennan C, Bryant M, Humphrey GC, Toronczak J, Schwartz T, Nguyen D, Ackermann G, D'Onofrio A, Strathdee SA, Schooley RT, Dorrestein PC, Knight R, Aslam S (2021) Assessment of the microbiome during bacteriophage therapy in combination with systemic antibiotics to treat a case of staphylococcal device infection. Microbiome 9(1). https://doi.org/10.1186/s40168-021-01026-9

Mukhtar S, Malik K, Mehnaz S (2019) Microbiome of halophytes: diversity and importance for plant health and productivity. Microbiol Biotechnol Lett 47(1):1–10. https://doi.org/10.4014/mbl.1804.04021

Nerva L, Sandrini M, Moffa L, Velasco R, Balestrini R, Chitarra W (2022) Breeding toward improved ecological plant–microbiome interactions. Trends Plant Sci 27(11):1134–1143. https://doi.org/10.1016/j.tplants.2022.06.004

Nobre T (2019) Symbiosis in sustainable agriculture: can olive fruit fly bacterial microbiome be useful in pest management? Microorganisms 7(8):238. https://doi.org/10.3390/microorganisms7080238

Ohore OE et al (2022) Ecological impact of antibiotics on bioremediation performance of constructed wetlands: microbial and plant dynamics, and potential antibiotic resistance genes hotspots. J Hazard Mater 424:127495. https://doi.org/10.1016/j.jhazmat.2021.127495

Orozco-Mosqueda MDC, Santoyo G, Glick BR (2023) Recent advances in the bacterial phytohormone modulation of plant growth. Plants 12(3):606

Pascale A, Proietti S, Pantelides IS, Stringlis IA (2020) Modulation of the root microbiome by plant molecules: the basis for targeted disease suppression and plant growth promotion. Front Plant Sci 10:1741. https://www.frontiersin.org/journals/plant-science/articles/10.3389/fpls.2019.01741/full

Pineda A, Kaplan I, Hannula SE, Ghanem W, Bezemer TM (2020) Conditioning the soil microbiome through plant-soil feedback suppresses an aboveground insect pest. New Phytol. https://www.ncbi.nlm.nih.gov/pmc/articles/PMC7155073/

Pretty J, Pervez Bharucha Z (2015) Integrated pest management for sustainable intensification of agriculture in Asia and Africa. Insects 6(1):152–182

Rai A, Sharma NK, Singh VK, Dwivedi BS, Singh JS, Rai PK (2023) Study of phosphate solubilizing fluorescent Pseudomonas recovered from rhizosphere and endorhizosphere of Aloe barbadensis (L.). Geomicrobiol J 40(4):347–359

Rändler-Kleine M et al (2020) How microbiome approaches can assist industrial development of biological control products. In: Prog Biol Control, pp 201–215. https://doi.org/10.1007/978-3-030-44838-7_13

Saleem B (2021) Phyllospheremicrobiome: plant defense strategies. In: Microbiomes and the global climate change, pp 173–201. https://doi.org/10.1007/978-981-33-4508-9_11

Sharma A, Kumar V, Shahzad B, Tanveer M, Sidhu GPS, Handa N et al (2019) Worldwide pesticide usage and its impacts on ecosystem. SN Appl Sci 1(11):1446

Sharma RR, Singh D, Singh R (2009) Biological control of postharvest diseases of fruits and vegetables by microbial antagonists: a review. Biol Control 50(3):205–221. https://doi.org/10.1016/j.biocontrol.2009.05.001

Shayanthan A, Ordoñez PAC, Oresnik IJ (2022) The role of synthetic microbial communities (Syncom) in sustainable agriculture. Frontiers. https://www.frontiersin.org/articles/10.3389/fagro.2022.896307/full

Shiny E, Mathew J, Jacob A, Rajashree HS (2024) Microbiome engineering to improve biocontrol and plant growth-promoting mechanisms. In Microbiome engineering. CRC Press, pp 149–155

Tan J, Kerstetter JE, Turcotte MM (2021) Eco-evolutionary interaction between microbiome presence and rapid biofilm evolution determines plant host fitness. Nat Ecol Evol 5(5):670–676. https://doi.org/10.1038/s41559-021-01406-2

Tkacz A, Poole P (2020) The plant microbiome: the dark and dirty secrets of plant growth. Plants People Planet 3(2):124–129. https://doi.org/10.1002/ppp3.10167

Trivedi P, Batista BD, Bazany KE, Singh BK (2022) Plant–microbiome interactions under a changing world: responses, consequences, and perspectives. New Phytol 234(6):1951–1959. https://doi.org/10.1111/nph.18016

Trivedi P, Leach JE, Tringe SG, Sa T, Singh BK (2020) Plant–microbiome interactions: from community assembly to plant health. Nat Rev Microbiol 18(11):607–621. https://www.nature.com/articles/s41579-020-0412-1

Trognitz F et al (2016) The role of plant–microbiome interactions in weed establishment and control. FEMS Microbiol Ecol 92(10). https://doi.org/10.1093/femsec/fiw138

Wassermann B, Abdelfattah A, Cernava T, Wicaksono W, Berg G (2022) Microbiome-based biotechnology for reducing food loss post harvest. Curr Opin Biotechnol 78:102808

Wisniewski M, Droby S (2019) The postharvest microbiome: the other half of sustainability. Biol Control 137:104025. https://doi.org/10.1016/j.biocontrol.2019.104025

Xiong C, Singh BK, He JZ, Han YL, Li PP, Wan LH et al (2021) Plant developmental stage drives the differentiation in ecological role of the maize microbiome. Microbiome 9(1):171. https://doi.org/10.1186/s40168-021-01118-6

Yadav AN (2020) Plant microbiomes for sustainable agriculture: current research and future challenges. In: Plant microbiomes for sustainable agriculture, pp 475–482. https://link.springer.com/chapter/10.1007/978-3-030-38453-1_16

Unlocking the Potential of Algal-Mediated Nanoparticles in Plant Science: Synthesis, Challenges, and Applications

9

Arya Gupta, Jyotsna Bora, Niharika V. Rao, Monika Joshi, and Subhasha Nigam

Abstract

Algal-mediated Nanoparticles (NP) offer a highly promising, sustainable and eco-friendly method for diverse applications. The present review explores the confluence of algae, nanotechnology, and environmental sustainability. A comprehensive method of green synthesis of NPs using different classes of algae, characterization techniques along with their detailed applications in various fields including agriculture, bioremediation, nanocoatings, renewable energy, and carbon capture is discussed. The mechanisms (intracellular and extracellular) involved in the NP synthesis focusing on the unique ability of algae to reduce metal ions into NPs are explained. Furthermore, current limitations and challenges associated with algal NP, scale-up challenges, and variability in the number of algae are critically discussed. Additionally, it also provides insights about the use of algal NPs in real-life examples and highlights the potential solutions to enhance the efficiencies and curb the limitations to address the associated challenges. Collaboration between regulatory bodies and researchers in the area of nanotechnology is also mentioned to maximize the benefits of NPS toward sustainable development.

Keywords

Algae · Nanoparticles · Green technology · Synthesis mechanism · Eco-friendly

A. Gupta · J. Bora · S. Nigam (✉)
Amity Institute of Biotechnology, Amity University, Noida, Uttar Pradesh, India
e-mail: gupta1@s.amity.edu; snigam@amity.edu

N. V. Rao · M. Joshi (✉)
Amity Institute of Nanotechnology, Amity University, Noida, Uttar Pradesh, India
e-mail: niharika.rao@s.amity.edu; mjoshi@amity.edu

R. E. Krishnankutty et al. (eds.), *Smart Agriculture*,
https://doi.org/10.1007/978-981-95-8785-8_9

1 Introduction

The modern era has witnessed technological advancement and nanotechnology has emerged as a scientific triumph for that advancement. The small sizes, high surface area, and tunable morphologies, nanomaterials show distinctive thermal, physical, biological, optical, chemical, mechanical, and electrical properties that vary greatly from their bulk material counterparts (Mandal and Ray Banerjee 2020). Nanomaterials can be synthesized through physical, chemical, and green synthesis methods. Physical and chemical methods generally utilize sophisticated instruments, harmful chemicals, toxic reagents, and consume excessive energy (Baig et al. 2021). Apart from that, by-products are harmful to the environment. However, excessive use of chemicals, nonpolar solvents, and toxic reagents during chemical synthesis has created a derogatory impact on the environment as well as human health. Considering this, exploring green methods and implementing natural and inherent molecules for the development of nanomaterials is an emerging alternative over the above-mentioned methods due to their safe protocol, low cost, and nontoxicity (Ijaz et al. 2020). Various bacteria, plant extracts, and algal extracts are the major precursors that act as reducing and capping agents for the development of these nanoparticles (Gnanasangeetha and Suresh 2020).

Among various bioagents, algae are significant producers of organic compounds and play a crucial role as the foundation of the food chain in aquatic systems. They are specially adapted to searching for resources in their environment through structural changes, storage, or increased resource efficiency and can excrete substances to improve nutrient availability. The properties of algae make them an intriguing part of the green synthesis of NPs. Algal biomass is used for NP synthesis (live or dead). Algae can accumulate or assemble heavy metal ions and have excellent potential to modify them into more malleable forms (Chaudhary et al. 2020). These enticing abilities are the reason algae have become the model organisms for synthesizing various classes of nanomaterials, mainly metallic NPs. Green nanomaterials differ from traditional materials because they are eco-friendly, stronger, lighter, and require less energy (Madani et al. 2022). Algal NPs obey the laws of green chemistry by reducing the dependence on toxic chemicals and avoiding their damaging consequences on the environment. The significant differences between biological synthesis and chemical synthesis are shown in Table 9.1.

Different strains of algal species (*Calothrix pulvinate*, *Chlorella vulgaris*, *Anabaena flos-aquae, Spirulina platensis, Plectonema boryanum*) are used for the biosynthesis of metallic and bimetallic NPs. Algal NPs have distinctive characteristics like low toxicity level, high biocompatibility, and large surface area. The various applications of algal NPs include drug delivery, ecological remediation, biomedical imaging, etc., which are further elucidated in the article (Bhardwaj et al. 2020). Among these applications, algal NPs show immense potential toward agro-industry and environmental remediation. The major factor favoring this, is due to its sustainable synthesis process and eco-friendliness. The article also encompasses the different synthesis techniques for algal NPs and provides an overview of NPs fabricated from different classes of algae. The general mechanism occurring during the

Table 9.1 Difference between biological and chemical synthesis of NPs

Parameters	Biological synthesis	Chemical synthesis	References
Design optimization	Optimal reaction condition such as selection of suitable organism, growth conditions, utilization of biocatalyst	Diverse physiochemical conditions such as requirements for a reaction, reactants concentration, catalyst	Álvarez-Chimal and Ángel Arenas-Alatorre (2023)
Synthesis mechanism	Enzymatic and nonenzymatic alterations using cell extract or whole cells	Electrolysis, chemical reduction, sol-gel processing, irradiation	Kulkarni et al. (2023)
Duration of synthesis	Longer duration due to subsequent NP and its microbial growth	Short duration	Kulkarni et al. (2023)
Usage	Diagnostic field (biosensors, targeted drug delivery, antimicrobial *agents),* therapeutic, clinical	Limited to clinical and biomedical areas due to the presence of toxins	Panchatcharam (2024)
Environment concerns	Do not harm the environment	Soil and water pollution because of the expenditure of energy and capital as well as toxic chemicals	Panchatcharam (2024), González-Almécija et al. (2024)

fabrication of these NPs, the challenges and limitations of fabrication and utilization of these NPs, and an overview of their scale up for industrial use are elucidated in the article.

2 Algae-Mediated Synthesis of Nanomaterials

Algal nanomaterials are derived from natural and synthetic sources. Some algae naturally cultivate inorganic NPs as part of their life cycle, while others can be synthesized from different algal elements to produce nanostructures with distinctive properties (Holyk et al. 2023). Algal NPs are cost effective, eco-friendly, and have diverse applications. Algae are rich in compounds like bioflavonoids, carbohydrates, terpenoids, aliphatic amines, sulfated polysaccharides, as well as other biological compounds that act as capping and reducing agents during NP synthesis (Behera et al. 2023).

Algal nanomaterials demonstrate a wide spectrum of pore dimensions categorized as microporous (pores <2 nm), mesoporous (pores between 2 and 50 nm), or macroporous (pores >50 nm), relying on the specific configuration and composition of the material.

However, Algal nanomaterials, particularly those in fiber-like or biopersistent granular type, and chemically modified algal residues (CMARs), can pose probable toxicity due to proportions, shape, and chemical arrangement. These factors can

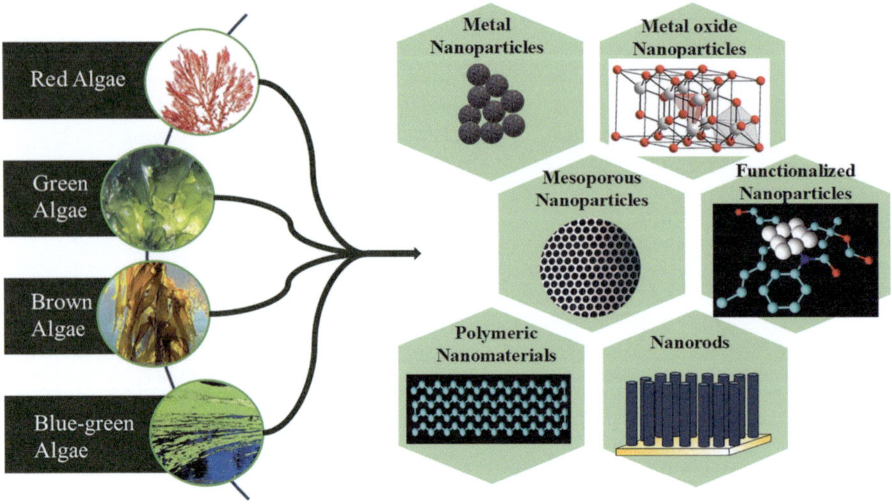

Fig. 9.1 An overview of NPs synthesized from different classes of algae

impact their interaction with biological systems, potentially leading to harm (Al-Maliki et al. 2022).

There are three main techniques for algal NP synthesis.

(i) Natural exploitation of live algal cells for the synthesis of NPs.
(ii) Disruption of algal cells accompanied by extraction through various down-stream-processing techniques like filtration and centrifugation.
(iii) Harvesting of NPs from the supernatants collected from the algal broth (Dubey et al. 2023; Huang et al. 2022a, b).

2.1 Synthesizing Nanomaterials Using Different Classes of Algae

Algae can be classified into different classes such as blue green, green, brown, and red algae depending on pigments, food reserves, and structural components. Figure 9.1 depicts the nanoparticles synthesis from different types of algae.

2.1.1 Brown Algae-Mediated Biosynthesis

Brown Algae are seaweed that belong to the family Sargassaceae and order Fucales. As of now, various species of brown algae are used to synthesize many metallic and metal oxide nanomaterials like silver, gold, titanium oxide, and zinc oxide. Nanoparticles (NPs) that are most commonly biosynthesized using brown algae include Ag, Cu, and Au NPs. Many NPs are synthesized from brown algae as shown in Table 9.2.

Table 9.2 NPs synthesized from different classes of algae

Algae	Class	Location	Size	Shape	NPs	Application	References
Sargassum muticum	Brown Algae	Extracellular	30–57 nm	Hexagonal	ZnO	Antiangiogenic and antiapoptotic effects	Hameed et al. (2023)
Sargassum muticum	Brown Algae	Extracellular	41.0 ± 5.7 nm, 0.4 ± 1.2 nm	Spherical	Ag, Au	Ovicidal and ovideterrent activity, antibacterial activity	González-Ballesteros et al. (2020), Khanna et al. (2019)
Kappaphycus alvarezii	Red Algae	Extracellular	10–40 nm	Polydisperse	Au	Agriculture and pharmaceutical industry	Gallego (2023)
Halymenia porphyriformis and *Solieria robusta*	Red Algae	Extracellular	–	–	Ag	Antibacterial activity	Menaa et al. (2021)
Cyanobacterium Oscillatoria limnetica	Blue Green Algae	–	30–17.97 nm	Quasi spherical	Ag	Antimicrobial activity	Hamouda and Aljohani (2024)
Cholera vulgaris	Green Microalgae	Extracellular	–	Spherical	Si	Anti-reflection coating	Dhaka et al. (2023)
Ulva intestinalis	Green Macroalgae	Extracellular	–	–	Ag and Au	Immunostimulant	González-Ballesteros et al. (2019)

2.1.2 Red Algae-Mediated Biosynthesis

Red algae are eukaryotic algae of the Rhodophyta family and are used as food as they contain many important minerals and vitamins. *Porphyra vietnamensis* has a strong reducing agent making it the most widely used species for the biosynthesis for many types of NPs (Palanisamy et al. 2024). Red algae-mediated Ag NPs compared to the physiochemical method are eco-friendly, more efficient, cost-effective, and less time-consuming (Chaudhary et al. 2020). Many NPs are synthesized from red algae as shown in Table 9.2.

2.1.3 Blue-Green Algae-Mediated Biosynthesis

Blue-green algae are prokaryotic algae. They are also known as Cyanobacteria and are a part of order Chroococcales. It has two well-defined families of Entophysalidaceae and Chroococcaceae. The members of both these families have different growth habitats forming colonies. They are found on moist rocks as parenchymatous cell masses, growing in dense patterns. Ag NPs synthesis by blue-green algae is mainly by the species *Spirulina platensis*. It is filamentous cyanobacterium that is free floating and has multicellular trichomes with one open end and left-handed helix (Chaudhary et al. 2020). Many NPs are synthesized from blue-green algae as shown in Table 9.2.

2.1.4 Green Algae-Mediated Biosynthesis

Green algae are eukaryotic organisms that are divided into micro and macro algae based on their habitat. Various species of micro green algae are used to generate Ag NPs. The most beneficial green macroalgae species is *Ulva fasciata* for the synthesis of NPs. Numerous microalgal species such as Chlorella, Chlamydomonas, etc., have been reported to be successfully used in the synthesis of green NPs (Menaa et al. 2021). Green algae possess inherent capabilities to uptake and accumulate metal ions from the environment, making them ideal candidates for the synthesis of metal oxide and metallic NPs. Additionally, the metabolic activities of green algae contribute to the reduction and stabilization of NPs, resulting in the formation of well-defined and stable nanomaterials. This approach offers several advantages such as cost-effectiveness, scalability, and reduced environmental impact compared to traditional chemical synthesis methods. Moreover, the use of green algae-mediated biosynthesis enables the production of NPs with controlled surface properties, shape, and size which are important for various applications in catalysis, sensing, drug delivery, and environmental remediation. Current research in this area aims to further understand the underlying mechanisms of NP formation by green algae and optimize the synthesis process to enhance the efficiency and yield of nanomaterials production. Green microalgae- and macroalgae-mediated biosynthesis and application of various NPs are depicted in Table 9.2.

Table 9.3 Characterization techniques used to analyze algal NPs

Algal species	Type of NPs	Imaging techniques	Applications	References
Portieria hornemannii	Ag	SEM, TEM, DLS	Antimicrobial, environmental applications	Uzair et al. (2020)
Ulva fasciata	ZnO	SEM, TEM, XRD	Antioxidant, environmental applications	Fouda et al. (2022)
Sargassum plagiophyllum	AgCl	SEM, TEM, DLS	Antibacterial, drug delivery	González-Ballesteros et al. (2020)
Bifurcaria bifurcate	CuO	SEM, TEM, XRD	Antimicrobial, environmental remediation	González-Ballesteros et al. (2020)
Jania rubens	Fe_3O_4	SEM, TEM, XRD	Magnetic applications, drug delivery	Maghraby et al. (2021)

2.2 Major Characterization Techniques for Algae Nanoparticles

Algal NPs are characterized by using various advanced analytical techniques to assess their biological, physical, and chemical properties. These techniques help us to understand the size, morphology, composition, and potential applications of the NPs synthesized from algae (Freeland et al. 2025; Holyk et al. 2023). The details of these techniques are shown in Table 9.3.

2.3 Factors Affecting Algal NP Synthesis

The algae-mediated NPs are converted by various factors, incorporating the algal species type, synthesis mechanisms (physical, chemical, biological), and growth conditions such as pH, temperature, reaction time, availability of nutrients, and light intensity as shown in Fig. 9.2. They affect the quantity and quality of algal biomass and the biochemical design which in turn affects the synthesis of NPs. The shape, size, and functionality of the NPs are also affected by the different cellular structures and different metabolic routes. All these factors cumulatively impact the stability, efficacy, and applications of algae-mediated NPs in various fields of nanocoatings, bioremediation, agriculture, and carbon utilization (Al-Maliki et al. 2022).

3 Mechanism Involved in the Synthesis of Algae-Mediated Nanomaterials

Biosynthesis methods are divided into extracellular and intracellular methods based on the location of NPs produced as shown in Fig. 9.3. Synthesis of nanomaterials using algae is done using three methods: NPs can be collected from supernatants of

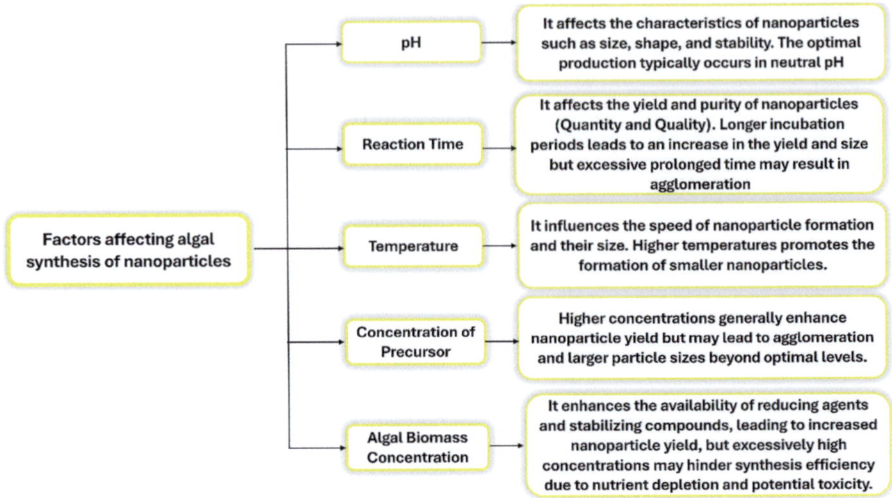

Fig. 9.2 Factors affecting synthesis of algal NPs

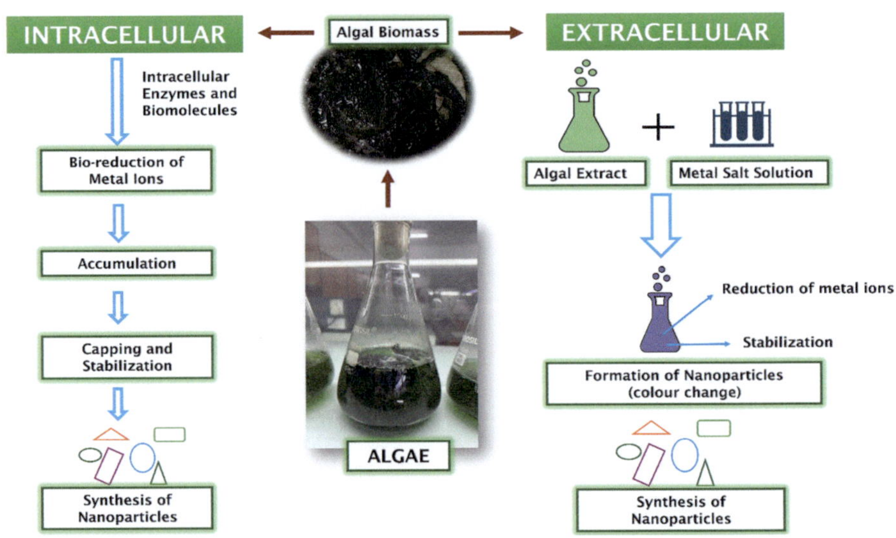

Fig. 9.3 Intracellular and extracellular mechanism for NPs synthesis from algae

algal broth, direct utilization of living algae cells, and destruction of algal cells and then their extraction through different downstream process techniques such as filtration and centrifugation (Jayaprakash and Kannappan 2022).

Algae can accumulate and modify heavy metal ions into more malleable forms. These enticing abilities are the reason algae have become the leading organisms for synthesizing different types of nanomaterials, mainly metallic NPs (Srivastava and Bhargava 2022). Microalgae have now become a good source for synthesizing nanomaterials. The synthesis of algae-mediated NMs includes the preparation of (i) algal extract, (ii) metal precursor solution, and (iii) incubation of algal extract with precursor metal. When the algal extract is incubated with a metal precursor solution, the biosynthesis of NPs is initiated. Algal extracts consist of biochemical compounds like fatty acids, proteins, carbohydrates, minerals, oils, chlorophyll, antioxidants, phycobilins, pigments, etc., which act as reducing agents. They reduce the charge of metal ions to a zero valent state. The visible color change of the reaction mixture illustrates nucleation. Bioreduction occurs in three phases—activation, growth, and termination phase. In the activation phase, the metal ions are reduced and the enzymes that are secreted by the algal cells lead to nucleation which is visible through color change of the solution. In the growth phase, metal elements that have been nucleated combine, forming thermodynamically stable NPs of different shapes and sizes. In the termination phase, the NPs acquire their final shape (Chaudhary et al. 2020). Many factors like temperature, pH, reaction time, exposure time, extract concentration, etc., can affect the formation and stabilization of NPs.

3.1 Intracellular Synthesis

Intracellular means the reactions or processes that take place inside the cell the NP synthesis occurs within the algal cell. No pretreatment of microalgae is required for intracellular synthesis as the procedure mainly depends on metabolic pathways NADPH or NADPH reductases that are generated during photosynthesis and respiration through the ETC (electron transport chain) or respiratory ETS at the cell wall or at the thylakoid membranes (Chaudhary et al. 2020). Studies have shown that Au NPs were biosynthesized intracellularly in an experiment where *Ulva intestinalis* and *Rhizoclonium fontinale* algal species were incubated with chloroauric acid at 20 °C for 72 h. Formation of Au NPs was indicated through visible color change of the thallus from green to purple. The fact that no color change was observed when biomass was incubated with gold metal solution further supported this, which led to the conclusion that this process of bioreduction is not linked with any metabolic pathways that involve intracellular enzymes or metabolites. Chloroplast of *Klebsormidium flaccidum* encapsulated in the silica gel suspension changed color from green to purple. This showed the potential of cells to reduce gold precursor (chloroauric acid) (Khoshnazar et al. 2023).

Intracellular synthesis of algal NPs offers better control over the shape, composition, and size of NPs offering an opportunity for industrial scale production. Research is being done in this area to gain a better understanding of the mechanism of NP synthesis through algae and to optimize the process for industrial production.

3.2 Extracellular Synthesis

Extracellular means the processes that take place outside the cell. This process happens when metal ions that are attached to the surface of algal cells are reduced at the surface by metabolites like nonprotein RNA, proteins, DNA, pigment, enzymes, ions, and lipids. As opposed to intracellular synthesis, the algal biomass in extracellular synthesis requires pretreatments like blending and washing (Srivastava et al. 2023). A study showed that several cells and an optimum concentration of metal precursors are essential for synthesis. By varying the dose of Au (III) ions and the cell number, Gold NPs were customized. The presence of gold peak at 530 nm confirmed the extracellular synthesis on the surface of *Spirulina platensis*, assisted by enzymes, biomolecules, and proteins (Hameed et al. 2023). One advantage of extracellular synthesis is the potential for tunable NP properties by manipulating the growth conditions and composition of the growth medium. Additionally, extracellular synthesis can be easily scaled up for industrial production compared to intracellular synthesis, which may require additional steps for NP extraction from the cells.

4 Applications of Algae-Mediated Nanomaterials for Environmental Sustainability

Environmental sustainability of nanoparticles is one of the crucial parameters for various applications including bioremediation and agriculture practices. In this context, the biological green route synthesis is an upcoming and reliable approach. Figure 9.4 and Table 9.4 highlight various applications of algae-synthesized nanoparticles.

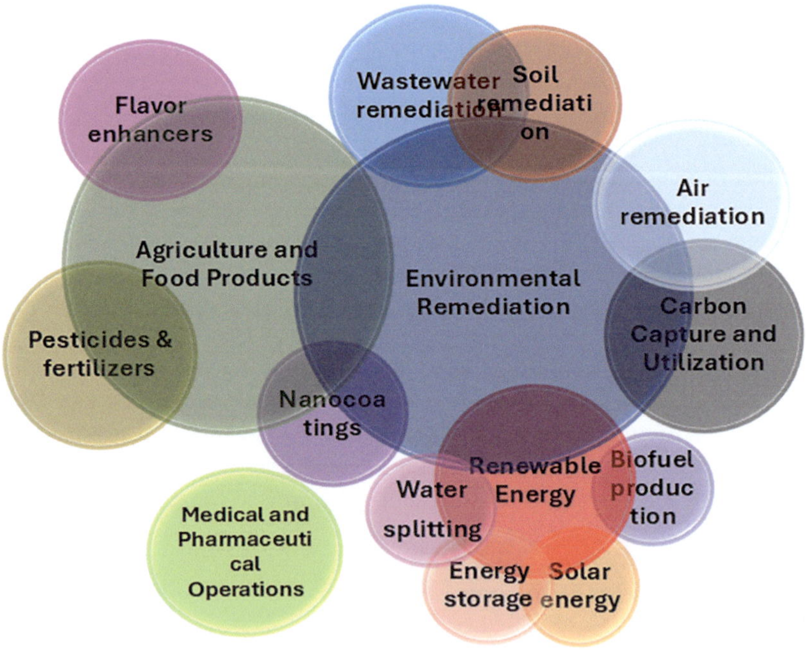

Fig. 9.4 Applications of algal NPs

Table 9.4 Different applications of Algal-based NPs

Algae	NPs	Applications	References
Chlorella ellipsoidea	Ag	Disintegration of the toxicant pollutant dyes methylene blue and methyl orange	Borah et al. (2020)
Sargassum myriocystum	Ag	Photocatalytic activity against methylene blue	Hamouda and Aljohani (2024)
Microchaete	Ag	Capability to decolorize against methyl red azo dyes	Husain et al. (2019)
Ulva latica	Ag	Effective in inhibiting the growth of *C.albicans*, *A.niger*, and *Candida parapsilosis* fungal strains	Chugh et al. (2021)
Sargassam vulgare	Ag	Anticancerous activity against human myeloblastic leukemic cells HL60 and HeLa cells	Chaudhary et al. (2020)
Noctiluca scintillans	Ag	Biosensing—Colorimetric sensing of hydrogen peroxide; used as an antiseptic and is an indicator for small dermal scratches such as in mouth, gum and teeth.	Elgamouz et al. (2020)
Acanthophora spicifera	Au	Strong anticancerous activities against the colorectal adenocarcinoma HT-29 cell line	Babu et al. (2020)
Sargassum myriocystum	Pt	Acts as a biosensors for detecting the adrenaline level in the body	Chaudhary et al. (2020)
Padina boergesenii	Ag	Nanofungicides	Vicente et al. (2021)
Chlorosacinopsis sp.	Microalgae-ZnO-NPs	Efficient sewage wastewater treatment and improved biodiesel production	Vasistha et al. (2021)
Oedogonium sp.	ZnO	Removal of organic and inorganic pollutants in wastewater by Adsorption Technique	Alprol et al. (2023)

4.1 Agriculture and Food Products

Algal nanotechnology can enhance agrarian practices by perfecting nutrient vacancies, crop growth, and pest control. The use of algae-synthesized NPs in agriculture and food products has the potential to revolutionize various aspects of food production, preservation, and safety. One application of algae-synthesized NPs is in crop cultivation and protection. Nanoparticles (NPs) such as zinc oxide and copper oxide, synthesized by algae, can be used as nano fertilizers or nano pesticides to improve crop yields and protect plants from pests and diseases (El-Refaey and Salem 2023). These NPs can enhance nutrient uptake, promote plant growth, and provide systemic protection against pathogens, reducing the need for traditional chemical inputs and minimizing environmental impacts. Furthermore, algae-synthesized NPs can be utilized for food packaging and preservation. Nanoparticles (NPs) with antimicrobial properties, such as silver NPs, can be incorporated into packaging materials to inhibit the growth of spoilage microbes and increase the shelf life of perishable food products. Additionally, NPs with barrier properties, such as graphene oxide, can be used to create impermeable coatings that prevent moisture loss and maintain the freshness and standard of packaged foods (Kokabi and Yousefzadi 2022). In food processing and manufacturing, algae-synthesized NPs can be employed for various purposes, including food fortification, flavor enhancement, and color stabilization. Nanoparticles (NPs) such as iron oxide and calcium carbonate, synthesized by algae, can be used to enrich foods with required minerals and nutrients, which helps address nutritional insufficiency in populations. Additionally, NPs with sensory properties, such as gold NPs, can be used as flavor enhancers or colorants in food formulations, enhancing the overall sensory experience for consumers. Moreover, algae-synthesized NPs have potential for improving food safety and quality control. Nanoparticles (NPs) with detection capabilities, such as quantum dots, can be used as nanosensors to detect contaminants, allergens, or pathogens in food products, enabling rapid and sensitive testing methods. These NPs can help ensure compliance with food safety regulations and prevent foodborne illnesses, enhancing customer's trust in the integrity of food products (Mangena 2025).

4.2 Environmental Remediation

Algae-intermediate nanotechnology can be employed for environmental remediation, including the decline of pollutants and the restoration of defiled ecosystems. Nanoparticles (NPs) deduced from algae can be used for the effective junking of heavy metals and oil painting tumbles (Aalam et al. 2024). Microalgae would be considered an ideal choice for the synthesis of NPs owing to their innate overaccumulation of heavy metals and high capacity for CO_2 capture and storage. Environmental remediation using algal NPs is usually accompanied by ecologically safe processes such as adsorption and benign conversions without the production of harmful by-products, making it an optimal choice (Khan et al. 2022). Many studies show promising applications of algal NPs such as gold, silver, and iron capable of targeting chemically diverse pollutants in wastewater, soil, and air (Tonelli et al. 2023).

4.2.1 Bioremediation of Wastewater

Nanomaterials have many properties like high sensitivity and reactivity, ease of functionalization, high surface-to-volume ratio, and high adsorption capacity making them suitable for use in wastewater treatment (Mangalam et al. 2024). These NPs, synthesized intracellularly by algae through eco-friendly and sustainable processes, offer unique properties that can enhance wastewater treatment performance. For instance, NPs such as silver NPs synthesized by algae have demonstrated strong antimicrobial properties, making them effective in disinfection and pathogen removal from wastewater. For the efficient removal of contaminants, algae-mediated iron NPs act as catalysts that with the help of oxidation process degrade the organic pollutants. Moreover, the use of algae-synthesized NPs in wastewater treatment promotes the use of renewable resources reducing the environmental impact of wastewater treatment processes (Sharma et al. 2023). Overall, the application of algae-synthesized NPs holds great promise for improving the efficiency and sustainability of wastewater treatment systems, contributing to the protection of water resources and human health.

4.2.2 Source of Renewable Energy

(i) Solar Energy: Algal NPs are incorporated into solar cells to enhance their efficacy by converting more sunlight into electricity. The NPs help in generating higher electrical current (Xu et al. 2023).

(ii) Biofuel Production: Algae have high lipid content that converts into biofuels such as biodiesel. Algal NPs can enhance the efficiency of lipid derivation for biofuel production (Mehariya et al. 2021). These NPs help to break down cell walls and discharge lipids more efficiently which increases the long-term oil yield. A lot of research has been done on green synthesized iron-based nanoparticles, since iron plays a highly efficient role as a catalyst in the enhancement of the reaction rate and the overall efficiency of the process. In addition to this, employment of green synthesized NPs has been revealed to lower the manufacturing costs (Srivastava et al. 2021).

(iii) Energy Storage: Algal NPs utilize the properties of specific algal species that can be developed for use in leading battery technologies. For example, some algae contain polysaccharides that display excellent charge storage capabilities, potentially facilitating the growth of exceptionally good and eco-friendly energy storage devices (Mehariya et al. 2021).

(iv) Water Splitting: Algal NPs are used in generating hydrogen as a renewable energy source by using them as agitators to intensify the competency of water-splitting reactions. Nanoparticles (NPs) absorb solar energy which boosts the breakdown of H_2O molecules into O_2 and H_2 (Parray et al. 2024).

4.2.3 Carbon Capture and Utilization

Algae can capture and convert carbon dioxide into biomass, reducing greenhouse gases. Nanotechnology can contribute to developing nanocomposite accouterments, similar to enhanced carbon nanotubes or NPs, for effective carbon capturing and application by algae. The biosequestration of CO_2 using microalgae cell factories is a sustainable, economical, and eco-friendly way to tackle the problem of increased CO_2 emissions (Onyeaka et al. 2021). Additionally, algae-mediated NPs can be used in biological processes for CO_2 fixation and biomass production. Algae cells naturally sequester CO_2 through photosynthesis, converting it into organic biomass and oxygen. Nanoparticles (NPs) synthesized by algae, such as iron NPs or quantum dots, can enhance photosynthetic efficiency and biomass productivity in algae cultures, leading to increased CO_2 sequestration rates and biomass yields. The biomass produced by algae can be used for various applications, including biofuels, animal feed, and bioplastics, further contributing to carbon mitigation efforts. Moreover, algae-synthesized NPs can be integrated into novel materials and systems for CO_2 capture and utilization. For example, NPs can be immobilized within porous materials or membranes to enhance their CO_2 adsorption capacity and selectivity. These

NP-based materials can be deployed in carbon capture and storage (CCS) systems or integrated into bioreactors for CO_2 utilization, offering scalable and cost-effective solutions for mitigating greenhouse gas emissions.

4.3 Medical and Pharmaceutical Applications

Algae-synthesized NPs hold significance due to their distinctive characteristics and biocompatibility. One area of application is drug delivery, where algae-synthesized NPs serve as carriers for therapeutic agents such as drugs, proteins, and nucleic acids. Nanoparticles (NPs) synthesized by algae, such as chitosan or alginate-based NPs, offer biocompatible and biodegradable platforms for the targeted delivery of drugs to targeted tissues or cells. These NPs can be made functional with targeting ligands or stimuli-responsive coatings to increase drug accumulation at disease sites and reduce off-target effects, improving patient outcomes and therapeutic efficacy. Furthermore, algae-synthesized NPs can be utilized for imaging and diagnostic purposes in medicine. Nanoparticles (NPs) such as gold NPs or quantum dots synthesized by algae exhibit unique optical properties that can be exploited for various imaging modalities, including fluorescence imaging, photoacoustic imaging, and computed tomography (CT). These NPs serve as contrast agents to visualize biological structures, monitor disease progression, and guide therapeutic interventions, facilitating early detection and personalized treatment strategies. In addition to drug delivery and imaging, algae-synthesized NPs hold promises for regenerative medicine and tissue engineering applications. Moreover, algae-synthesized NPs have been investigated for their antimicrobial properties, which can be incorporated in the development of novel antimicrobial agents and wound dressings.

4.4 Nanocoatings

The plating of a thin subcaste of NPs onto an object's surface to enhance its performance and characteristics is known as nanocoating. Nanocoating can give several benefits which include advanced erosion resistance, increased continuity, increased conductivity, better optical characteristics, and tone-drawing capabilities (Gu et al. 2020).

One potential application of nanocoatings by algae-synthesized NPs is in the field of corrosion protection. Nanoparticles (NPs) such as silver, zinc oxide, and iron oxide, synthesized by algae, can be incorporated into coatings to provide enhanced corrosion resistance to metals and alloys. These NPs can act as barriers against corrosive agents, inhibit the formation of corrosion products, and promote self-healing mechanisms, thereby extending the lifespan of coated materials and structures. Another application area is the development of antimicrobial coatings for various surfaces, including medical devices, food packaging, and household appliances. Algae-synthesized NPs, particularly silver NPs, exhibit strong antimicrobial characteristics that can effectively stop or reduce the growth of bacteria, fungi, and viruses on coated surfaces. This can help mitigate the spread of infectious diseases and reduce the need for traditional chemical disinfectants, promoting a safer and healthier environment.

Furthermore, nanocoatings produced using algae-synthesized NPs can be utilized for environmental remediation purposes. For example, NPs such as titanium dioxide and zinc oxide synthesized by algae have photocatalytic properties that can degrade organic pollutants and purify air and water when incorporated into coatings applied to building facades, wastewater treatment facilities, and air purification systems (Gaur et al. 2024).

5 Scaling Up Algae-Mediated Metal NP Synthesis for Industrial Use

Industrial production of algal-mediate NPs requires various considerations. A few of them are discussed below.

5.1 Algal Strain Selection and Growth Optimization

Algal strains with high biomass yield, fast growth rate, and capable of tolerating high concentrations of metal ions are preferred. The NP synthesis should be high and fast and environmental concerns should be considered while selecting a suitable strain. The conditions like temperature, pH, light availability, and other factors should be optimized to produce maximum yield (Srinivasan and Kulshreshtha 2020; Sarwer et al. 2022).

5.2 Bioreactor Design and Sterilization

Efficient bioreactors should be designed for efficient growth and production. The cost should be kept in mind while finding the most effective process for better productivity. The bioreactor should be operated by professionals and strict measures should be taken to prevent contamination and ensure sterilization (Benner et al. 2022).

5.3 Downstream Processing

Effective techniques to purify NPs from the algal biomass should be developed to reduce waste and save costs. The size of NPs can be controlled to achieve desired results and efficient methods should be developed to incorporate NPs into products to make them into a useful product (Uzair et al. 2020).

5.4 Waste Management and Environmental Safety

Waste generated should be recycled, treated, or disposed of in a way that does not cause harm to the environment. The production process should be evaluated to check for any environmental concerns. Safety assessment of the process and NPs and their impact on environment are important for regulatory approval.

6 Real World Examples of NPs Being Used Worldwide

Algal mediated NPs have noteworthy commercial applications in areas like pharmaceutical industries, agriculture, food industry etc., as mentioned above. The use of NPs in fertilizers and pesticides has always been questionable even though its usage improves the quality of the fertilizers. However, due to their nontoxic nature, algal NPs have demonstrated potential to be utilized in the agro-industry (Mangena 2025). Table 9.5 summarizes specific device applications of algal-synthesized NPs globally.

Table 9.5 Real World Examples of NPs being Used Worldwide

Industry/Research lab	Country	Work done	References
School of Science, Hainan University; State Key Laboratory of Metastable Materials Science and Technology, Yanshan University	China	Utilizing algal synthesized NPs for Dopamine Detection	Huang et al. (2022b)
Department of Analytical Chemistry and Department of Cell and Molecular Biology, University of Kashan	Iran	Carbonaceous sensor synthesized using microalgae *Chlorella vulgaris* for enzymatic determination of glucose in blood serum	Jafari et al. (2021)
Algenol	USA	Personal care ingredients and natural colorants	Algenol Biotech (2025)
School of Mechanical Engineering and Automation; Beijing Advanced Innovation Center for Biomedical Engineering, and School of Biological Science and Medical Engineering, Beihang University	China	Mechanical Microrobots were prepared for targeted delivery, drug loading and chemo–photothermal therapy using *Spirulina* cells as biotemplate	Wang et al. (2019)
Diatom Research Laboratory, Amity University; Department of Chemistry, Hindu College, University of Delhi; Department of Biotechnology, North Bengal University	India	*T. weissflogii* was used for delivery of curcumin loaded drugs	Saxena et al. (2022)
Genetic Engineering and Biotechnology Research Institute, SRTA City; Indiana University; Tanta University; Roswell Park Comprehensive Cancer Center; The University of Jordan	Egypt, USA, Jordan	*Arthrospira* sp. Polysaccharides were use to synthesize AgNPs and checked for their antibacterial efficacy	El-Deeb et al. (2020)

7 Challenges and Solutions

The major limitations and challenges of algal NPs are discussed below.

7.1 Challenges

(i) Restricted applications: Algae-based nanotechnology mainly focuses on energy production, waste treatment, and biofuel production.
(ii) Scale-up challenges: Expanding up algal nanotechnology from the laboratory to industrial production is challenging due to higher costs, lower yields and technical issues in maintaining stable experimental conditions.
(iii) Variations of algae: Algae quality differs since they are living organisms that have distinct compositions, conditions for development, and genetic characteristics.
(iv) Cost considerations: As compared to conventional resources or other nanomaterials, algae-based nanotechnology can still be a bit costly. The high expenses of algae cultivation, extraction, and purification might restrict its economic viability.
(v) Limited Knowledge: Despite ongoing research, there is still an inadequate understanding of the full potential of algae-based nanotechnology. Further research is needed to explore its applications, enhance production processes, and analyze long-term human and environmental consequences.

7.2 Potential Solutions

(i) Standardization and Optimization of Cultivation Conditions—The conditions in which algae are grown can greatly affect NP yield and quality. Optimizing factors like temperature, light, nutrient availability can improve algal biomass production. Following standard SOPs can provide consistency and safety.
(ii) Regulatory Guidelines and Risk Assessment—Establishing proper safety and regulatory guidelines to produce algal NPs on industrial scale and conducting thorough risk assessments of NPs can help address environmental and health concerns.
(iii) Characterization Techniques like TEM, SEM, DLS, XRD, etc., are important to better understand the physical, chemical and biological features of algal NPs which helps in their effective application.
(iv) Transparency—Maintaining transparency in the production techniques and research findings can help in gaining public and government trust and can also help in commercialization of the products.
(v) Spreading awareness about phyconanotechnology through educational camps or workshops can help the public build interest in this field and minimize any concerns related to using products containing algal NPs.

7.3 Collaboration Between Regulatory Bodies and Researchers in the Area of Nanotechnology

Collaboration between regulatory bodies and researchers in the area of nanotechnology is essential to maximize the benefits this field has to offer toward sustainable development. It is also important to address the unique challenges posed by this rapidly evolving field. This collaboration must take care of various aspects like safety assessments, regulatory frameworks, and commercialization strategies. Certain examples of such collaborations are the ISO Standards developed for nanotechnology, including terminology, measurement, and safety practices. These standards were created through extensive participation between scientists, industry experts, and governing bodies worldwide. The U.S. Food and Drug Administration (FDA) has collaborated extensively with researchers to develop guidelines for the safe use of nanomaterials in medicine. This collaboration has led to the approval of several nanomedicine products, such as Doxil, a NP-based cancer treatment (U.S. Food and Drug Administration 2022). The European Union's NanoSafety Cluster fosters collaboration between researchers, regulators, and industry to address nanotechnology-related challenges. These examples show that effective collaboration between researchers and regulatory bodies is crucial for realizing the full potential of nanotechnology while mitigating associated risks.

7.4 Future Prospects

Tests on algae-derived NPs (NPs) assess their efficacy in eliminating various contaminants such as organic/aromatic compounds and heavy metals. These NPs have shown encouraging results in catalyzing dye breakdown, lowering the population of chloroquine-resistant species, and displaying photocatalytic activity against harmful contaminants. They effectively prevent bacterial adherence and biofilm formation on surfaces (Chaudhary et al. 2020).

The production of Ag NPs (SNPs) making use of *Fusarium oxysporum* includes checking their antibacterial activity, antioxidant ability, hydrogen peroxide detection, and photocatalytic performance (Jayaprakash and Kannappan 2022). The copper oxide (CuO) NPs demonstrated photocatalytic activity. This suggests that they can remove organic contaminants from water or wastewater (Sorekine et al. 2022). The utilization of brown algae as a biocatalyst in the preparation of magnesium oxide (MgO) NPs prevents the requirement for harmful substances and energy-intensive procedures. The most effective and safe method for antibacterial activity and dye removal is photocatalysis with semiconductor-based zinc oxide (ZnO) NPs and an aqueous extract of Punica granatum peel (Ye et al. 2024).

8 Conclusion

In conclusion, algae-synthesized NPs represent a promising avenue for diverse applications across various fields. These NPs, synthesized through eco-friendly and sustainable processes within algae cells, offer unique properties that can be tailored for specific applications. In biomedicine, algae-synthesized NPs act as a promising means of diagnosis, drug delivery vehicles, and antimicrobial agents due to their intrinsic properties and biocompatibility. In the bioremediation of environment, these NPs can be utilized for wastewater analysis, pollutant removal, and soil remediation, contributing to the preservation of water reserves and ecosystem health. Furthermore, the use of algae-mediated NPs as antibacterials agents can give a cost-effective result for managing multidrug-resistant bacteria. They are environment-friendly conflation, strong antibacterial parcels, and have the capacity to fight biofilms against contagious conditions. Overall, ongoing research and innovation in the synthesis and application of algae-synthesized NPs continue to expand their potential impact across diverse fields, offering sustainable solutions to global challenges while minimizing environmental harm.

Funding Nil

Compliance with Ethical Standards

Conflicts of Interest The authors declare that they have no conflict of interest in the publications.

Informed Consent Not applicable.

Research Involving Human Participants and/or Animals This article does not contain any studies of human or animal subjects performed by any of the authors.

References

Aalam G, Ahmad S, Amir M et al (2024) Future perspectives of nanoparticles in plants and algae. In: Molecular impacts of nanoparticles on plants and algae. Elsevier, pp 425–444. https://doi.org/10.1016/B978-0-323-95721-2.00019-1

Algenol Biotech (2025) The Algenol advantage. https://www.algenol.com/. https://www.algenol.com/. Accessed 15 Feb 2025

Al-Maliki RM, Alsalhy QF, Al-Jubouri S et al (2022) Classification of nanomaterials and the effect of Graphene Oxide (GO) and recently developed nanoparticles on the ultrafiltration membrane and their applications: a review. Membranes (Basel) 12:1043. https://doi.org/10.3390/membranes12111043

Alprol AE, Mansour AT, El-Beltagi HS, Ashour M (2023) Algal extracts for green synthesis of zinc oxide nanoparticles: promising approach for algae bioremediation. Materials 16:2819. https://doi.org/10.3390/ma16072819

Álvarez-Chimal R, Ángel Arenas-Alatorre J (2023) Green synthesis of nanoparticles: a biological approach. In: Green chemistry for environmental sustainability – Prevention-Assurance-Sustainability (P-A-S) approach. IntechOpen

Babu B, Palanisamy S, Vinosha M et al (2020) Bioengineered gold nanoparticles from marine seaweed Acanthophora spicifera for pharmaceutical uses: antioxidant, antibacterial, and anticancer activities. Bioprocess Biosyst Eng 43:2231–2242. https://doi.org/10.1007/s00449-020-02408-3

Baig N, Kammakakam I, Falath W (2021) Nanomaterials: a review of synthesis methods, properties, recent progress, and challenges. Mater Adv 2:1821–1871. https://doi.org/10.1039/D0MA00807A

Behera M, Behera PR, Bhuyan PP et al (2023) Algal nanoparticles and their antibacterial activity: current research status and future prospectives. Drugs Drug Candidates 2:554–570. https://doi.org/10.3390/ddc2030029

Benner P, Meier L, Pfeffer A et al (2022) Lab-scale photobioreactor systems: principles, applications, and scalability. Bioprocess Biosyst Eng 45:791–813. https://doi.org/10.1007/s00449-022-02711-1

Bhardwaj B, Singh P, Kumar A et al (2020) Eco-friendly greener synthesis of nanoparticles. Adv Pharm Bull 10:566–576. https://doi.org/10.34172/apb.2020.067

Borah D, Das N, Das N et al (2020) Alga-mediated facile green synthesis of silver nanoparticles: photophysical, catalytic and antibacterial activity. Appl Organomet Chem 34. https://doi.org/10.1002/aoc.5597

Chaudhary R, Nawaz K, Khan AK et al (2020) An overview of the algae-mediated biosynthesis of nanoparticles and their biomedical applications. Biomolecules 10:1498. https://doi.org/10.3390/biom10111498

Chugh D, Viswamalya VS, Das B (2021) Green synthesis of silver nanoparticles with algae and the importance of capping agents in the process. J Genet Eng Biotechnol 19:126. https://doi.org/10.1186/s43141-021-00228-w

Dhaka A, Chand Mali S, Sharma S, Trivedi R (2023) A review on biological synthesis of silver nanoparticles and their potential applications. Results Chem 6:101108. https://doi.org/10.1016/j.rechem.2023.101108

Dubey V, Parikh KD, Lin K-YA et al (2023) Algae-based synthesis to generate nanomaterials for nanoremediation. In: Green nanoremediation. Springer International Publishing, Cham, pp 109–126

El-Deeb NM, Abo-Eleneen MA, Al-Madboly LA et al (2020) Biogenically synthesized polysaccharides-capped silver nanoparticles: immunomodulatory and antibacterial potentialities against resistant pseudomonas aeruginosa. Front Bioeng Biotechnol 8. https://doi.org/10.3389/fbioe.2020.00643

Elgamouz A, Idriss H, Nassab C et al (2020) Green synthesis, characterization, antimicrobial, anticancer, and optimization of colorimetric sensing of hydrogen peroxide of algae extract capped silver nanoparticles. Nanomaterials 10:1861. https://doi.org/10.3390/nano10091861

El-Refaey AA, Salem SS (2023) Algae materials for bionanopesticides: nanoparticles and composites. In: Algae Materials. Academic, pp 219–230. https://doi.org/10.1016/B978-0-443-18816-9.00004-6

Fouda A, Eid AM, Abdelkareem A et al (2022) Phyco-synthesized zinc oxide nanoparticles using marine macroalgae, *Ulva fasciata* delile, characterization, antibacterial activity, photocatalysis, and tanning wastewater treatment. Catalysts 12:756. https://doi.org/10.3390/catal12070756

Freeland B, Ahad IU, Foley G, Brabazon D (2025) Advanced characterization techniques for nanostructures. In: Micro and nanomanufacturing volume II. Springer Nature Switzerland, Cham, pp 53–89

Gallego GR (2023) Potential of Kappaphycus alvarezii (seaweed) extract as biostimulant on the growth of mangrove propagules (Rhizophora mangle L.) through foliar application. Int J Fish Aquat Stud 11:50–57. https://doi.org/10.22271/fish.2023.v11.i2a.2790

Gaur J, Kumar S, Pal M et al (2024) Current trends: Zinc oxide nanoparticles preparation via chemical and green method for the photocatalytic degradation of various organic dyes. Hybrid Adv 5:100128. https://doi.org/10.1016/j.hybadv.2023.100128

Gnanasangeetha D, Suresh M (2020) A review on green synthesis of metal and metal oxide nanoparticles. Nat Environ Pollut Technol 19:1789–1800. https://doi.org/10.46488/NEPT.2020.v19i05.002

González-Almécija B, Manuel López R, Asensio-Calavia P et al (2024) Harnessing the power of metallic nanoparticles: antimicrobial peptide functionalization with gold and silver nanoparticles. In: Nanomaterials for biomedical and bioengineering applications. Springer Nature Singapore, Singapore, pp 57–78

González-Ballesteros N, Diego-González L, Lastra-Valdor M et al (2019) Immunostimulant and biocompatible gold and silver nanoparticles synthesized using the Ulva intestinalis L. aqueous extract. J Mater Chem B 7:4677–4691. https://doi.org/10.1039/C9TB00215D

González-Ballesteros N, Rodríguez-Argüelles MC, Lastra-Valdor M et al (2020) Synthesis of silver and gold nanoparticles by Sargassum muticum biomolecules and evaluation of their antioxidant activity and antibacterial properties. J Nanostructure Chem 10:317–330. https://doi.org/10.1007/s40097-020-00352-y

Gu Y, Xia K, Wu D et al (2020) Technical characteristics and wear-resistant mechanism of nano coatings: a review. Coatings 10:233. https://doi.org/10.3390/coatings10030233

Hameed H, Waheed A, Sharif MS et al (2023) Green Synthesis of Zinc Oxide (ZnO) nanoparticles from green algae and their assessment in various biological applications. Micromachines (Basel) 14:928. https://doi.org/10.3390/mi14050928

Hamouda RA, Aljohani ES (2024) Assessment of silver nanoparticles derived from brown algae Sargassum vulgare: insight into antioxidants, anticancer, antibacterial and hepatoprotective effect. Mar Drugs 22:154. https://doi.org/10.3390/md22040154

Holyk J, Kajnts D, Vantyukh D (2023) Classification of nanomaterials and the opportunities of their use in civil engineering. In: Urban development and spatial planning, pp 95–113. https://doi.org/10.32347/2076-815x.2023.82.95-113

Huang G, Chen X, Li N et al (2022a) A convenient synthesis of gold nanoparticles in Spirulina extract for rapid visual detection of dopamine in human urine. Colloids Surf A Physicochem Eng Asp 650:129675. https://doi.org/10.1016/j.colsurfa.2022.129675

Huang Y, Gao M, Wang W et al (2022b) Effects of manufactured nanomaterials on algae: Implications and applications. Front Environ Sci Eng 16:122. https://doi.org/10.1007/s11783-022-1554-3

Husain S, Afreen S, Hemlata et al (2019) Cyanobacteria as a bioreactor for synthesis of silver nanoparticles-an effect of different reaction conditions on the size of nanoparticles and their dye decolorization ability. J Microbiol Methods 162:77–82. https://doi.org/10.1016/j.mimet.2019.05.011

Ijaz I, Gilani E, Nazir A, Bukhari A (2020) Detail review on chemical, physical and green synthesis, classification, characterizations and applications of nanoparticles. Green Chem Lett Rev 13:223–245. https://doi.org/10.1080/17518253.2020.1802517

Jafari SM, Masoum S, Tafreshi SAH (2021) A microlagal-based carbonaceous sensor for enzymatic determination of glucose in blood serum. J Ind Eng Chem 101:195–204. https://doi.org/10.1016/j.jiec.2021.06.012

Jayaprakash M, Kannappan S (2022) An overview of a sustainable approach to the biosynthesis of AgNPs for electrochemical sensors. Arab J Chem 15:104324. https://doi.org/10.1016/j.arabjc.2022.104324

Khan F, Shahid A, Zhu H, Wang N, Javed MR, Ahmad N, Xu J, Alam MA, Mehmood MA (2022) Prospects of algae-based green synthesis of nanoparticles for environmental applications. Chemosphere 293:133571. https://doi.org/10.1016/j.chemosphere.2022.133571

Khanna P, Kaur A, Goyal D (2019) Algae-based metallic nanoparticles: synthesis, characterization and applications. J Microbiol Methods 163:105656. https://doi.org/10.1016/j.mimet.2019.105656

Khoshnazar SM, Asadi A, Holghoomi R, Abdolmaleki A (2023) Green synthesis of silica nanoparticles/nanocomposites for biomedical applications: a narraitive review. Biochem Mosc Suppl B Biomed Chem 17:41–49. https://doi.org/10.1134/S1990750823600085

Kokabi M, Yousefzadi M (2022) Algal nanoparticles and their potential application in agriculture. In: Nano-enabled agrochemicals in agriculture. Academic, pp 189–198. https://doi.org/10.1016/B978-0-323-91009-5.00021-5

Kulkarni D, Sherkar R, Shirsathe C et al (2023) Biofabrication of nanoparticles: sources, synthesis, and biomedical applications. Front Bioeng Biotechnol 11. https://doi.org/10.3389/fbioe.2023.1159193
Madani M, Hosny S, Alshangiti DM et al (2022) Green synthesis of nanoparticles for varied applications: Green renewable resources and energy-efficient synthetic routes. Nanotechnol Rev 11:731–759. https://doi.org/10.1515/ntrev-2022-0034
Maghraby YR, Farag MA, Ramadan AR (2021) Protective action of Jania rubens nanoencapsulated algal extract in controlling vegetable oils' rancidity. ACS Omega 6:5642–5652. https://doi.org/10.1021/acsomega.0c06069
Mandal A, Ray Banerjee E (2020) Introduction to nanoscience, nanotechnology and nanoparticles. In: Nanomaterials and biomedicine. Springer Singapore, Singapore, pp 1–39
Mangalam J, Sharma J, Sharma M et al (2024) Recent advances of nanomaterials for non-steroidal anti-inflammatory drugs removal from wastewater: a critical review. Clean Techn Environ Policy 26:1705–1729. https://doi.org/10.1007/s10098-023-02701-1
Mangena P (2025) A review on the role of algae-nanoparticles as a sustainable agricultural strategy to improve resilience against abiotic stresses in leguminous crops. Plant Nano Biol 11:100125. https://doi.org/10.1016/j.plana.2024.100125
Mehariya S, Goswami RK, Karthikeysan OP, Verma P (2021) Microalgae for high-value products: A way towards green nutraceutical and pharmaceutical compounds. Chemosphere 280:130553. https://doi.org/10.1016/j.chemosphere.2021.130553
Menaa F, Wijesinghe U, Thiripuranathar G et al (2021) Marine algae-derived bioactive compounds: a new wave of nanodrugs? Mar Drugs 19:484. https://doi.org/10.3390/md19090484
Onyeaka H, Miri T, Obileke K et al (2021) Minimizing carbon footprint via microalgae as a biological capture. Carbon Capture Sci Technol 1:100007. https://doi.org/10.1016/j.ccst.2021.100007
Palanisamy J, Palanichamy VS, Vellaichamy G et al (2024) A comprehensive review on the green synthesis of silver nanoparticles from marine sources. Naunyn Schmiedeberg's Arch Pharmacol. https://doi.org/10.1007/s00210-024-03547-0
Panchatcharam P (2024) Recent advancements in the green synthesis of bioactive metallic nanoparticles from biological entities and their biomedical applications. In: Biogenic wastes-enabled nanomaterial synthesis. Springer Nature Switzerland, Cham, pp 239–256
Parray JA, Singh N, Haghi AK (2024) Algal biology and biomass characteristics. In: Microalgae as a sustainable source of green energy generation and bioeconomy, pp 11–39
Sarwer A, Hamed SM, Osman AI et al (2022) Algal biomass valorization for biofuel production and carbon sequestration: a review. Environ Chem Lett 20:2797–2851. https://doi.org/10.1007/s10311-022-01458-1
Saxena A, Dutta A, Kapoor N et al (2022) Envisaging marine diatom Thalassiosira weissflogii as a "SMART" drug delivery system for insoluble drugs. J Drug Deliv Sci Technol 68:102983. https://doi.org/10.1016/j.jddst.2021.102983
Sharma A, Pal K, Saini N, Kumar S, Bansal D, Mona S (2023) Remediation of contaminants from wastewater using algal nanoparticles via green chemistry approach: an organized review. Nanotechnology 34(35):352001. https://doi.org/10.1088/1361-6528/acd45a
Sorekine G, Anduwan G, Waimbo MN et al (2022) Photocatalytic studies of copper oxide nanostructures for the degradation of methylene blue under visible light. J Mol Struct 1248:131487. https://doi.org/10.1016/j.molstruc.2021.131487
Srinivasan B, Kulshreshtha G (2020) Algal biomass for biofuels and bioproducts. In: Algal biomass for biofuels and bioproducts, pp 139–160
Srivastava S, Bhargava A (2022) Biological synthesis of nanoparticles: algae. In: Green nanoparticles: the future of nanobiotechnology. Springer Singapore, Singapore, pp 139–171
Srivastava M, Srivastava N, Saeed M, Mishra PK, Saeed A, Gupta VK, Malhotra BD (2021) Bioinspired synthesis of iron-based nanomaterials for application in biofuels production: a new in-sight. Renew Sust Energ Rev 147:111206. https://doi.org/10.1016/j.rser.2021.111206
Srivastava S, Fuentes F, Bhargava A (2023) Algal synthesis of gold nanoparticles: applications in bioenergy. In: Technological advancement in algal biofuels production, pp 109–127

Tonelli F, Silva C, Delgado V, Tonelli F (2023) Algae-based green AgNPs, AuNPs, and FeNPs as potential nanoremediators. Green Process Synth 12(1):20230008. https://doi.org/10.1515/gps-2023-0008

U.S. Food and Drug Administration (2022) Drug products, including biological products, that contain nanomaterials – guidance for industry. https://www.fda.gov/regulatory-information/search-fda-guidance-documents/drug-products-including-biological-products-contain-nanomaterials-guidance-industry. https://www.fda.gov/regulatory-information/search-fda-guidance-documents/drug-products-including-biological-products-contain-nanomaterials-guidance-industry. Accessed 15 Feb 2025

Uzair B, Liaqat A, Iqbal H et al (2020) Green and cost-effective synthesis of metallic nanoparticles by algae: safe methods for translational medicine. Bioengineering 7:129. https://doi.org/10.3390/bioengineering7040129

Vasistha S, Khanra A, Rai MP (2021) Influence of microalgae-ZnO nanoparticle association on sewage wastewater towards efficient nutrient removal and improved biodiesel application: an integrated approach. J Water Process Eng 39:101711. https://doi.org/10.1016/j.jwpe.2020.101711

Vicente TFL, Lemos MFL, Félix R et al (2021) Marine Macroalgae, a source of natural inhibitors of fungal phytopathogens. J Fungi 7:1006. https://doi.org/10.3390/jof7121006

Wang X, Cai J, Sun L et al (2019) Facile fabrication of magnetic microrobots based on spirulina templates for targeted delivery and synergistic chemo-photothermal therapy. ACS Appl Mater Interfaces 11:4745–4756. https://doi.org/10.1021/acsami.8b15586

Xu B, Li S, Shi R, Liu H (2023) Multifunctional mesoporous silica nanoparticles for biomedical applications. Signal Transduct Target Ther 8:435. https://doi.org/10.1038/s41392-023-01654-7

Ye S, Xie C, Agar OT et al (2024) Alginates from brown seaweeds as a promising natural source: a review of its properties and health benefits. Food Rev Intl 40:2682–2710. https://doi.org/10.1080/87559129.2023.2279583

Integrated Metagenomic Insight into Microbial Dye Degradation

10

Bishwarup Sarkar and Sougata Ghosh

Abstract

Bioremediation is a promising technology that can be established for the removal of toxic dyes from the contaminated environment. However, the cultivation of microbes from the toxic environment, followed by their evaluation of dye-degradation ability, is a cumbersome process. Culture-dependent microbial dye degradation often is unable to elucidate information about the microbial dynamics involved in the degradation of pollutants in the environment. Integrated metagenomics, on the other hand, can provide useful insights into the role of microorganisms that are responsible for dye degradation. Hence, this chapter provides useful insights into the integrated metagenomics approach for unravelling the complexity and diversity of the microbial communities that are capable of dye degradation. Degradation of toxic dyes such as reactive black 5, direct black G, malachite green, and others by various microflora is investigated by metagenomics for understanding the mechanisms of remediation. Further, the functional and structural characteristics of these microbial consortia in response to dye degradation are also assessed. Hence, further consideration of the suitable and adapted microbial biomass waste from the evaluation of metagenomics-based microflora can provide an efficient way to fabricate unique microbial consortia that could help in enhanced dye degradation.

Keywords

Microorganisms · Metagenomics · Dye Degradation · Detoxification · Metabolic pathways · Enzymes

B. Sarkar
College of Science, Northeastern University, Boston, MA, USA

S. Ghosh (✉)
Department of Microbiology, School of Science, RK University, Rajkot, Gujarat, India

R. E. Krishnankutty et al. (eds.), *Smart Agriculture*,
https://doi.org/10.1007/978-981-95-8785-8_10

1 Introduction

Environmental pollution due to the release of dye-contaminated effluents from various industries associated with paper, textiles, paint, cosmetics, leather, and food has risked all forms of life either directly or indirectly (Bencheqroun et al. 2019). Precursors such as naphthalene and benzene are used to develop more than a thousand of industrially important dyes that are used on a large scale throughout the world and enter the environment every day during the disposal of industrial or municipal effluent into the water bodies (Bloch et al. 2022). Although several conventional methods such as adsorption, flocculation, membrane separations, precipitation, and wet oxidation are used for dye removal, they are often inefficient, labor-intensive, and time-consuming (Wang et al. 2020). Thus, the residual dyes with high solubilizing properties in water can inhibit the penetration and diffusion of the sunlight in the water, affecting the metabolic activities of the aquatic flora and fauna. A decrease in photosynthesis and dissolved oxygen levels is extremely hazardous for the aquatic ecosystem (Ajaz et al. 2020).

In view of the background, several biological methods for the removal of synthetic dyes from the water are being developed that are not only economical but also rapid, efficient, simple, and environmentally benign (Crini and Lichtfouse 2018). Bacteria, cyanobacteria, fungi, yeast, algae, and even plants are used for the adsorption, degradation, and mineralization of various dyes, which are mainly attributed to their surface properties and enzymatic activities (Tochhawng et al. 2019). However, there is a threshold limit for each microorganism for dye removal. Hence, alternative approaches are continuously being investigated and developed to enhance the dye removal capacity of the microbes. Compared to pure microbial cultures, mixed cultures or microbial consortia are more efficient in dye removal due to their metabolic synergy (Kalyani et al. 2009). Recombinant DNA technology is the most modern approach where a multidisciplinary strategy is employed for successful genetic manipulation of microbes for the enhancement of dye removal by them (Kumar et al. 2020). Hence, this chapter gives an elaborate account of various integrated metagenomic approaches for microbe-mediated dye degradation that are listed in Table 10.1. Further, rational optimization of various parameters such as microbial load, initial dye concentration, pH, temperature, nutrients, and heavy metals can lead to designing of efficient process for genetically modified microbe-associated dye degradation.

2 Integrated Metagenomic Approach

There are various reports on the usefulness of integrated metagenomics for evaluating the dye-degradation potential of various microbial consortia. For example, Dafale et al. (2010) reported the metagenomic approach for studying azo dye degradation from a two-stage anoxic–oxic bioreactor system. The bacterial biomass was obtained from a textile effluent treatment plant that was gradually exposed to remazole black-5 dye. Under anoxic reactor conditions, more than 90% of 100 mg/L

Table 10.1 Metagenomic approach for microbial dye degradation

Microbial sample	Dye degraded	Metagenomic analysis	References
Bacterial biomass from textile effluent sample	Remazole black 5	Enrichment of *Thauera* sp., *Pseudoxanthomonas* sp., *Desulfomicrobium* sp., *Acidovorax* sp., and *Ottowia* sp. in oxic reactor	Dafale et al. (2010)
CES-1 microbial composite culture	–	Bioaugmentation of culture increased the abundance of *Acinetobacter* sp., *Glucanacetobacter* sp., *Arcobacter* sp., and *Marinobacter manganoxydans*	Santhanarajan et al. (2021)
DDMZ1 bacterial flora	Reactive black 5	Upregulation of genes that expressed NADH: Ubiquinone reductase, NADPH: Quinone reductase, and NAD($^+$)-dependent glutamate dehydrogenase	Zhang et al. (2020)
HT1 halo-thermophilic bacterial consortium	Metanil yellow G	Identification of genes expressing FMN-dependent NADH-azoreductase, Lac-domain containing protein, and catechol dioxygenases	Guo et al. (2021)
Mangrove sediment microflora	Malachite green	Expression of three recombinant proteins namely Mgv-rPOD, Mgv-rLACC, and Mgv-rCYP	Qu et al. (2018)
Microbial consortium from textile effluent	Congo red	16 s rDNA metagenomic analysis highlighted the abundance of *Acinetobacter*, *Comamonas*, *Erwinia*, *Trichococcus*, *Dysgonomonas*, and *Citrobacter* sp.	Neihsial et al. (2022)
Microbiome collected from a seawater sample	Reactive deep blue M-2GE, reactive brilliant orange K-7R, reactive red KM-8B, and KD-8B	Identification of Lac15 enzyme	Fang et al. (2011)
Microbiome from a polluted river sample	Deep blue 35	Identification of genes expressing catalase-peroxidase, laccase, and polyphenol oxidase	Sánchez-Reyes et al. (2022)
Mixed bacterial culture collected from contaminated soil samples and sludge sites	Direct blue 71	Identified absolute abundance of *Acinetobacter*, *Comamonas*, *Aeromonadaceae*, *Pseudomonas*, *Flavobacterium*, *Porphyromonadaceae*, and *Enterobacteriaceae*	Zin et al. (2020)
Thermophilic microflora	Direct black G	Upregulation of 16 differently expressed genes that were translated into FAD/NAD(P)-binding protein, catechol-2-3-dioxygenase, 4-oxalocrotonate tautomerase, nicotinamide adenine dinucleotide (NADH) -ubiquinone reductase, and NADH-quinone oxidoreductase	An et al. (2020)

(continued)

Table 10.1 (continued)

Microbial sample	Dye degraded	Metagenomic analysis	References
Wastewater bacterial consortium	Reactive red 2	Electron transfer to azoreductase by flavin mononucleotide (FMN) and flavin protein for dye degradation	Gu et al. (2021)
ZW1 bacterial consortium	Metanil yellow G	Enhanced expression of laccase, azoreductase, FMN-dependent reductase, and NADPH-/NADH-quinone oxidoreductases	Guo et al. (2020)

of remazole black-5 dye decolorization was observed in the first 48 h of incubation. The rate of the feeding solution of 2 mL/min was initially fed into the anoxic bioreactor, wherein the effluent was then collected from the oxic bioreactor. The metabolite concentration was decreased by 73% in the oxic bioreactor. The toxic effect of the wastewater on microbial diversity was also analyzed, which showed a 62% decrease in the *Escherichia coli* colony-forming units (CFUs) in the presence of the anoxic effluent. Thereafter, the diversity of the two-stage bioreactor was calculated on the basis of sequence analysis of the 16S ribosomal DNA (rDNA) metagenomic library screening. A phylogenetic tree was constructed using the 16S rDNA sequence thus obtained from the oxic bioreactor. *Thauera* sp., *Pseudoxanthomonas* sp., *Desulfomicrobium* sp., *Acidovorax* sp., and *Ottowia* sp. were the primary genera that were detected using a culture-independent tool. Likewise, *Clostridium* sp., *Klebsiella* sp., and *Bacteroides bacterium* were the primary bacterial species present in the anoxic reactor.

Santhanarajan et al. (2021) also studied the metagenomic insights for textile dye wastewater treatment after the bioaugmentation of a CES-1 microbial composite culture which is shown in Fig. 10.1. The bioaugmentation resulted in the improvement of removal efficiencies for COD, total nitrogen, total phosphorus, and suspended solids along with the reduction in the intensity of the color after 300–531 days of incubation. The considerable decrease in sludge formation corresponding to the total nitrogen removal as well as chemical oxygen demand (COD) removal was noted. In addition, the diversity of the composite culture was also improved after bioaugmentation with *Acinetobacter* sp., *Glucanacetobacter* sp., *Arcobacter* sp., and *Marinobacter manganoxydans* as the primary denitrifiers.

A cometabolic degradation of reactive black 5 dye (RB5) was also investigated by Zhang et al. (2020) in a metagenomic and metaproteomic study. A natural bacterial flora named DDMZ1 was used in this study for the RB5 dye degradation that was isolated and enriched from an activated sludge sample obtained from a sewage treatment plant. A maximum decolorization efficiency of 98% was achieved for 200 mg/L of RB5 dye after 48 h of incubation in the presence of 3 g/L of fructose in the yeast extract medium. Hence, fructose was considered a cometabolic substrate along with the yeast extract medium that enhanced the DDMZ1 flora performance effectively. Moreover, the biomass also increased considerably after fructose supplementation in the yeast extract medium. Later, azoreductase assays were performed as these enzymes are known for their application in decolorization of azo

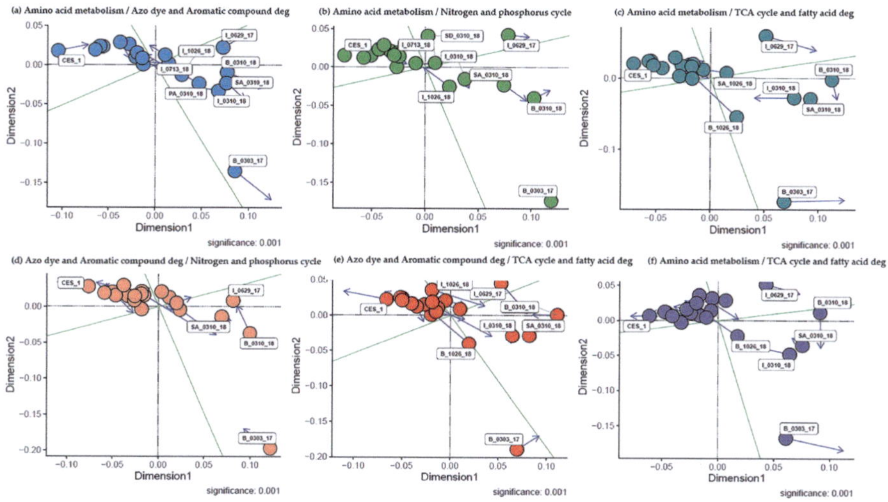

Fig. 10.1 Procrustes analysis of the major potential degradation pathways in textile wastewater treatment system before and after bioaugmentation of CES-1 over the monitoring period (531 days). Blue arrow indicates the significant positive correlation between the two pathways ($p < 0.05$). (Reprinted from Santhanarajan et al. 2021)

dyes such as RB5 wherein the enzyme is responsible for the azo bond cleavage. Both the intracellular and extracellular activities of the azoreductase enzyme were increased after the addition of RB5 dye. The highest intracellular and extracellular azoreductase activity of 18.2 and 29.9 U/mg of protein was attained in the presence of yeast extract medium and the addition of fructose. Thereafter, metagenomic shotgun sequencing of the bacterial flora was carried out to analyze the metabolic functional profiles. The genes identified from the sample cultivated in yeast extract medium and fructose were classified within 38 functional groups according to gene ontology (GO) functional annotation. On the basis of clusters of orthologous groups of proteins (COGs) analysis, the gene sequences obtained in all different samples were classified into about 23 functional groups. The carbohydrate-active enzymes (CAZy) database highlighted six main enzyme classes that were active in carbohydrate degradation namely, glycoside hydrolases (GH), glycosyl transferases (GT), carbohydrate esterases (CE), auxiliary activities (AA), carbohydrate-binding modules (CBM), and polysaccharide lyases (PL). The gene sequences classified into GH and GT were most abundant when cultivated in the presence of yeast extract medium with fructose. Pathway annotations to the microbial genomes also revealed the possible involvement of around 261 KEGG pathways wherein the pathways primarily involved in the degradation of benzoate, aminobenzoate, naphthalene, xylene, polycyclic aromatic hydrocarbon, nitrotoluene, ethylbenzene, and aromatic compounds were considerably enriched. Hence, the molecular mechanism of RB5 dye degradation was hypothesized based on the metagenomic insights wherein the yeast extract was proposed to provide the necessary energy to upregulate the expression of candidate genes in *Pseudomonas* sp. expressing functional proteins such as NADH:

ubiquinone reductase that is involved in H^+-translocation along with upregulated expression of other enzymes such as NADH: ubiquinone reductase, NADPH: quinone reductase and NAD(+)-dependent glutamate dehydrogenase as shown in Fig. 10.2. Likewise, the yeast extract medium and fructose were suggested to provide the necessary energy and carbon source to *Burkholderia* sp. to increase the production of similar functional proteins that are involved in the TCA cycle and glycolysis.

In another similar study, Guo et al. (2021) characterized a halo-thermophilic bacterial consortium (HT1) for azo dye decolorization. The enrichment of the HT1 consortium isolated from a textile wastewater sludge sample was carried out in a liquid basal medium supplemented with 100 mg/L of MYG dye that was incubated at 50°C under static conditions. A dye degradation efficiency of 92.1% was observed after 36 h of enrichment at 50°C and 10% salinity. Next, the metagenomic analysis of the HT1 consortium revealed *Bacillus* as the dominant genus that represented 92.4% of the consortium. Several factors showed their effect on MYG decolorization. For instance, the optimum temperature for maximum dye degradation was 50°C. Similarly, the optimum pH range of 6.0–8.0 demonstrated more than 90% dye degradation after a sufficient time of incubation. The initial dye concentration also played an important role in the determination of dye-degradation efficiency. Later, metagenomic sequencing was performed to obtain metabolic functional profiles of genes. COG analysis showed 22 various functional categories whereas CAZymes function was classified into six primary categories as evident from Fig. 10.3. Moreover, the KEGG database revealed several categories of annotated pathways some of which included pyruvate metabolism, dioxin degradation, xylene degradation, toluene degradation, polycyclic aromatic hydrocarbon degradation, naphthalene degradation, amino-benzoate degradation, benzoate degradation, and fluorobenzoate degradation. Additionally, FMN-dependent NADH-azoreductase, Lac-domain containing protein, catechol dioxygenases along with other hydrolases were identified in the metagenome as well which are known to reduce aromatic amines of the azo dyes.

In another study, Qu et al. (2018) reported the metagenomic-based elucidation of gene families responsible for the degradation of malachite green (MG) dye the pathway for which is demonstrated in Fig. 10.4. Sediment samples from a mangrove forest were used in this study from which the environmental DNA (eDNA) was extracted and analyzed. Metagenome sequencing of the small subunit of ribosomal RNA (SSU rRNA) then revealed 50.195% and 17.03% abundance of *Proteobacteria* and *Actinobacteria*, respectively. Around 44 putative genes responsible for MG dye degradation were identified in the nonredundant database. Around 93.54% of these genes were categorized in the peroxidase gene family, followed by 3.40% and 3.06% in the laccase (LACC) and cytochrome p450 gene families, respectively. A phylogenetic tree was constructed based on the amino acid sequences wherein the *mgv-laccase*, *mgv-p450*, and *mgv-peroxidase* were gathered within the LACC of *Pelobacter seleniigenes*, CYP of *Salinisphaera shabanensis*, and POD from *Salinibacter ruber*, respectively. Various MG-degrading potential was observed for

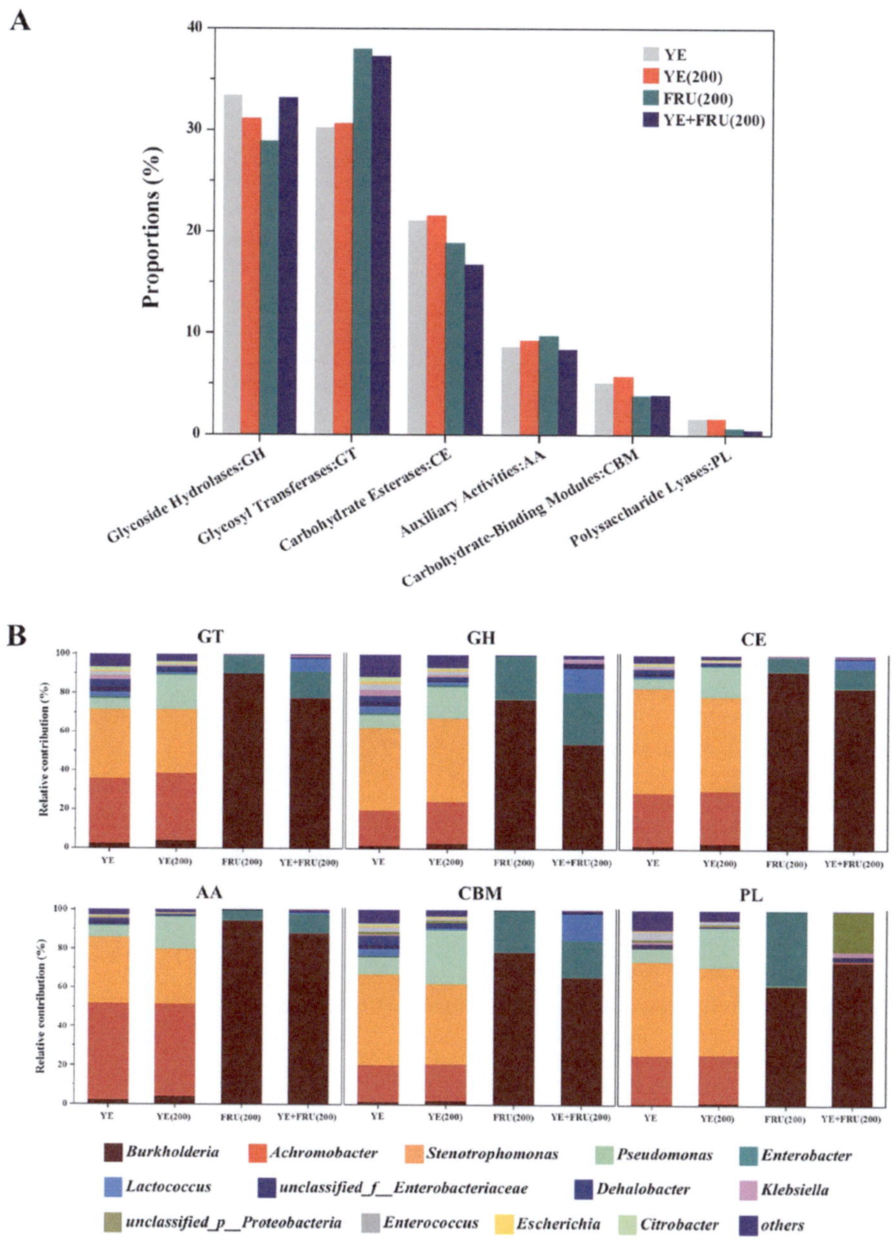

Fig. 10.2 CAZy functional annotation of gene sequences from four assays (**a**). Barplot of species and CAZy functional contribution analysis (**b**). YE- YE medium + DDMZ1; YE(200)- YE medium + DDMZ1 + RB5 (200 mg/L); FRU(200)- FRU medium + DDMZ1 + RB5 (200 mg/L); YE + FRU(200)- YE + FRU medium + DDMZ1 + RB5 (200 mg/L). GT- Glycosyl Transferases; GH- Glycoside Hydrolases; CE- Carbohydrate Esterases; AA- Auxiliary Activities; CBM- Carbohydrate-Binding Modules; PL- Polysaccharide Lyases. (Reprinted with permission from Zhang et al. 2020. Copyright © 2019 Elsevier Ltd.)

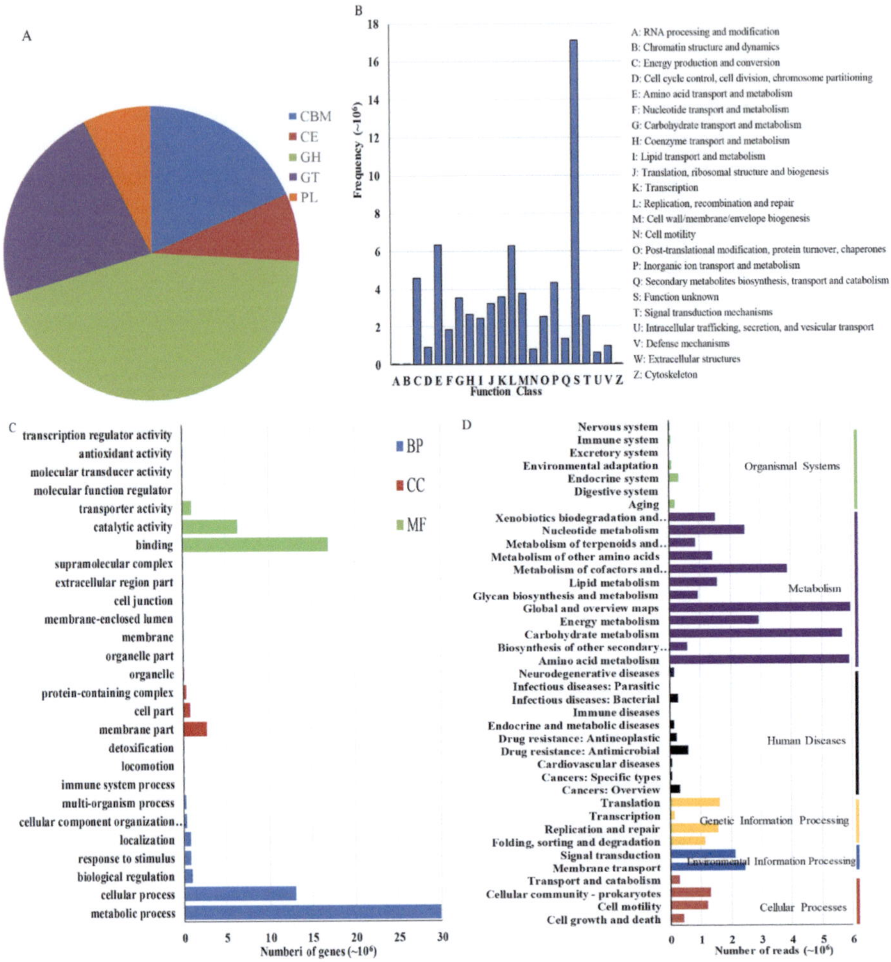

Fig. 10.3 (**a**) Distribution scale diagram of differential genes annotated to CAZymes; (**b**) COG function classification; (**c**) Gene function classification in the consortium HT1, BP biological process, CC cellular components, MF Molecular function; (**d**) KEGG classification map. (Reprinted with permission from Guo et al. 2021. Copyright © 2021 Published by Elsevier Ltd.)

three recombinant proteins wherein Mgv-rPOD exhibited the highest decolorization activity of 97.3% within 40 min in the presence of 300 mg/L of MG dye.

Recently, Neihsial et al. (2022) also demonstrated the diversity and dye degradation activity of a microbial consortium obtained from textile effluent. The soil sediment sample contaminated with dyes was cultivated in Bushnell Hass broth medium that was supplemented with yeast extract, glycerol, and 500 mg/L of Congo red dye. Later, the dye decolorization efficiency was improved wherein up to 2500 mg/L of Congo red was effectively decolorized. A maximum decolorization efficiency of 99% was achieved in the presence of 2000 mg/L of Congo red dye after 120 h of

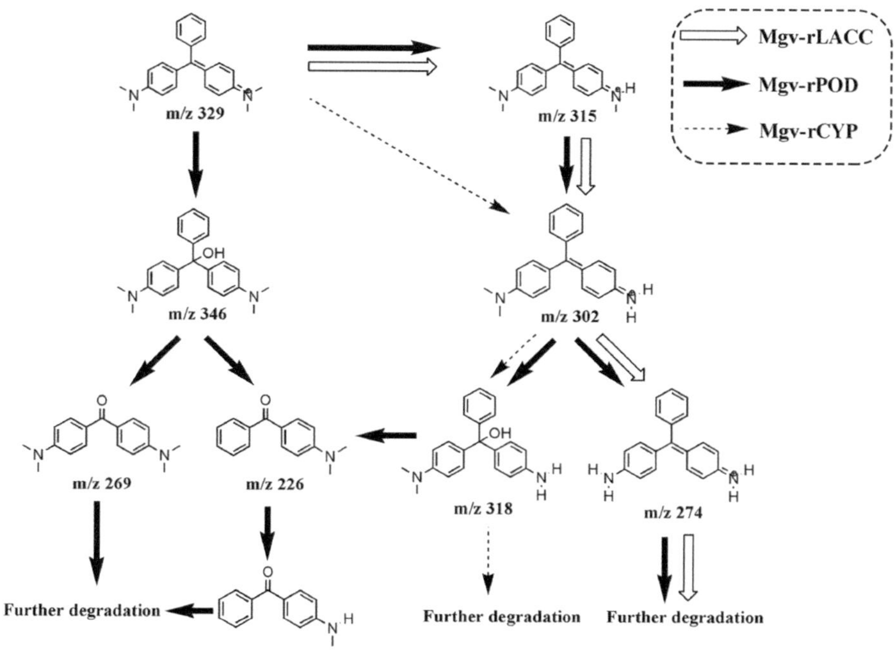

Fig. 10.4 The putative MG degradation pathway deduced from the products of Mgv-rPOD, Mgv-rLACC, and Mgv-rCYP. The intermediates connected by the black, white and dashed arrows indicated the MG-degrading pathway of Mgv-rPOD, Mgv-rLACC, and Mgv-rCYP, respectively. (Reprinted from Qu et al. 2018)

incubation. The effect of heavy metals along with changes in the pH and temperature on dye decolorization activity was also studied wherein 0.5 mM of nickel and copper reduced the decolorization efficiency to 92% and 70%, respectively after 72 h of incubation. Effective decolorization activity was observed within a wide pH range of 6.0–9.0 after incubation for 24 h. An optimal temperature range of 24–40°C showed almost 97% decolorization in 24 h incubation. The molecular profiling of the bacterial community was then performed using next-generation sequencing. The 16S rDNA metagenomic operational taxonomic units (OTUs) demonstrated the dominance of various phyla such as Proteobacteria, Bacteroidetes, and Firmicutes. Likewise, *Acinetobacter*, *Comamonas*, *Erwinia*, *Trichococcus*, *Dysgonomonas*, and *Citrobacter* were the abundant bacterial genera that were present in the consortium. The functional annotations of identified genes were then also utilized for predicting a degradation pathway that is schematically represented in Fig. 10.5. The genes responsible for reductive deamination of the two azo bonds present in the Congo red dye followed by conversion of the product into biphenyl were present in the metagenome dataset. Further, it was proposed that the biphenyl was mineralized to form catechol.

A laccase enzyme characterized by Fang et al. (2011) was identified from a marine microbial metagenome that demonstrated dye decolorization properties. The

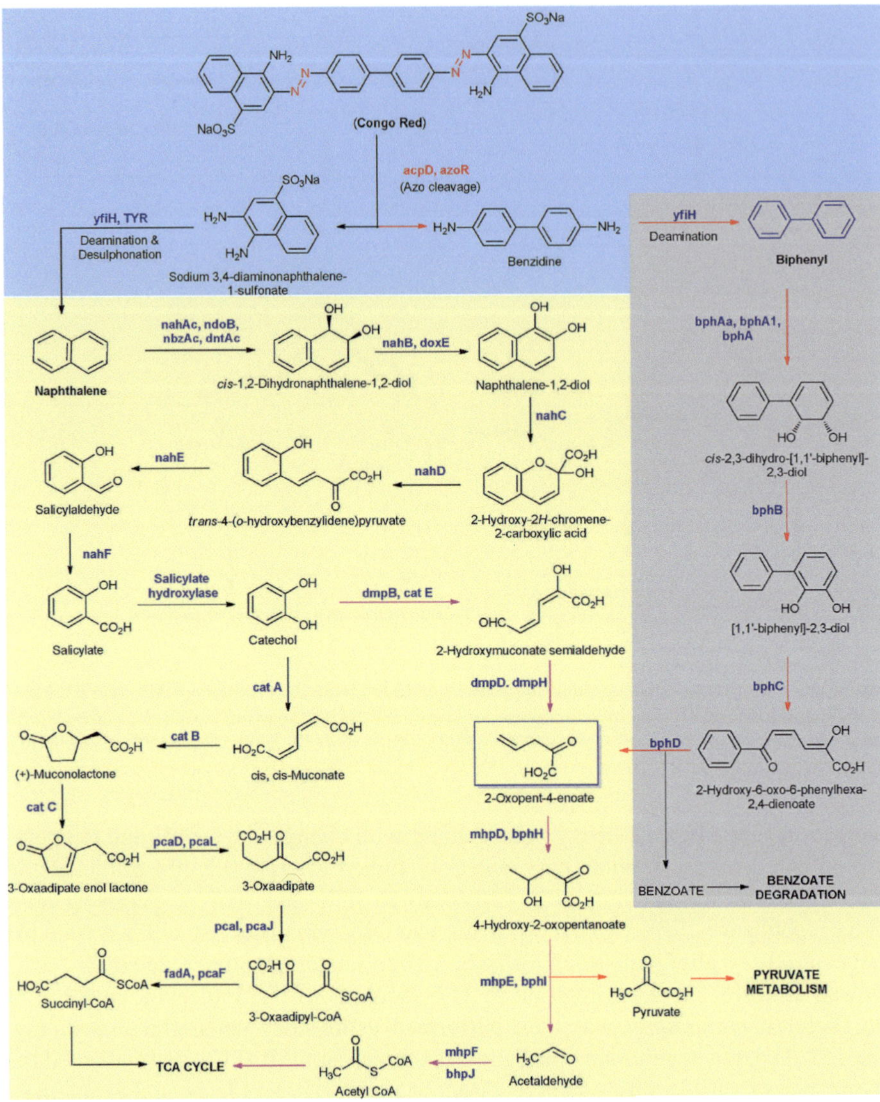

Fig. 10.5 Schematic representation of proposed CR degradation pathways generated by mapping predicted genes in KEGG pathway. (Reprinted with permission from Neihsial et al. 2022. Copyright © 2022 Elsevier Ltd.)

microbes were collected from seawater samples in China. A marine microbial metagenome library was then constructed that was comprised of 1.4 Gb of DNA. An inverse polymerase chain reaction (PCR) was carried out to investigate novel laccase enzymes which then highlighted the presence of the *lac15p* gene that encoded a 439 amino acid protein having three copper binding domains that were conserved. The Lac15 protein was recombinantly expressed in *E. coli* that showed considerable activity toward syringaldazine at an optimal pH and temperature of 7.5 and 45 °C. The identified laccase in this study also displayed significant decolorization of four azo dyes namely, reactive deep blue M-2GE, reactive brilliant orange K-7R, reactive red KM-8B, and KD-8B. The addition of mediators such as 2,2′-azino-bis(3-ethylbenzothiazoline-6-sulfonic acid) (ABTS) resulted in a further increase in the decolorization efficiency.

In a similar recent study, Sánchez-Reyes et al. (2022) reported Hi-C metagenomic deconvolution of a microbiome which provided novel genome composites having textile dye degradation ability. The microbiome sample was collected from a polluted river and was cultured in a growth medium that was supplemented with anthraquinone Deep Blue 35 dye as the primary carbon source. Thereafter, a Hi-C proximity-ligation deconvolution method was performed which showed around 100 genome composites out of which 80% species were potentially novel. *Methanobacterium*, *Clostridium*, and *Cupriavidus* genera comprised 50, 22, and 11% of the overall microbiome community. A decolorization efficiency of 23% along with 30% COD removal was achieved after 30 days of dye enrichment. Annotations of the metagenomes using the KEGG orthology system revealed various functional profiles that corresponded with carbohydrate, energy, and lipid metabolism, glycan biosynthesis, xenobiotic biodegradation, and membrane transport. Further analysis of the functional annotations identified the genes responsible for the expression of catalase-peroxidase, laccase, and polyphenol oxidase in the genome composites.

Zin et al. (2020) also demonstrated the decolorization of direct blue 71 (DB71) dye using a mixed bacterial culture that was collected from contaminated soil samples and sludge sites. The bacterial culture was cultivated in a minimal salt medium containing glucose as the carbon source along with ammonium sulphate and 50 mg/L of direct blue 71 dye. Thereafter, the metagenomic analysis of the mixed culture was performed wherein 30%, 11%, 10%, 10%, 8%, 6%, and 4% abundance of *Acinetobacter*, *Comamonas*, *Aeromonadaceae*, *Pseudomonas*, *Flavobacterium*, *Porphyromonadaceae*, and *Enterobacteriaceae* was observed in the microbial community. The dye decolorization optimization was carried out using response surface methodology (RSM) and artificial neural network (ANN) in which the ANN model exhibited a higher R^2 value and a minimum average absolute deviation of 0.99 and 0.04, respectively.

Likewise, An et al. (2020) demonstrated the application of integrated metagenomic and metaproteomic analyses for elucidating the functional potential of genes and their modes of action on direct black G (DBG) dye. A thermophilic microflora sample was used in this study that was obtained from soil samples contaminated near a dye factory drain outlet. The decolorization rate of DBG dye was increased progressively with successive increases in the incubation time. A maximum DBG dye degradation rate of 85% was achieved after 40 h of incubation. Moreover, the supplementation of beef extract, peptone, and glucose in the growth medium along with DBG dye increased the dye-degradation rate to 97% within 28 h of incubation. The metagenomic analysis of the microflora was carried out. Illumina MiSeq sequencing platform and quantitative reverse transcriptase polymerase chain reaction (qRT-PCR) were used for metagenomic analyses wherein a total of 1266 genes were differentially expressed. The downregulation of gene expression was proposed to be due to the toxic effect of DBG dye which facilitated reduced cellular metabolism. The dilution effect was also speculated to cause downregulation of the gene expression as the dye degradation efficiency reduced from 93% to 7% after diluting the microbial community from 10^{-3} to 10^{-11}, respectively. In addition, functional annotations to the microbial genome also revealed the presence of seven glycosyltransferases, 3 types of carbohydrate-binding molecules (CBM), 3 kinds of carbohydrate esterases, and 3 types of glycoside hydrolases. A total of 682 differently expressed genes (DEGs) were annotated to molecular function, biological processes, and cell-component categories, respectively. On the basis of the Kyoto Encyclopedia of Genes and Genomes (KEGG) database, around 562 DEGs were involved in metabolism. The involvement of genes responsible for chlorocyclohexane, chlorobenzene, toluene, benzoic acid, aromatic acid, xylene, dioxin, and sulphur degradation was considerably enhanced during pathway annotations. Thereafter, quantitative-reverse transcriptase polymerase chain reaction (qRT-PCR) validated upregulation in 16 key DEGs in the presence of DBG dye. The gene encoding a FAD/NAD(P)-binding protein was maximally upregulated among all the genes. Other genes such as catechol-2-3-dioxygenase and 4-oxalocrotonate tautomerase were also considerably enhanced. Based on the analysis of the function of proteins, a detailed molecular mechanism for DBG degradation was proposed. The pathway suggested the involvement of enzymes such as nicotinamide adenine dinucleotide (NADH) -ubiquinone reductase and NADH-quinone oxidoreductase subunit related to the citric acid (TCA) cycle and glycolysis.

Similarly, Gu et al. (2021) also recently reported metagenomic insights on the reactive red 2 (RR2) dye degradation in the presence of varying concentrations of ferroferric oxide. The dye degradation was studied using the hydrolysis/acidification process in 2-liter sequencing batch reactors with the addition of ferroferric oxide that was operated at 30°C for 6 h. Three different organic carbon sources namely, peptone, starch, and sodium acetate were added with the chemical oxygen demand (COD) ratio of 6:3:1 in the synthetic wastewater samples. A maximum dye decolorization efficiency of 76.44% was achieved after supplementation of 15 g/L of Fe_3O_4 in the reactor. The abundance of microbes expressing azoreductase enzyme was also drastically increased with the successive increase in the Fe_3O_4 concentration up to 15 g/L. In addition, the dominant azo dye-degrading bacteria was altered from *Bacillus*, *Aeromonas*, and *Lysobacter* to *Acinetobacter* and *Desulfovibrio* after increasing the Fe_3O_4 concentration. Likewise, the abundance of *Geobacter* also increased with a concomitant increase in the Fe_3O_4 content. Hence, based on the observed results, Fe_3O_4 was proposed to facilitate extracellular electron transfer which promoted the degradation of RR2 dye. Additionally, *Geobacter* and *Staphylococcus* abundance was enriched after Fe_3O_4 addition which expressed enzymes responsible for the generation and dehydrogenation of NADH and succinate along with the degradation of azo dyes. Thus, these bacterial genera were suggested to generate electrons through carbon utilization which were then transferred to azoreductase enzyme by flavin mononucleotide (FMN) and flavin protein after which the reduced enzyme was proposed to facilitate dye degradation.

The metagenomic analysis of a halo-alkaliphilic bacteria consortium for the metanil yellow G (MY-G) dye degradation was also reported by Guo et al. (2020). The bacterial consortium ZW1 was isolated from a sludge sample collected from a wastewater treatment plant wherein the consortium was enriched in a thermostatic incubator with a growth medium containing basic nutrients. The microbial community structure of the enriched consortium was then evaluated using Illumina sequencing which showed 49.83%, 30.73%, and 19.22% abundance of *Halomonas*, *Marinobacter*, and *Clostridiisalibacter*, respectively. Thereafter, decolorization efficiency of the consortium ZW1 was studied which showed only 0.98% MY-G decolorization after 16 h of incubation. However, supplementation of yeast extract as both primary carbon and organic nitrogen source resulted in 93.27% decolorization efficiency of MY-G dye within 16 h of incubation. A pH value of 8.0, 10% salinity, incubation temperature of 40°C, and initial dye concentration of 200 mg/L showed maximum dye decolorization by the bacterial consortia. The ZW1 consortium also showed effective decolorization of other dyes such as direct black-15, chrysophenine G, acid black ATT, and acid red GR. Later, the degradation mechanism of the MY-G by the ZW1 consortium was investigated by metagenomic analysis. Around 102 genes from the metagenomic analysis corresponded with the CAZy database, which were majorly carbohydrate-binding modules that are known for the degradation of complex recalcitrant substrates. Likewise, the sequences were divided into 23 different functional groups in accordance with the COG database that showed the majority of the genes to be involved in amino acid transport and metabolism, energy production and conversion, inorganic ion transport, and signal transduction mechanisms. In addition, pathway annotations of genes responsible for MY-G degradation corresponded with the genes that are known for chlorocyclohexane, chlorobenzene, benzoate, fluorobenzoate, dioxin, xylene, and toluene degradation. Hence, various enzymes responsible for refractory pollutant degradation were highlighted through metagenomic analysis that primarily included laccase, azoreductase, FMN-dependent reductase, and NADPH-/NADH-quinone oxidoreductases.

3 Conclusion

Microbe-mediated dye degradation is an eco-friendly method for removal of hazardous dyes from the environment. Various microbes are isolated and identified that can play promising role in the process of dye adsorption, degradation and mineralization. Metagenomic approaches are employed to identify the mechanism of dye degradation where several metabolic enzymes such as 4-oxalocrotonate tautomerase, azoreductase, catalase-peroxidase, catechol dioxygenases, FMN-dependent reductase, laccase, $NAD(^{+})$-dependent glutamate dehydrogenase, NADPH-/NADH-quinone oxidoreductases, and polyphenol oxidase are identified to play a significant role. Thus, there is a wide potential to use the identified gene coding the aforementioned enzymes to develop genetically modified microbes with higher dye-degrading potential employing recombinant DNA technology. However, for getting maximum dye degradation, various process parameters such as microbial load, time, temperature, pH, initial dye concentration, heavy metals, and nutrients should be carefully optimized. In view of the background, it can be concluded that genetically modified microorganisms can help to develop highly efficient water-treatment process for dye removal in order to ensure clean environment.

References

Ajaz M, Shakeel S, Rehman A (2020) Microbial use for azo dye degradation-a strategy for dye bioremediation. Int Microbiol 23(2):149–159

An X, Chen Y, Chen G, Feng L, Zhang Q (2020) Integrated metagenomic and metaproteomic analyses reveal potential degradation mechanism of azo dye-Direct Black G by thermophilic microflora. Ecotoxicol Environ Saf 196:110557

Bencheqroun Z, Mrabet IE, Kachabi M, Nawdali M, Neves I, Zaitan H (2019) Removal of basic dyes from aqueous solutions by adsorption onto Moroccan clay (fez city). Mediterr J Chem 8(2):158–167

Bloch K, Mohammed SM, Karmakar S, Shukla S, Asok A, Banerjee K, Patil-Sawant R, Mohd Kaus NH, Thongmee S, Ghosh S (2022) Catalytic dye degradation by novel phytofabricated silver/zinc oxide composites. Front Chem 10:1013077

Crini G, Lichtfouse E (2018) Advantages and disadvantages of techniques used for wastewater treatment. Environ Chem Lett 17:145–155

Dafale N, Agrawal L, Kapley A, Meshram S, Purohit H, Wate S (2010) Selection of indicator bacteria based on screening of 16S rDNA metagenomic library from a two-stage anoxic–oxic bioreactor system degrading azo dyes. Bioresour Technol 101(2):476–484

Fang Z, Li T, Wang Q, Zhang X, Peng H, Fang W, Hong Y, Ge H, Xiao Y (2011) A bacterial laccase from marine microbial metagenome exhibiting chloride tolerance and dye decolorization ability. Appl Microbiol Biotechnol 89(4):1103–1110

Gu M, Yin Q, Wu G (2021) Metagenomic analysis of facilitation mechanism for azo dye reactive red 2 degradation with the dosage of ferroferric oxide. J Water Process Eng 41:102010

Guo G, Hao J, Tian F, Liu C, Ding K, Xu J, Zhou W, Guan Z (2020) Decolorization and detoxification of azo dye by halo-alkaliphilic bacterial consortium: systematic investigations of performance, pathway and metagenome. Ecotoxicol Environ Saf 204:111073

Guo G, Liu C, Hao J, Tian F, Ding K, Zhang C, Yang F, Liu T, Xu J, Guan Z (2021) Development and characterization of a halo-thermophilic bacterial consortium for decolorization of azo dye. Chemosphere 272:129916

Kalyani DC, Telke AA, Govindwar SP, Jadhav JP (2009) Biodegradation and detoxification of reactive textile dye by isolated Pseudomonas sp. SUK1. Water Environ Res 81(3):298–307

Kumar V, Chandra R, Thakur IS, Saxena G, Shah MP (2020) Recent advances in physicochemical and biological treatment approaches for distillery wastewater. In: Shah M, Banerjee A (eds) Combined application of physico-chemical and microbiological processes for industrial effluent treatment plant. Springer, Singapore, pp 79–118

Neihsial R, Singha NA, Singh AK (2022) Taxonomic diversity and predictive metabolic functions of a heavy metal tolerant multiple azo dye degrading bacterial consortium from textile effluents. Int Biodeterior Biodegr 171:105421

Qu W, Liu T, Wang D, Hong G, Zhao J (2018) Metagenomics-based discovery of malachite green-degradation gene families and enzymes from mangrove sediment. Front Microbiol 9:2187

Sánchez-Reyes A, Bretón-Deval L, Mangelson H, Salinas-Peralta I, Sanchez-Flores A (2022) Hi-C deconvolution of a textile dye–related microbiome reveals novel taxonomic landscapes and links phenotypic potential to individual genomes. Int Microbiol 25(1):99–110

Santhanarajan AE, Sul WJ, Yoo KJ, Seong HJ, Kim HG, Koh SC (2021) Metagenomic insight of a full scale eco-friendly treatment system of textile dye wastewater using bioaugmentation of the composite culture CES-1. Microorganisms 9(7):1503

Tochhawng L, Mishra VK, Passari AK, Singh BP (2019) Endophytic fungi: role in dye decolorization. In: Singh B (ed) Advances in endophytic fungal research. Fungal biology. Springer, Cham, pp 1–15

Wang Y, Jiang L, Shang H, Li Q, Zhou W (2020) Treatment of azo dye wastewater by the self-flocculating marine bacterium *Aliiglaciecola lipolytica*. Environ Technol Innovation 19:100810. https://doi.org/10.1016/j.eti.2020.100810

Zhang Q, Xie X, Liu Y, Zheng X, Wang Y, Cong J, Yu C, Liu N, Sand W, Liu J (2020) Co-metabolic degradation of refractory dye: a metagenomic and metaproteomic study. Environ Pollut 256:113456

Zin KM, Effendi Halmi MI, Abd Gani SS, Zaidan UH, Samsuri AW, Abd Shukor MY (2020) Microbial decolorization of triazo dye, direct blue 71: an optimization approach using response surface methodology (RSM) and artificial neural network (ANN). Biomed Res Int 2020:2734135

Current Trends in Agriculture Research Based on AI and Machine Learning

11

Saumya Dwivedi, R. Rajesh, and Manickavelu Alagu

Abstract

From manual subsistence systems to modern, digitally driven ecosystems, agriculture has undergone a series of transformative shifts over the centuries. Today's convergence of Artificial Intelligence (AI) and Machine Learning (ML) with traditional farming is marked to drive a food systems revolution worldwide. This chapter provides a comprehensive review of the latest trends in agriculture research based on AI and ML. It highlights how crucial agriculture is in ensuring global food security and walks us through the evolution from traditional farming practices to the digital revolution (Agriculture 4.0 and 5.0). The chapter also explores how new and emerging technologies are being implemented across key areas like crop management, pest detection, yield prediction, and livestock monitoring. It also looks at how different parts of the world are embracing AI adoption, especially how countries like India are performing compared to developed economies. Additionally, the chapter examines the challenges in acquiring and managing agricultural data, summarizing current research patterns, mapping out future areas of focus, and discussing potential strategies and policies to help support the transition toward an efficient, sustainable, and resilient food system globally.

Keywords

Agriculture · Artificial Intelligence · Machine Learning · Precision Farming · Smart Agriculture · Agrifood Systems · Sustainability

S. Dwivedi · R. Rajesh
Department of Computer Science, Central University of Kerala, Kasaragod, Kerala, India
e-mail: saumya.2600707001@cukerala.ac.in; rajeshr@cukerala.ac.in

M. Alagu (✉)
Department of Genomic Science, Central University of Kerala, Kasaragod, Kerala, India
e-mail: amanicks@cukerala.ac.in

R. E. Krishnankutty et al. (eds.), *Smart Agriculture*,
https://doi.org/10.1007/978-981-95-8785-8_11

1 Introduction

Agriculture stands as the fundamental pillar of global food security. The widely accepted definition by the Food and Agriculture Organization (FAO 2009) states that food security is said to be achieved "when all people, at all times, have physical, social, and economic access to sufficient, safe, and nutritious food". The role of agriculture is essential in achieving sustainable development goals aimed at eliminating world hunger and malnutrition (FAO et al. 2023). The urgency for agricultural advancement comes from the demographic projections indicating that the global population is expected to hit the 9 billion mark by 2050 (Godfray et al. 2010). This necessitates not only an increase in food production but also a paradigm shift toward highly optimized resource utilization (Calicioglu et al. 2019).

However, achieving these goals is increasingly complex due to a formidable array of interconnected challenges. Modern agriculture must contend with the following:

- *Climate Change*: Altered precipitation patterns, increased frequency and intensity of extreme weather events (like droughts, floods, and heatwaves), and shifting agro-ecological zones (El Bilali et al. 2020).
- *Resource Depletion and Degradation*: Including progressive soil degradation (erosion, nutrient depletion, salinization) (Hossain et al. 2020), dwindling freshwater availability for irrigation (Khondoker et al. 2023), and loss of agricultural biodiversity (Ebel et al. 2021).
- *Biotic Stresses*: Such as the emergence of novel pests and diseases, increased pathogen resistance, and invasive species impacting crop yields (Fried et al. 2017).
- *Socioeconomic Pressures*: Including labor shortages in many agricultural regions (Bousmah and Grenier 2022), market volatility (Mustafa et al. 2024), supply chain disruptions (as evidenced by the COVID-19 pandemic) (Mahajan and Tomar 2021), and the need for equitable access to technology and resources, especially for smallholder farmers (Abraham and Pingali 2020).
- *Sustainability Imperatives*: Growing pressure to reduce agriculture's environmental footprint, concerning greenhouse gas emissions, water pollution from runoff, and impacts on ecosystems (Adedibu 2023).

This complex interplay of pressures facing modern agriculture, from demographic shifts to environmental concerns, is visually summarized in Fig. 11.1. Beyond small fixes, there is a need for a transformative innovation to tackle this complex web of challenges. While foundational, traditional methods lack the capacity to effectively address this scale of predicament. In these trying times, AI and ML emerge as a unique and powerful ally of agricultural revolution (Dawn et al. 2023).

Then the question arises, how exactly are AI and ML relevant here? The answer is that these technologies possess the distinct ability as follows:

- *Process Unprecedented Data Complexity*: Agriculture generates vast, heterogeneous datasets from diverse sources (sensors, satellites, drones, genomics,

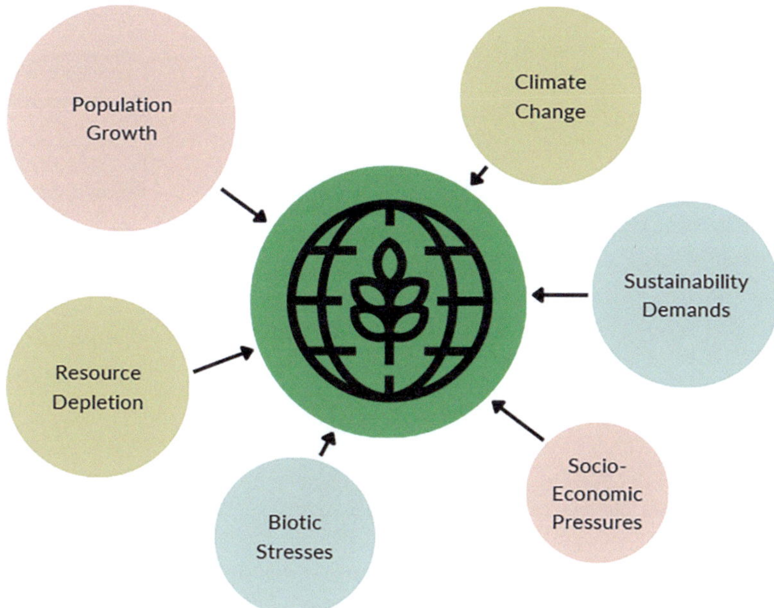

Fig. 11.1 The agricultural challenge nexus

weather stations, market data, etc.). Artificial intelligence/Machine learning (AI/ML) excels at integrating and analyzing this complex information in ways not feasible through conventional statistics or human analysis alone (Elbasi et al. 2024).

- *Identify Nonlinear Patterns and Predict Outcomes*: AI and ML algorithms can uncover complex relationships between datasets to enable highly accurate predictions for yield forecasting, disease outbreak detection, optimal planting times, and resource needs (Mishra and Mishra 2023).
- *Enable Precision and Automation at Scale*: By analyzing real-time data obtained by IoT sensors, AI/ML facilitates precision agriculture. By applying the right inputs (water, fertilizer, pesticides) at the right place, time, and quantity, resource use can be optimized and waste minimized (Ramirez 2025). Furthermore, it drives automation for labor-intensive tasks, addressing workforce shortages (Saleem et al. 2021).
- *Facilitate Adaptive Management*: ML models can learn and adapt to changing environmental conditions and new data, offering dynamic decision support crucial for resilience in the face of climate variability and other unforeseen disruptions (Kameswari 2023).

The fusion of AI and ML with agriculture will pave the way toward a new era of data-driven agricultural intelligence (Sinha et al. 2023). By leveraging insights derived from intricate data analysis, smart agriculture powered by AI/ML offers a pathway to significantly enhance productivity, optimize resource efficiency, bolster

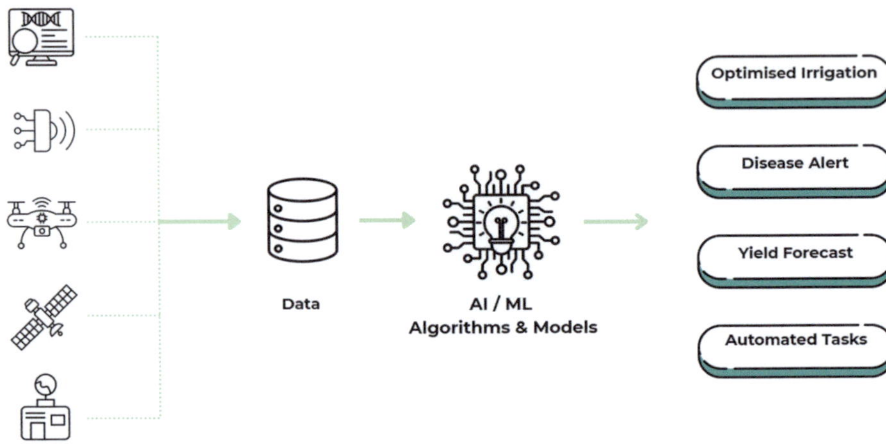

Fig. 11.2 The flow of information in AI-powered agriculture: from diverse data sources through analysis to actionable intelligence and optimized outcomes

supply chain resilience, and inherently, improve the overall sustainability of our global food systems (Dhal and Kar 2024). Figure 11.2 provides a schematic overview of the information flow in AI-driven agriculture, from multisource data collection through algorithmic analysis to practical applications.

2 The Evolution of Agriculture: From Traditional Practices to Agriculture 5.0

Agriculture has undergone significant transformations throughout history. The initial early phase (Agriculture 1.0) was manual work and subsistence farming practices. When the mechanization era came, it marked a significant revolutionary shift (Agriculture 2.0) with innovations like tractors, fertilizers, and pesticides. The subsequent Green Revolution certainly enhanced yield, but also introduced environmental concerns (Pingali 2012).

The late twentieth century witnessed the advent of Precision Agriculture (Agriculture 3.0) which introduced electronic systems, automation, and information technology into farming practices. This era witnessed the emergence of sophisticated technologies like GPS (Global Positioning System), yield-monitoring systems, and variable rate technology. These advancements enabled farmers to optimize inputs on a site-specific basis (Zhang et al. 2002; Lowenberg-DeBoer and Erickson 2019).

Agriculture 4.0 came with the convergence of advanced digital technologies, comprising Internet of Things (IoT), cloud computing, Big Data analytics, and especially AI/ML. This convergence led to an interconnected farm management ecosystem that optimizes agricultural operations (Saiz-Rubio and Rovira-Más 2020; Zhai et al. 2020). Today's smart farming frameworks leverage a network of

Table 11.1 Different agricultural eras across key dimensions

Era	Primary focus	Key technologies/methods	Key Outcome/shift
Ag 1.0	Subsistence	Manual Labor, Basic Tools	Basic food production
Ag 2.0	Increased Yield (Mechanization)	Tractors, Fertilizers, Pesticides	Green Revolution, Higher output
Ag 3.0	Optimization (Precision Ag)	GPS, Sensors, VRT, IT, Automation	Site-specific management, Input saving
Ag 4.0	Interconnected Efficiency	IoT, Cloud, Big Data, AI/ML, Real-time Data	Smart Farming, Data-driven decisions
Ag 5.0	Sustainability and Resilience	Advanced Robotics, Human-Robot Collab, XR, 6G, AI/ML	Holistic goals beyond productivity

sensors and real-time data analysis to optimize efficiency, maximize productivity, and ensure traceability across the agri-food supply chain (Kamilaris et al. 2017).

With the previous advancements as foundation, Agriculture 5.0 represents the next evolutionary phase in agricultural advancement. It incorporates human–robot collaboration, advanced robotics, extended reality (XR), and highly anticipated next-gen networks like 6G. This phase aims to go beyond productivity and efficiency, and encompass broader societal and environmental objectives (Fountas et al. 2024). The framework of Agriculture 5.0 not only focuses on enhanced yield, but also sustainability and resilience, which ensures that technological advancements contribute to food security while preserving natural resources and promoting societal equity. To better understand this trajectory, Table 11.1 provides a concise comparison of the different agricultural eras based on their primary objectives, core methodologies, and defining outcomes.

3 The Engines of Modern Agriculture: AI, ML, and Deep Learning

At the heart of the ongoing agricultural transformation lies three interconnected concepts: Artificial Intelligence (AI), Machine Learning (ML), and Deep Learning (DL). While often used interchangeably, it is important to understand their hierarchy and specific contributions (Dokic et al. 2020).

3.1 Artificial Intelligence (AI)

Artificial intelligence (AI) is an umbrella term which generally refers to the field of Computer Science dedicated to creating systems capable of performing tasks that typically require human intelligence (Jaboob et al. 2024). This term includes a wide range of capabilities like reasoning, knowledge representation, planning, learning, natural language processing, perception, and motion control. In terms of

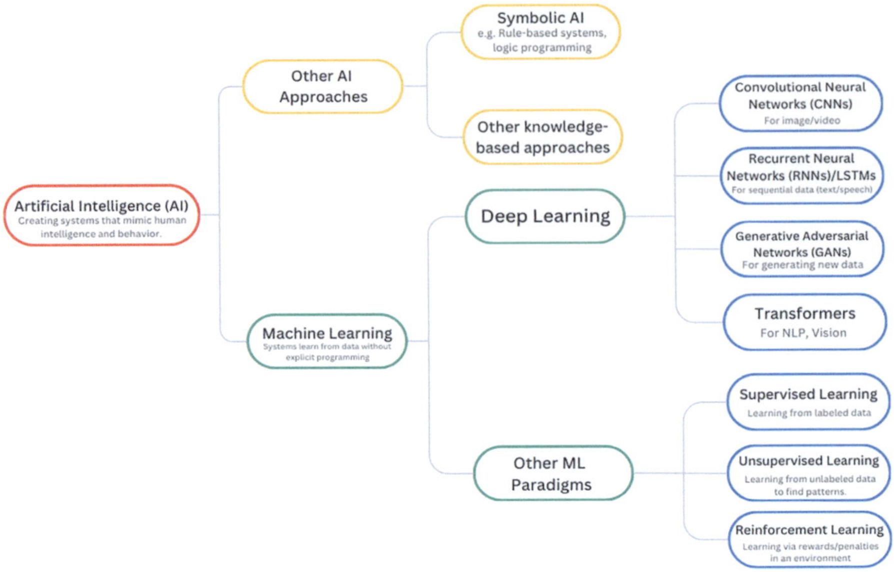

Fig. 11.3 Common classification within the continuously expanding field of Artificial Intelligence

agricultural advancement, AI is generally used to represent the ambition of building intelligent machines (Russell and Norvig 2016). To better understand the structure of this rapidly evolving domain, Fig. 11.3 presents a common classification within Artificial Intelligence, outlining its major approaches and the specific methodologies nested within Machine Learning and Deep Learning.

3.2 Machine Learning (ML)

A subset of AI, Machine Learning focuses mainly on the ability of systems to learn from data and make decisions without being explicitly programmed for (Russell and Norvig 2016). Instead of taking fixed instructions, ML algorithms analyze datasets, identify key patterns, and build a statistical model capable of making predictions (or decisions) on the new data (Mitchell 1997). The core components of ML are Data, Algorithm, and the Model.

ML algorithms are often categorized by the way they learn:

(a) *Supervised Learning*: These algorithms (models) learn from labeled data (generally input–output pairs) to predict the output for new inputs. An agricultural application includes predicting yield based on labeled historical data (Chlingaryan et al. 2018).
(b) *Unsupervised Learning*: In this type, the model tries to uncover hidden structure or patterns in unlabeled data. For example, grouping of fields with similar soil properties (Ließ et al. 2021).

(c) *Reinforcement Learning*: In this type of learning, the model learns optimal strategies by interacting with an environment and receiving rewards or penalties. An example is training a robot for autonomous harvesting (Kolhalkar et al. 2025).

Training ML models particularly in supervised learning, involves an iterative optimization process to find the model parameters ('θ') that minimizes a loss function ('$J(\theta)$'). This loss function quantifies the discrepancy between the model's predictions and the actual target values in the training data. A foundational and widely used algorithm for this optimization is Gradient Descent (Ruder 2016).

The core idea behind Gradient Descent is to iteratively adjust the model parameters in the direction opposite to the gradient of the loss function with respect to those parameters. The update rule for Gradient Descent can be expressed as follows:

$$\theta := \theta - \alpha \nabla J\left(\theta\right)$$

where:

- 'θ' represents the set of model parameters (e.g., weights and biases in a neural network).
- '=' denotes an update operation.
- 'α' is the learning rate.
- '$\nabla J(\theta)$' is the gradient of the loss function 'J' with respect to the parameters 'θ'.

This iterative process is repeated until the loss converges to a minimum or a predefined number of iterations is reached. Variants like Stochastic Gradient Descent (SGD) and Mini-batch Gradient Descent are commonly used in practice, especially for large datasets, as they update parameters based on smaller subsets of data, leading to a faster and more robust training. Gradient Descent is fundamental in training most ML models and is the core mechanism behind backpropagation in Deep Learning (Ruder 2016).

3.3 Deep Learning

Deep Learning (DL) is a specialization of Machine Learning that leverages Artificial Neural Networks (ANNs) which are multilayered computational models inspired by neural structures to automatically learn hierarchical representations from data (Goodfellow et al. 2016). This process allows DL models to discover intricate patterns by progressively building complex features from simpler ones (e.g., from image edges to object parts), thereby minimizing the need for manual feature engineering (Goodfellow et al. 2016).

Information propagates through the network as neurons in each layer compute a weighted sum of inputs from the preceding layer, add a bias, and apply a nonlinear activation function. The training of these networks, typically via backpropagation and optimization algorithms like Gradient Descent, involves iteratively adjusting the network's weights and biases to minimize a loss function that quantifies the discrepancy between predicted and actual outputs (Nielsen 2015). While powerful, the efficiency of DL models with their numerous parameters is heavily reliant on substantial training datasets to achieve generalization and avoid overfitting (LeCun et al. 2015).

In many agricultural applications, such as classifying different crop types, identifying multiple disease categories, or grading produce, the task is a multiclass classification. For such tasks, the output layer of a DL network often employs the Softmax function. The Softmax function takes a vector of arbitrary real-valued scores (logits) from the final layer and transforms them into a probability distribution over the "*K*" possible classes. The output probabilities sum to 1, representing the model's confidence for each class (Franke and Degen 2023).

The Softmax function for the *i*-th class $\sigma(z_i)$ given a vector of logits z is defined as follows:

$$\sigma\left(z_i\right) = \frac{e^{z_i}}{\sum_{k=1}^{K} e^{z_k}}$$

Where:

- 'z_i' is the logit for class 'i'.
- 'e^{z_i}' is the exponential of the logit, ensuring nonnegativity.
- '$\sum_{k=1}^{K} e^{z_k}$' is the normalization term, which is the sum of the exponentials of all logits for all K classes.

3.3.1 Key Deep Learning Architectures

Deep learning consists of various network architectures. Each of them is designed to excel at specific types of data and tasks. Some prominent architectures relevant to agriculture are as follows:

(a) *Convolutional Neural Networks (CNNs)*
   - *Architecture*: Specifically designed to process grid-like data such as images, CNNs consist of convolutional layers (which apply learnable filters to detect spatial hierarchies of features like edges, textures, and motifs), pooling layers (for reducing dimensionality and providing translational invariance), and fully connected layers (for classification or regression based on learned features) (Goodfellow et al. 2016).
   - *Relevance in Agriculture*: CNNs are extensively used for image-based tasks like plant disease detection/classification, weed identification, fruit counting/grading, land cover mapping from satellite/drone imagery, and livestock monitoring (Kamilaris and Prenafeta-Boldú 2018).

(b) *Recurrent Neural Networks (RNNs)*
   - *Architecture*: RNNs are designed to process sequential data where the order of information matters (like time series or natural language). Recurrent

Neural Networks (RNNs) possess memory by having connections that loop back from a layer's output to its input, which allows information from the previous step to influence the current step. Variants like Long Short-Term Memory (LSTM) and Gated Recurrent Unit (GRU) networks are used to tackle challenges like the vanishing gradient problem, enabling RNNs to learn long-range dependencies effectively (Noh 2021).

- *Relevance in Agriculture*: Analyzing time-series sensor data (like soil moisture, temperature over time for crop growth modeling, and irrigation scheduling), weather forecasting, and predicting commodity prices.

(c) *Transformers*

- *Architecture*: Transformers were originally developed for Natural Language Processing (NLP) tasks, but now they have revolutionized the field and are increasingly applied to other domains including computer vision. Their architecture is based primarily on a self-attention mechanism which allows the model to weigh the importance of different parts of the input sequence (or image patches) when processing information that helps in capturing contextual relationships effectively. Transformers can be processed in a highly parallel manner, leading to efficient training on large datasets (Vaswani et al. 2017).
- *Relevance in Agriculture*: Advanced NLP tasks in agriculture (like summarizing research, extracting information from farm logs), analyzing complex sequential data (like genomic sequences for crop breeding), interpreting multitemporal satellite imagery for change detection, and potentially in advanced robotic control systems (Lu et al. 2024).

A core component of Transformers is Scaled Dot-Product Attention, defined as follows:

$$\text{Attention}\left(Q,K,V\right)=\text{softmax}\left(\frac{QK^T}{\sqrt{d_k}}\right)V$$

Where:

- 'Q' (Queries), 'K' (Keys), and 'V' (Values) are matrices derived from the input embeddings. The attention mechanism calculates how much focus each element (Query) should place on other elements (Keys) to compute an output representation based on the associated Values.
- 'QK^T' represents the dot product of Queries and Keys, measuring their similarity or compatibility.
- '$\sqrt{d_k}$' is a scaling factor, where 'd_k' is the dimension of the keys. The scaling prevents the dot products from becoming too large, which could lead to vanishing gradients in the Softmax function.
- "*Softmax*" is applied to the scaled scores to obtain attention weights, which sum to 1.

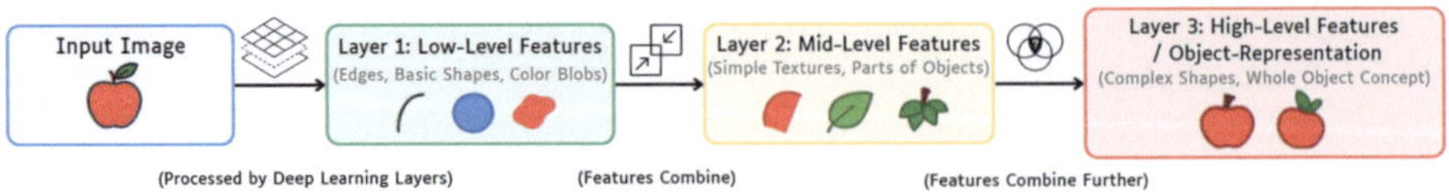

Fig. 11.4 Conceptual illustration of hierarchical feature learning in a Deep Learning model

- These weights are then multiplied by the "*V*" (Values) matrix to produce the output of the attention layer, which is a weighted representation of the input sequence where important elements receive higher weights (Vaswani et al. 2017).

While AI provides the vision, ML provides the general methods for learning from data, and DL offers a powerful set of techniques, particularly adept at handling complex patterns and raw data. This makes it a cornerstone of many advanced agricultural AI applications today (Kamilaris and Prenafeta-Boldú 2018). Figure 11.4 shows how the process of hierarchical feature learning within a Deep Learning model involves layers progressively extracting more complex representations.

4 The Role of Technology in Addressing Global Agricultural Challenges

Agriculture today faces unprecedented challenges: rapid population growth, climate change, and increasing resource constraints. Research by Godfray et al. (2010) indicates that, to feed about 9 billion people by 2050, a tremendous pressure will be exerted on existing agricultural systems, which are already falling behind. To address these issues, the agricultural sector is increasingly embracing technological solutions. Artificial intelligence (AI) and ML are now being used to not only improve operational efficiency and productivity, but also to reduce environmental impact through optimized resource usage.

For instance, AI-powered predictive models demonstrate significant results in forecasting crop yields and understanding the impact of weather, enabling farmers to modify their response in response to climate variations (Premachandra and Kumara 2021). Machine Learning algorithms have shown considerable efficiency in crop monitoring, disease detection, and resource management, directly contributing to improved agricultural productivity (Liakos et al. 2018). These technological advancements are instrumental in ensuring an equilibrium between increased production and environmental sustainability.

5 Use Cases of AI and ML in Agricultural Research

Artificial intelligence (AI) and ML have been the most significant technological advancements in agriculture by far. These technologies function as robust analytical tools, possessing the capability to process diverse and voluminous datasets encompassing satellite imagery, drone footage, sensor measurements, and historical records. By implementing supervised learning, unsupervised learning, and reinforcement learning methodologies, these models address a wide range of agricultural tasks, ranging from crop yield prediction and pest detection to optimizing robotic control strategies (Kamilaris and Prenafeta-Boldú 2018).

Deep learning models, particularly CNNs and Recurrent Neural Networks (RNNs), have been prominently used due to their exceptional capacity in capturing spatial and temporal patterns in data (Benos et al. 2021; Jha et al. 2019). Also, development efforts in explainable AI (XAI) are underway to enhance the transparency and interpretability of complex decision-making processes, thereby fostering confidence among farmers and agronomists (Martin et al. 2024; Mohan et al. 2025).

Let us explore some concrete examples of how these technologies are being applied to address critical challenges in agriculture today. The following use cases illustrate the practical and diverse application of AI/ML across different stages of the agricultural value chain:

Case 1: Early Detection of Plant Diseases Using Deep Learning and Computer Vision

- *Problem*: Plant diseases are a major threat to crop yields and quality worldwide. Traditional methods can only do so much, like manual scouting, which is time-consuming, costly and often late in controlling the spread. This leads to broader, often less effective, application of pesticides, thereby increasing costs and environmental impact (Mahlein 2016).
- *Artificial Intelligence (AI)/Machine Learning (ML) Approach*: This problem is well suited for Deep Learning, particularly CNN, which is a model highly effective for image analysis. This approach involves supervised learning.
  - Training: A large dataset containing labeled images of plant/leaves/stems is collected. The labels are generally "healthy" or with specific types of disease along with their severity (like "Wheat Rust—Severity level 2," "Tomato Blight—Early stage").
  - Model: CNN model is trained on the labeled dataset. Having multiple layers, the network automatically learns to identify the intricate visual features indicative of the different diseases, even at stages which are sometimes hard for human eyes to detect.
- *Outcome/Impact*:
  - Early detection: The model can detect diseases early, often days or weeks before, compared to traditional methods.
  - Accuracy: The models provide high classification accuracy in their results for various diseases.
  - Accessibility: Mobile app versions of the models allow farmers to quickly diagnose their crops by just taking a picture from their smartphones.
  - Precision treatment: The accuracy of the model and early detection lead to early prevention and enable targeted treatments to locations where it is truly

needed, which in turn reduces chemical usage, costs, and environmental impacts.

- *Example Systems/Research*: Deep learning and feature extraction using PlantVillage dataset have been applied for plant disease detection (Mohameth et al. 2020). A variety of these diseases are visually represented in Fig. 11.5 (Mohanty et al. 2016).

Case 2: Optimizing Irrigation Through Sensor Fusion and Machine Learning

- *Problem*: Water scarcity is a major global challenge, and agriculture happens to be the largest consumer of freshwater resources. Traditional irrigation methods with fixed schedules often lead to significant water waste due to overirrigation or

Fig. 11.5 Sample images from PlantVillage dataset for different types of leaf diseases (Mohanty et al. 2016)

yield loss due to underirrigation. Optimizing water usage is crucial for sustainability (FAO 2022).

- *Artificial Intelligence/Machine Learning (AI/ML) Approach*: ML techniques involving Supervised Learning and Reinforcement Learning (RL) can be combined and fused with sensors to predict water needs based on past data and learning optimal irrigation strategies over time.
  - Data Integration: Data collected from in-field sensors (soil, moisture, humidity, temperature), weather stations (current and forecast), satellite/drone imagery, and crop-specific physiological models can be integrated.
  - Model: ML models like Regression Models, Support Vector Machines (SVMs), Decision trees, and potential RL agents are trained to analyze the fused data. The models learn complex relationships between environmental conditions, soil moisture levels, crop water requirements, and weather.
- *Outcome/Impact*:
  - Water Savings: Significant reductions in water consumption compared to traditional methods.
  - Improved Yield/Quality: Maintaining optimal soil moisture lets the crops grow without stress leading to better yield and crop quality.
  - Resource Efficiency: Reduced energy consumption associated with water pumping.
  - Automated Decision Support: Provides farmers with precise recommendations on when and how much to irrigate and even this can be automated using IoT devices on irrigation system valves.
- *Example System/Research*: Artificial intelligence has been used to optimize water consumption in agriculture through a predictive algorithm-based irrigation management system (Preite and Vignali 2024).

Case 3: Enhancing Crop Yield Prediction with ML and Remote Sensing

- *Problem*: Accurately predicting the crop yield before harvest is crucial for farmers (because it involves management decisions and securing loans), policymakers (for food-security planning and import/export decisions), and community traders (for market stability). Traditional methods rely on historical averages and limited field surveys which are not always accurate especially with the increasing climate variability (Lobell et al. 2009).
- *Artificial Intelligence/Machine Learning (AI/ML) Approach*: ML models including the classical ones and sometimes DL models like Recurrent Neural Networks (RNNs) or Long Short-Term Memory (LSTMs) are used in supervised learning especially for time-series data.
  - Feature Engineering: Multiple factors that influence yield are identified through feature selection. This often involves combining data from different sources over the growing season. For deeper research, genomic data (genotypic and phenotypic) can be integrated and the genotypic data that is highly

associated to the phenotype (often yield) can be selected through rigorous feature selection.
  - Model: Algorithms like Multiple Linear Regression, Random Forests, Gradient Boosting Machines, or ANNs/RNNs are trained on both the historical data and the genomic data corresponding to actual yields. The models learn complex, often nonlinear relationships between these inputs and the final yield.
- *Outcome/Impact*:
  - Improved accuracy: ML models have consistently demonstrated higher accuracy in yield prediction compared to traditional methods.
  - Timeliness: Provides earlier and more frequent yield estimates throughout the year.
  - Granularity: Allows yield prediction at different scales, from individual small landholders to large regions.
  - Informed decision making: Enables better planning for harvesting, storage, logistics, and market positioning. Helps the government anticipate potential food shortages or surpluses.
- *Example System/Research*: Our team is currently working on exploring genomic data of rice landraces using machine learning methods where a part of our research involves using genotypic data of rice varieties to predict the phenotypic data (like yield).

6 Applications of AI and ML in Agriculture

The implementation of AI and ML methodologies in agriculture span a diverse and continually expanding area. The key areas are the following:

6.1 Crop Management and Disease Detection

Artificial intelligence (AI)-driven agricultural methodologies have revolutionized crop management, enabling early disease detection, optimized crop yields, and precise application of inputs. Deep learning models like CNNs have shown exceptional accuracy (>95%) in plant disease identification through image analysis, which has facilitated timely and economically efficient interventions (Kamilaris and Prenafeta-Boldú 2018; Mohanty et al. 2016; Atila et al. 2021; Saleem at al. 2020). These advancements help protect yields and mitigate losses occurring from pets and diseases.

Similarly, hyperspectral imaging technology integrated with hierarchical self-organizing classification systems has demonstrated remarkable efficacy in differentiating biotic and abiotic stress conditions in agricultural settings. Field trials conducted at Rothamsted Research utilized a ground-based Specim V10 system to acquire 12,120 canopy spectra across five strategically selected wavebands, each spanning 20 nm. Three sophisticated neural network architectures (counter propagation ANNs, supervised Kohenen networks, and XY-fusion systems) were used to analyze the resulting spectral data and each model achieved cross-validated accuracy rates exceeding 95% in distinguishing between healthy wheat specimens, yellow rust infections, and nitrogen-deficiency conditions (Pantazi et al. 2017).

6.2 Precision Agriculture and Resource Optimization

Precision agriculture methodologies utilize AI/ML to analyze sensor data (e.g., soil moisture and nutrient levels) and remote sensing imagery to facilitate variable rate technology (VRT) implementation for targeted application of fertilizers and pesticides. This approach results in enhanced resource-use efficiency while simultaneously reducing environmental impact (Schieffer and Dillon 2015). Furthermore, smart irrigation systems incorporating Machine Learning models to optimize water distribution based on real-time data, achieve substantial water conservation. This is a significant achievement, especially in arid and water-stressed regions (Goap et al. 2018; Bwambale et al. 2023).

In modern precision agriculture, an Internet-of-Things layer of both analogue and digital transducers (e.g., soil-moisture probes, electrochemical nutrient sensors and optical canopy scanners) continuously feeds real-time data into AI-driven decision engines. By embedding basic pre-processing and anomaly detection at the edge, these systems cut redundant transmissions and curb node power draw, enabling, for instance, an LSTM-based scheduler to sustain 180-day autonomous operation on a single Li-ion cell (Shadrin et al. 2020).

Another study by Munir et al. (2021) introduces a system incorporating a multidimensional approach to determining precise crop water requirements. This advanced framework integrates structured domain knowledge, encoded within an ontological framework with real-time sensor network data. The system employs a KNN algorithm to process these integrated inputs, generating precise predictions of field-specific irrigation requirements. A significant architectural feature involves the strategic deployment of an edge server positioned between the sensor network and primary IoT infrastructure, facilitating localized data processing. There are multiple operational benefits to this design, like reduced central server processing demands, minimized communication latency, and enhanced accuracy in crop-specific water-requirement predictions.

6.3 Autonomous Machinery and Robotics

Artificial intelligence (AI)-powered automations are being deployed extensively for labor-intensive tasks like planting, weeding, harvesting, and sorting. Autonomous tractors and robotic systems integrated with advanced computer-vision capabilities are already showing improvements in labor efficiency and operational productivity (Bergerman et al. 2016; Jha et al. 2019). These methodologies address labor shortage challenges and also contribute to production cost reduction, making a significant advancement in modern agricultural systems.

A striking example comes from recent work on autonomous strawberry pickers, where AI-driven perception and manipulation have pushed robotic harvesting into real-world polytunnels. The system combines an Intel R200 RGB-D camera, Hokuyo LiDAR for navigation and two low-cost grippers mounted on a horizontal dual-arm manipulator, all orchestrated under ROS and powered by a 48-hour battery pack. A novel obstacle-separation algorithm computes pushing trajectories that clear leaves, stems, and nearby berries, allowing the robot to pluck fruit clusters and deposit each berry directly into a market punnet, eliminating any post-harvest repacking. In field trials, first-attempt pick success ranged from 50% (highly occluded fruit) to 97.1% (minimally obstructed), rising to 75–100% after a second pass with updated vision feedback. Cycle-time benchmarks showed an average of 6.1 s per strawberry in single-arm mode and just 4.6 s when both arms operate in tandem. By automating one of the most labor-intensive tasks on the farm, such systems demonstrate how intelligent robotics can both alleviate workforce shortages and cut harvesting costs (Xiong et al. 2020).

6.4 Livestock Management

In the livestock sector, AI enabled wearable devices are emerging as essential instruments for monitoring animal health and behavior. This implementation with continuous monitoring helps detect diseases early which in turn improves animal welfare and productivity (Neethirajan et al. 2024). For example, sensor-based systems track vital signs and activity levels to facilitate timely intervention, consequently minimizing the spread of disease and improving overall herd management (Mishra and Mishra 2023; Lin 2025).

Artificial intelligence (AI)-enabled monitoring and ML techniques are increasingly being employed for optimizing livestock production and management. Machine learning (ML) models have been efficient in predicting diseases like Sub-Clinical Mastitis (SCM), with one study using Random Forest achieving an accuracy of 90% for prediction independent of Somatic Cell Count (SCC) (Ebrahimie et al. 2018), and another one employing Gradient Boosting Machine (GBM) that reached 84.9% accuracy in predicting subclinical bovine mastitis (Ebrahimi et al. 2019).

Beyond that, ML is also employed in health management tasks like classifying germ-transmission routes. A Random Forest model achieved 98% accuracy in distinguishing contagious and environmental types (Hyde et al. 2020). A deep learning model Mask R-CNN was used for cattle-image segmentation with a mean pixel accuracy of 0.92 (Qiao et al. 2019), and another CNN model successfully identified pig faces with an accuracy of 96.7% (Hansen et al. 2018). Predictive ML systems also contribute to production outcomes and welfare, forecasting poultry egg production issues a day in advance with an estimation accuracy of 0.9854 (Morales et al.

2016) and classifying calving difficulties with up to 75% accuracy using models like Neural Networks or multimodal regression (Fenlon et al. 2017).

6.5 Yield Prediction and Decision Support Systems

The integration of diverse data streams from satellites, on-field sensors, and historical records enables the development of robust ML models for predicting crop yield. These analytical models generate actionable insights and yield forecasts, which aid in making decisions for stakeholders across the agri-food supply chain (Crane-Droesch 2018). In many instances, these predictive systems have improved decision accuracy by up to 15%, depending on the crop and climate conditions (Lowenberg-DeBoer and Erickson 2019).

A study by Khaki and Wang (2019) presented a model that achieved a root-mean-square-error (RMSE) of 12% of the average yield and 50% of the standard deviation for the validation dataset when using predicted weather data, with performance improving slightly to an RMSE of 11% of the average yield and 46% of the standard deviation with perfect weather data. Computational results confirmed that this DNN approach significantly outperformed other popular prediction methods, including Lasso, shallow neural networks, and regression trees. Furthermore, the analysis revealed the significant finding that environmental factors exerted a greater influence on the crop yield than genotype within the context of the analyzed data.

Artificial intelligence (AI) and ML-driven decision support systems are also being specifically tailored for controlled environments like greenhouses (Rayhana et al. 2020). Kocian et al. (2020) utilized this approach by implementing a system with IoT smart sensors installed within greenhouses to monitor diverse parameters in real-time, including environmental conditions, soil properties, and critical plant growth indicators such as leaf area index (LAI), accumulated dry weight (DW), and evapotranspiration (ET). These real-time data streams are sent to the cloud via IoT devices, enabling the implementation of an agriculture decision-support system aimed at predicting crop growth and informing management practices in these specific settings.

A brief overview of AI and ML research trends in agriculture, including key applications and representative studies, is presented in Table 11.2.

7 Comparative Analysis: India Versus Developed Countries

The adoption of AI and ML models across the globe exhibits notable disparities, especially between developed economies and developing nations such as India. The developed countries typically benefit from their large-scale, consolidated farming operations that facilitate substantial investments in expensive and cutting-edge AI applications like autonomous machinery and comprehensive sensor networks. On

Table 11.2 Compilation of AI and ML research trends in agriculture

Research domain	Key trends and applications	Representative studies
Crop disease detection	Development of CNN models for leaf image classification;	Mohanty et al. (2016)
	EfficientNet models achieving >95% accuracy	Atila et al. (2021)
Yield prediction	Use of LSTM, Random Forests, and deep neural networks integrating satellite images, weather data, and soil information	Crane-Droesch (2018), Nevavuori et al. (2020)
	Deep neural networks outperform Lasso and regression trees; RMSE $\approx$ 11–12% of mean yield; environmental factors shown more influential than genotype	Khaki and Wang (2019)
Precision Agriculture and Resource Management	Variable rate technology guided by IoT sensors and remote sensed data;	Schieffer and Dillon (2015)
	Smart irrigation systems reducing water usage	Goap et al. (2018)
	Advanced control strategies for precision irrigation	Abioye et al. (2020)
	Ontology-driven irrigation requirement estimation; KNN-based field watering prediction; edge server for local preprocessing, reduced latency and central load	Shadrin et al. (2020), Munir et al. (2021)
Autonomous Machinery and Robotics	AI-powered robotics for planting, weeding, and harvesting;	Bergerman et al. (2016)
	Computer vision-assisted navigation	Jha et al. (2019)
	Dual-arm strawberry picker with RGB-D and LiDAR guidance; first-pass success rates 50–97%, rising to 75–100% on a second pass; cycle times 4.6–6.1 s per fruit	Xiong et al. (2020)
Livestock Management	Wearable sensor devices monitoring animal health and behavior;	Mishra and Mishra (2023)
	AI-powered wearable devices to detect bird flu and other illnesses;	Lin (2025)
	ML for sub-clinical mastitis prediction and germ transmission classification;	Ebrahimie et al. (2018), Ebrahimi et al. (2019), Hyde et al. (2020)
	Deep learning for cattle segmentation (Mask R-CNN) and pig face identification (CNN);	Qiao et al. (2019), Hansen et al. (2018)
	Forecasting egg production and calving difficulty.	Morales et al. (2016), Fenlon et al. (2017)
Explainable AI (XAI)	Development of transparent AI decision frameworks;	Martin et al. (2024)
	AgroXAI system providing farmers with tailored crop recommendations	Turgut et al. (2024)

(continued)

Table 11.2 (continued)

Research domain	Key trends and applications	Representative studies
Stress Detection & Spectral Analysis	Hyperspectral imaging combined with hierarchical self-organizing classifiers for biotic vs. abiotic stress; cross-validated accuracies >95%	Pantazi et al. (2017)
Decision-Support Systems	IoT-enabled real-time monitoring (LAI, DW, ET) with cloud analytics; models to predict growth and inform management in controlled environments	Kocian et al. (2020)

the other hand, India's agricultural sector is characterized by millions of smallholder farms with limited capital and fragmented landholdings (Aguilar 2024).

India's digital initiatives, such as the "Saagu Baagu" project and the Agri Stack platform, represent significant efforts to address the technological gaps by providing farmers with AI-driven advisory services, soil testing, and digital connectivity. Initiatives like these have not only improved crop yields by an estimated 21% per acre but have also doubled the net income of participating farmers (Aguilar 2024; Jurgens and Kaushik 2024). Nevertheless, challenges still persist, including limited digital literacy, inadequate technological infrastructure, and the necessity for localized AI solutions that address diverse agro-ecological conditions. Recent studies have also highlighted concerns about credit accessibility and the socioeconomic implications of labor displacement, which are challenges that manifest distinctly from those seen in developed countries (Garcia-Murillo and MacInnes 2023).

8 Current Utilization and Emerging Research Trends

Recent scientific advancements have demonstrated a robust implementation of AI and ML applications in modern agriculture. The emergence of advanced disease detection models, high-accuracy yield-prediction systems, and autonomous farming equipment exemplifies the transition toward data-driven, precision agriculture. Notably, the integration of remote sensing with multitemporal datasets from unmanned aerial vehicles (UAVs) and satellites now supports detailed field-level analysis, while IoT devices collect critical environmental metrics in near real-time (Kamilaris et al. 2017; Weiss et al. 2020).

Other groundbreaking case studies have documented significant achievements in AI implementation, particularly in automated pest detection using deep neural network models on edge computing devices (Albanese et al. 2021). They further demonstrated how decision support systems can contribute to the transformation of operations on farms by integrating multimodal data streams (Zhai et al. 2020). Still, the transition from pilot initiatives to large-scale implementation continues to encounter substantial challenges like data quality, standardization, and most importantly, the heterogeneous nature of agricultural environments.

Research initiatives focusing on the integration of explainable AI methodologies demonstrate growing recognition of the crucial role of transparency in fostering farmer trust and facilitating broader adoption of AI enabled frameworks (Martin et al. 2024). Within this evolving landscape, substantial research continues to focus on critical issues like data heterogeneity, durability of IoT sensors in harsh field conditions, and overcoming infrastructural constraints in rural areas (Kamilaris et al. 2017; Lowenberg-DeBoer et al. 2020).

9 Data Requirements and Challenges

A wide array of sensors forms the foundations of data collection in modern agriculture, providing the raw inputs necessary for sophisticated AI and Ml analyses. Table 11.3 summarizes key sensor types and the data they collect (Mulla 2013; Pajares et al. 2013; Li et al. 2010).

The implementation of robust and efficient AI models in agriculture continues to rely on the availability of extensive, high-quality datasets. Data in this domain is characterized by the "Vs" of Big Data: volume, variety, velocity, and veracity. The collection of accurate, labeled data, especially for image-based disease detection presents a major challenge, as it typically requires domain expertise and substantial manual effort (Barbedo 2018).

Other technical challenges here are sensor calibration, noise in the data stream, and insufficient data transmission infrastructure in rural areas. The integration of heterogeneous datasets derived from IoT devices and sensors, UAVs, and satellite imagery requires sophisticated data fusion and preprocessing techniques. To address these issues, it is necessary to establish clear data governance frameworks that protect the privacy rights, delineate data ownership parameters, and address security concerns (Wiseman et al. 2019).

10 Future Outlook

The future of agriculture is increasingly aligned with AI and ML implementations, which also present a significant market potential. Market analysis indicates that AI in the agricultural sector is projected to reach USD 3.0 billion value by 2026 (Globe Newswire 2024). This dominant role of software solutions and IoT-driven farm management frameworks emphasize the ongoing digital transformation in this sector. Collaborations like the US–India Artificial Intelligence (USIAI) Initiative, established in April 2021, showcase a global commitment to leveraging AI for improved and optimized good grain production and sustainable farming practices (Mishra and Fanase 2025).

Emerging trends encompass the further integration of human–robot collaboration, expanded precision agriculture technologies into remote sensing and cloud-based data analytics, and the development of adaptive regulatory methodologies that facilitate innovation while also addressing potential risks (Duckett et al. 2018).

Table 11.3 Key sensors enabling AI/ML applications in agriculture

Sensor type	Measures	Data generated	Relevance to AI/ML in agriculture
Location and positioning			
GPS/GNSS Receiver	Geographic coordinates (Latitude, Longitude, Altitude)	Precise location data points	Essential for georeferencing all other sensor data, creating field maps, enabling site-specific management, guiding autonomous vehicles (path planning).
Imaging sensors			
RGB (Visible Light) Cameras	Reflectance in Red, Green, Blue light spectra	Standard digital images	Input for Computer Vision (DL) models to identify weeds, pests, diseases; perform plant counting, assess growth stages, detect visible stress symptoms.
Multispectral Cameras	Reflectance in specific visible & near-infrared (NIR) bands	Images with multiple layers (bands)	Data used to calculate Vegetation Indices (e.g., NDVI, NDRE) which are features for ML models predicting biomass, nitrogen status, water stress, yield.
Hyperspectral Cameras	Reflectance across hundreds of narrow, contiguous spectral bands	High-dimensional image data ("data cubes")	Input for advanced ML/DL models to detect subtle nutrient deficiencies, specific diseases, crop varieties, or soil properties with high specificity.
Thermal Infrared Cameras	Surface temperature based on emitted infrared radiation	Thermal map images	Data used by ML models to estimate canopy temperature, correlating with plant water stress (transpiration cooling) for irrigation management and stress detection.
LiDAR (Light Detection & Ranging)	Distance to objects using pulsed laser light	3D point clouds representing physical structures	Input for ML algorithms to estimate plant height, canopy volume, biomass; map field topography; obstacle detection for autonomous navigation.
Environmental sensors			
Weather Station Sensors	Temp, humidity, rainfall, wind speed/direction, solar radiation	Time-series data of atmospheric conditions	Critical input features for ML models predicting crop growth, yield, disease/pest risk, evapotranspiration rates for irrigation scheduling.

(continued)

Table 11.3 (continued)

Sensor type	Measures	Data generated	Relevance to AI/ML in agriculture
Soil Moisture Sensors	Volumetric water content or soil water potential	Point measurements or profile data, often time-series	Key input for ML models optimizing irrigation schedules, predicting water stress, validating water distribution models.
Soil Electrical Conductivity (EC)	Ability of soil to conduct electricity	EC measurements (point or mapped)	Feature used in ML models to infer soil texture variations, salinity levels, and potentially nutrient zones for variable rate applications.
Soil pH Sensors	Acidity or alkalinity of the soil	pH values (point or mapped)	Input feature for ML models recommending soil amendments or predicting nutrient availability.
Soil Nutrient Sensors (e.g., Nitrate, Potassium)	Concentration of specific ions in the soil solution	Nutrient concentration values (point or mapped, often requires sampling)	Direct input for ML models creating precise variable rate fertilization maps (often combined with other data like yield maps and EC).
Plant-specific sensors			
Leaf Wetness Sensors	Presence and duration of moisture on surfaces	Binary (wet/dry) or duration data, time-series	Input feature for ML models predicting fungal disease outbreaks (many require leaf wetness for spore germination).
Dendrometers	Micro-variations in stem or fruit diameter	High-resolution time-series data of growth/shrinkage	Data used by ML models to detect water stress (stem shrinkage) in real-time, particularly for tree/vine crops, informing precise irrigation decisions.
Chlorophyll Meters/SPAD	Relative chlorophyll content in leaves	Index values (point measurements)	Feature data for ML models assessing plant nitrogen status and photosynthetic potential, often correlated with yield or stress.
Other relevant sensors			
Flow Meters	Volume or rate of liquid flow (e.g., irrigation water)	Flow rate data, cumulative volume	Provides ground truth data for calibrating irrigation models, monitoring actual water applied vs. recommended by AI/ML systems.
Accelerometers / Gyroscopes	Motion, orientation, vibration of equipment	Inertial measurement unit (IMU) data	Used in ML models for equipment monitoring (e.g., detecting malfunctions), stabilizing sensor platforms (drones), or refining positioning data.

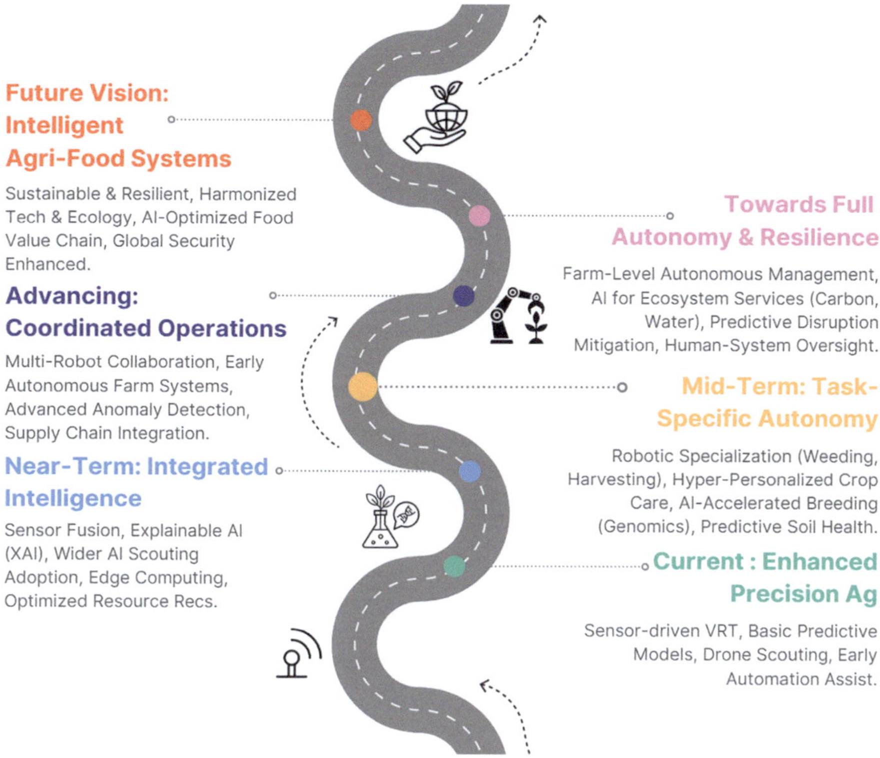

Fig. 11.6 The Future Roadmap for AI and ML in Agriculture

In line with these technological advancements, the research initiatives must continue to address the rising and existing fundamental challenges related to data standardization, model interpretability, and the adaptation of AI models to localized agricultural conditions (Gumbi et al. 2023); even more so in the regions characterized by small-scale farming operations, such as India. Figure 11.6 offers a conceptual journey into the evolving landscape of AI and ML in agriculture, from today's precision techniques to tomorrow's innovative agri-food systems.

11 Strategies for Adopting AI-Driven Agricultural Research

For a transformative AI integration strategy in agriculture, the combination of technological innovation with capacity building and supportive policy frameworks is a must. The key elements include the following:

- Artificial Intelligence (AI) Literacy in Agricultural Sciences: The development of a comprehensive understanding, encompassing both cognitive and psychological dimensions of AI learning (Kymäläinen et al. 2024) ensures that agricultural professionals understand AI content, applications, ethical implications, and practical evaluation measures.

- High-Quality Data Infrastructure: The establishment of federated and standardized database systems like India's upcoming IDEA platform is critical for accelerating AI research (Beriya 2022).
- Research Prioritization: Strategic focus on context-specific challenges, including the development of affordable AI solutions for smallholder farmers, optimization of climate resilience capabilities, and mitigating post-harvest losses remains essential for sustained research advancement (Chand et al. 2022).
- Ethical Guidelines and Stakeholder Engagement: Developing clear ethical frameworks for AI deployment will address biases and ensure equitable access across agricultural communities.
- Policy and Regulatory Support: Establishing data governance frameworks, regulatory sandboxes, and multistakeholder platforms is necessary to sustain innovation and foster collaboration between governments, industries, and research institutions (FAO 2022; Wolfert et al. 2017).

12 Policy Recommendations and Institutional Support for AI-Based Innovations

The optimization of AI's potential in agriculture necessitates comprehensive policy frameworks and coordinated institutional efforts. Policymakers must establish coherent and long-term strategic visions that facilitate digital transformation in agriculture while implementing secure data governance frameworks to protect privacy and ensure ethical use (Wolfert et al. 2017). The cultivation of public–private partnerships also represents a crucial mechanism for fostering innovation and enhancing the scalability of AI solutions. For example, the development of centers dedicated to AI-driven agricultural research, like the Center for Applied Artificial Intelligence in Agriculture at the University of Florida, is a promising step (UF/IFAS Center for Applied Artificial Intelligence in Agriculture 2025) .

Furthermore, the promotion of multistakeholder dialogue is essential to ensure that emerging technologies benefit all segments of the agricultural value chain. Innovative solutions such as AI-powered chatbots (like Kisan e-Mitra under the PM-KISAN scheme) are instrumental in enhancing communication and accessibility of service for farmers (Babu et al. 2024). Additionally, targeted educational programs and infrastructural investments supporting capacity building in rural regions are crucial for bridging the digital divide between developed nations and regions with smallholder farming populations like India.

13 Conclusion

The convergence of Artificial Intelligence and Machine Learning technologies with agricultural practices represents a fundamental transformation in the future of food systems. The integration of digital innovation with traditional agriculture, encompassing precision farming, crop disease detection, advanced yield prediction, and autonomous machinery, demonstrates significant potential for addressing global challenges. As outlined in this chapter, AI-driven agricultural research presents transformative benefits for increased productivity, optimized resource use, sustainability, and enhanced resilience in the face of existing and rising environmental challenges.

Nevertheless, it is equally important to acknowledge and address the significant challenges in this technological transformation, including data quality assurance, infrastructural constraints, and the need for context-specific solutions. The successful integration of these technologies necessitates a balanced approach combining technological innovation, systematic capacity building, robust policy frameworks, and stakeholder engagement. Through such coordinated efforts, AI and ML technologies can effectively empower farmers, bridge developmental gaps between regions, and contribute to a sustainable and secure global food system.

References

Abioye EA, Abidin MSZ, Mahmud MSA, Buyamin S, Ishak MHI, Abd Rahman MKI, Otuoze AO, Onotu P, Ramli MSA (2020) A review on monitoring and advanced control strategies for precision irrigation. Comput Electron Agric 173:105441. https://doi.org/10.1016/j.compag.2020.105441

Abraham M, Pingali P (2020) Transforming smallholder agriculture to achieve the SDGs. In: The role of smallholder farms in food and nutrition security. Springer International Publishing, Cham, pp 173–209

Adedibu PA (2023) Ecological problems of agriculture: impacts and sustainable solutions. ScienceOpen Preprints

Aguilar M (2024) AI for agriculture: how Indian farmers are harvesting innovation. World Economic Forum. Retrieved 14 April 2025, from https://www.weforum.org/impact/ai-for-agriculture-in-india/

Albanese A, Nardello M, Brunelli D (2021) Automated pest detection with DNN on the edge for precision agriculture. IEEE J Emerg Select Topics Circuits Syst 11(3):458–467. https://doi.org/10.1109/JETCAS.2021.3101740

Atila Ü, Uçar M, Akyol K, Uçar E (2021) Plant leaf disease classification using EfficientNet deep learning model. Eco Inform 61:101182. https://doi.org/10.1016/j.ecoinf.2020.101182

Babu M, Anand R, Deepanraj V (2024) Information and communication technology (ICT) applications in natural farming. Retrieved 14 April 2025

Barbedo JG (2018) Factors influencing the use of deep learning for plant disease recognition. Biosyst Eng 172:84–91. https://doi.org/10.1016/j.biosystemseng.2018.05.013

Benos L, Tagarakis AC, Dolias G, Berruto R, Bochtis D (2021) Machine learning in agriculture: a comprehensive updated review. Sensors 21(11):3758. https://doi.org/10.3390/s21113758

Bergerman M, Billingsley J, Reid J, van Henten E (2016) Robotics in agriculture and forestry. In: Siciliano B, Khatib O (eds) Springer handbook of robotics. Springer, Cham. https://doi.org/10.1007/978-3-319-32552-1_56

Beriya A (2022) India digital ecosystem of agriculture and agristack: an initial assessment. ICT India Working Paper No. 68

Bousmah I, Grenier G (2022) Labor shortages and immigration: the case of the Canadian agriculture sector. Agribusiness 38(1):220–235

Bwambale E, Abagale FK, Anornu GK (2023) Smart irrigation for climate change adaptation and improved food security. IntechOpen. https://doi.org/10.5772/intechopen.106628

Calicioglu O, Flammini A, Bracco S, Bellù L, Sims R (2019) The future challenges of food and agriculture: an integrated analysis of trends and solutions. Sustainability 11(1):222

Chand R, Joshi P, Khadka S (2022) Indian agriculture towards 2030: pathways for enhancing farmers' income, nutritional security and sustainable food and farm systems. Springer Nature, Singapore

Chlingaryan A, Sukkarieh S, Whelan B (2018) Machine learning approaches for crop yield prediction and nitrogen status estimation in precision agriculture: a review. Comput Electron Agric 151:61–69

Crane-Droesch A (2018) Machine learning methods for crop yield prediction and climate change impact assessment in agriculture. Environ Res Lett 13(11):114003. https://doi.org/10.1088/1748-9326/aae159

Dawn N, Ghosh T, Ghosh S, Saha A, Mukherjee P, Sarkar S et al (2023) Implementation of artificial intelligence, machine learning, and internet of things (IoT) in revolutionizing agriculture: a review on recent trends and challenges. Int J Exp Res Rev 30:190–218

Dhal SB, Kar D (2024) Transforming agricultural productivity with AI-driven forecasting: innovations in food security and supply chain optimization. Forecasting 6. INL/JOU-24-81560-Rev000

Dokic K, Blaskovic L, Mandusic D (2020) From machine learning to deep learning in agriculture–the quantitative review of trends. IOP Conference Series: Earth and Environmental Science 614:012138

Duckett T, Pearson S, Blackmore S, Grieve B, Chen WH, Cielniak G et al (2018) Agricultural robotics: the future of robotic agriculture. arXiv:1806.06762

Ebel R, Menalled F, Ahmed S, Gingrich S, Baldinelli GM, Félix GF (2021) How biodiversity loss affects society. In: Handbook on the human impact of agriculture. Edward Elgar Publishing, Cheltenham, pp 352–376

Ebrahimi M, Mohammadi-Dehcheshmeh M, Ebrahimie E, Petrovski KR (2019) Comprehensive analysis of machine learning models for prediction of sub-clinical mastitis: deep learning and gradient-boosted trees outperform other models. Comput Biol Med 114:103456

Ebrahimie E, Ebrahimi F, Ebrahimi M, Tomlinson S, Petrovski KR (2018) Hierarchical pattern recognition in milking parameters predicts mastitis prevalence. Comput Electron Agric 147:6–11

El Bilali H, Bassole IHN, Dambo L, Berjan S (2020) Climate change and food security. Agriculture & Forestry/Poljoprivreda i šumarstv 66(3)

Elbasi E, Mostafa N, Zaki C, AlArnaout Z, Topcu AE, Saker L (2024) Optimizing agricultural data analysis techniques through AI-powered decision-making processes. Appl Sci 14(17):8018

FAO (2009) Declaration of the World Summit on Food Security. FAO. Retrieved 13 April 2025, from https://www.fao.org/fileadmin/templates/wsfs/Summit/Docs/Final_Declaration/WSFS09_Declaration.pdf

FAO (2022) The state of the world's land and water resources for food and agriculture—systems at breaking point. Main report. FAO, Rome

FAO, IFAD, UNICEF, WFP, WHO (2023) The state of food security and nutrition in the world 2023: urbanization, agrifood systems transformation and healthy diets across the rural–urban continuum. FAO, Rome. https://doi.org/10.4060/cc3017en

Fenlon C, O'Grady L, Mee JF, Butler ST, Doherty ML, Dunnion J (2017) A comparison of 4 predictive models of calving assistance and difficulty in dairy heifers and cows. J Dairy Sci 100(12):9746–9758

Fountas S, Espejo-García B, Kasimati A, Gemtou M, Panoutsopoulos H, Anastasiou E (2024) Agriculture 5.0: cutting-edge technologies, trends, and challenges. IT Professional 26(1):40–47. https://doi.org/10.1109/MITP.2024.3358972

Franke M, Degen J (2023) The softmax function: properties, motivation, and interpretation. https://doi.org/10.31234/osf.Io/vsw47

Fried G, Chauvel B, Reynaud P, Sache I (2017) Decreases in crop production by non-native weeds, pests, and pathogens. In: Impact of biological invasions on ecosystem services. Springer International Publishing, Cham, pp 83–101

Garcia-Murillo M, MacInnes I (2023) The promise and perils of artificial intelligence: overcoming the odds. International Telecommunications Society (ITS). Retrieved from https://www.econstor.eu/bitstream/10419/277963/1/Garcia-Murillo-MacInnes.pdf

Globe Newswire (2024) AI in agriculture statistics: transforming farming practices for a sustainable future. GlobeNewswire News Room. Retrieved from https://www.globenewswire.com/

Goap A, Sharma D, Shukla AK, Krishna CR (2018) An IoT based smart irrigation management system using machine learning and open source technologies. Comput Electron Agric 155:41–49. https://doi.org/10.1016/j.compag.2018.09.040

Godfray HCJ, Beddington JR, Crute IR, Haddad L, Lawrence D, Muir JF, Pretty J, Robinson S, Thomas SM, Toulmin C (2010) Food security: the challenge of feeding 9 billion people. Science 327(5967):812–818

Goodfellow I, Bengio Y, Courville A (2016) Deep learning. MIT Press, Cambridge

Gumbi N, Gumbi L, Twinomurinzi H (2023) Towards sustainable digital agriculture for smallholder farmers: a systematic literature review. Sustainability 15(16):12530. https://doi.org/10.3390/su151612530

Hansen MF, Smith ML, Smith LN, Salter MG, Baxter EM, Farish M et al (2018) Towards on-farm pig face recognition using convolutional neural networks. Comput Ind 98:145–152

Hossain A, Krupnik TJ, Timsina J, Mahboob MG, Chaki AK, Farooq M et al (2020) Agricultural land degradation: processes and problems undermining future food security. In: Environment, climate, plant and vegetation growth. Springer International Publishing, Cham, pp 17–61

Hyde RM, Down PM, Bradley AJ, Breen JE, Hudson C, Leach KA et al (2020) Automated prediction of mastitis infection patterns in dairy herds using machine learning. Sci Rep 10(1):1–8

Jaboob A, Durrah O, Chakir A (2024) Artificial intelligence: an overview. Eng Appl Artif Intell:13–22

Jha K, Doshi A, Patel P, Shah M (2019) A comprehensive review on automation in agriculture using artificial intelligence. Artif Intell Agric 2:1–12. https://doi.org/10.1016/j.aiia.2019.05.004

Jurgens J, Kaushik P (2024) Farmers in India are using AI for agriculture—here's how they could inspire the world. World Economic Forum. Retrieved 14 April 2025, from https://www.weforum.org/stories/2024/01/how-indias-ai-agriculture-boom-could-inspire-the-world/

Kameswari L (2023) A review on application of machine learning techniques in climate change and disaster management. Technoarete Trans Clim Change Disaster Manage Res 2(1)

Kamilaris A, Prenafeta-Boldú FX (2018) Deep learning in agriculture: a survey. Comput Electron Agric 147:70–90

Kamilaris A, Kartakoullis A, Prenafeta-Boldú FX (2017) A review on the practice of big data analysis in agriculture. Comput Electron Agric 143:23–37. https://doi.org/10.1016/j.compag.2017.09.037

Khaki S, Wang L (2019) Crop yield prediction using deep neural networks. Front Plant Sci 10:621. https://doi.org/10.3389/fpls.2019.00621

Khondoker M, Mandal S, Gurav R, Hwang S (2023) Freshwater shortage, salinity increase, and global food production: a need for sustainable irrigation water desalination—a scoping review. Earth 4(2):223–240

Kocian A, Massa D, Cannazzaro S, Incrocci L, Lonardo SD, Milazzo P et al (2020) Dynamic Bayesian network for crop growth prediction in greenhouses. Comput Electron Agric 169

Kolhalkar NR, Pandit AA, Kedar SA, Yedukondalu G (2025) Artificial intelligence algorithms for robotic harvesting of agricultural produce. In: 2025 1st International Conference on AIML-Applications for Engineering & Technology (ICAET), pp 1–6

Kymäläinen HR, von Cräutlein M, Elo K, Galambosi S, Honkanen A, Pietikäinen J, Södervik I (2024) Integrating artificial intelligence (AI) literacy into curricula: the case of agricultural sciences at the University of Helsinki

LeCun Y, Bengio Y, Hinton G (2015) Deep learning. Nature 521(7553):436–444

Li S, Simonian A, Chin BA (2010) Sensors for agriculture and the food industry. Electrochem Soc Interface 19(4):41

Liakos KG, Busato P, Moshou D, Pearson S, Bochtis D (2018) Machine learning in agriculture: a review. Sensors 18(8):2674

Ließ M, Gebauer A, Don A (2021) Machine learning with GA optimization to model the agricultural soil-landscape of Germany: an approach involving soil functional types with their multivariate parameter distributions along the depth profile. Front Environ Sci 9:692959. https://doi.org/10.3389/fenvs.2021.692959

Lin B (2025) Cow wearables help America's dairy farmers detect bird flu and other illnesses. CIO J (Wall Street J). Retrieved 7 April 2025, from https://www.wsj.com/articles/cow-wearables-help-americas-dairy-farmers-detect-bird-flu-and-other-illnesses-57470646

Lobell DB, Cassman KG, Field CB (2009) Crop yield gaps: their importance, magnitudes, and causes. Annu Rev Environ Resour 34(1):179–204

Lowenberg-DeBoer J, Erickson B (2019) Setting the record straight on precision agriculture adoption. Agron J 111(4):1552–1569. https://doi.org/10.2134/agronj2018.12.0779

Lowenberg-DeBoer J, Huang IY, Grigoriadis V et al (2020) Economics of robots and automation in field crop production. Precis Agric 21:278–299. https://doi.org/10.1007/s11119-019-09667-5

Lu Y, Lu X, Zheng L, Sun M, Chen S, Chen B et al (2024) Application of multimodal transformer model in intelligent agricultural disease detection and question-answering systems. Plants 13(7):972. https://doi.org/10.3390/plants13070972

Mahajan K, Tomar S (2021) COVID-19 and supply chain disruption: evidence from food markets in India. Am J Agric Econ 103(1):35–52

Mahlein AK (2016) Plant disease detection by imaging sensors–parallels and specific demands for precision agriculture and plant phenotyping. Plant Dis 100(2):241–251

Martin RJ et al (2024) XAI-powered smart agriculture framework for enhancing food productivity and sustainability. IEEE Access 12:168412–168427. https://doi.org/10.1109/ACCESS.2024.3492973

Mishra V, Fanase P (2025) India, the United States, and the Global AI Order. In: India's AI imperative, p 10

Mishra H, Mishra D (2023) Artificial intelligence and machine learning in agriculture: transforming farming systems. Res Trends Agric Sci 1:1–16

Mitchell TM (1997) Machine learning. McGraw-Hill, New York

Mohameth F, Bingcai C, Sada KA (2020) Plant disease detection with deep learning and feature extraction using plant village. J Comput Commun 8(6):10–22

Mohan RNVJ, Rayanoothala PS, Sree RP (2025) Next-gen agriculture: integrating AI and XAI for precision crop yield predictions. Front Plant Sci 15:1451607

Mohanty SP, Hughes DP, Salathé M (2016) Using deep learning for image-based plant disease detection. Front Plant Sci 7:1419. https://doi.org/10.3389/fpls.2016.01419

Morales IR, Cebrián DR, Blanco EF, Sierra AP (2016) Early warning in EGG production curves from commercial hens: a SVM approach. Comput Electron Agric 121:169–179

Mulla DJ (2013) Twenty five years of remote sensing in precision agriculture: key advances and remaining knowledge gaps. Biosyst Eng 114(4):358–371

Munir MS, Bajwa IS, Ashraf A, Anwar W, Rashid R (2021) Intelligent and smart irrigation system using edge computing and IoT. Complexity 2021(1):6691571

Mustafa Z, Vitali G, Huffaker R, Canavari M (2024) A systematic review on price volatility in agriculture. J Econ Surv 38(1):268–294

Neethirajan S, Scott S, Mancini C, Boivin X, Strand E (2024) Human-computer interactions with farm animals—enhancing welfare through precision livestock farming and artificial intelligence. Front Vet Sci 11:1490851. https://doi.org/10.3389/fvets.2024.1490851

Nevavuori P, Narra N, Linna P, Lipping T (2020) Crop yield prediction using multitemporal UAV data and spatio-temporal deep learning models. Remote Sens 12(23):4000. https://doi.org/10.3390/rs12234000

Nielsen MA (2015) Neural networks and deep learning. Determination Press, San Francisco

Noh SH (2021) Analysis of gradient vanishing of RNNs and performance comparison. Information 12(11):442. https://doi.org/10.3390/info12110442
Pajares G, Peruzzi A, Gonzalez-de-Santos P (2013) Sensors in agriculture and forestry. Sensors 13(9):12132–12139. https://doi.org/10.3390/s130912132
Pantazi XE, Moshou D, Oberti R, West J, Mouazen AM, Bochtis D (2017) Detection of biotic and abiotic stresses in crops by using hierarchical self organizing classifiers. Precis Agric 18:383–393. https://doi.org/10.1007/s11119-017-9507-8
Pingali PL (2012) Green revolution: impacts, limits, and the path ahead. Proc Natl Acad Sci 109(31):12302–12308. https://doi.org/10.1073/pnas.0912953109
Preite L, Vignali G (2024) Artificial intelligence to optimize water consumption in agriculture: a predictive algorithm-based irrigation management system. Comput Electron Agric 223:109126
Premachandra JSANW, Kumara PPNV (2021) A novel approach for weather prediction for agriculture in Sri Lanka using Machine Learning techniques. In 2021 International Research Conference on Smart Computing and Systems Engineering (SCSE) 4:182–189. IEEE. https://doi.org/10.1109/SCSE53661.2021.9568319
Qiao Y, Truman M, Sukkarieh S (2019) Cattle segmentation and contour extraction based on mask R-CNN for precision livestock farming. Comput Electron Agric 165:104958
Ramirez JGC (2025) Optimizing water and fertilizer use in agriculture through AI-driven IoT networks: a comprehensive analysis. Artif Intell Mach Learn Rev 6(1):1–7
Rayhana R, Xiao G, Liu Z (2020) Internet of things empowered smart greenhouse farming. IEEE J Radio Freq Identif 4(3):195–211. https://doi.org/10.1109/JRFID.2020.2984391
Ruder S (2016) An overview of gradient descent optimization algorithms. arXiv:1609.04747
Russell SJ, Norvig P (2016) Artificial intelligence: a modern approach. Pearson
Saiz-Rubio V, Rovira-Más F (2020) From smart farming towards agriculture 5.0: a review on crop data management. Agron 10(2):207. https://doi.org/10.3390/agronomy10020207
Saleem MH, Khanchi S, Potgieter J, Arif KM (2020) Image-based plant disease identification by deep learning meta-architectures. Plants 9(11):1451. https://doi.org/10.3390/plants9111451
Saleem MH, Potgieter J, Arif KM (2021) Automation in agriculture by machine and deep learning techniques: a review of recent developments. Precis Agric 22(6):2053–2091
Schieffer J, Dillon C (2015) The economic and environmental impacts of precision agriculture and interactions with agro-environmental policy. Precis Agric 16:46–61. https://doi.org/10.1007/s11119-014-9382-5
Shadrin D, Menshchikov A, Somov A, Bornemann G, Hauslage J, Fedorov M (2020) Enabling precision agriculture through embedded sensing with artificial intelligence. IEEE Trans Instrum Meas 69(7):4103–4113
Sinha M, Dasgupta T, Mitra A (2023) AI and data-driven approaches for agriculture. In: Proceedings of the 15th Annual Meeting of the Forum for Information Retrieval Evaluation, pp 158–159
Turgut Ö, Kök İ, Özdemir S (2024) AgroXAI: explainable AI-driven crop recommendation system for agriculture 4.0. In: 2024 IEEE International Conference on Big Data (BigData). IEEE, pp 7208–7217
UF/IFAS Gulf Coast Research and Education Center (2025) Center for artificial intelligence. University of Florida Institute of Food and Agricultural Sciences. Retrieved from https://gcrec.ifas.ufl.edu/research-programs/center-for-artificial-intelligence/
Vaswani A, Shazeer N, Parmar N, Uszkoreit J, Jones L, Gomez AN et al (2017) Attention is all you need. In: Advances in neural information processing systems 30, Long Beach
Weiss M, Jacob F, Duveiller G (2020) Remote sensing for agricultural applications: a meta-review. Remote Sens Environ 236:111402. https://doi.org/10.1016/j.rse.2019.111402
Wiseman L, Sanderson J, Zhang A, Jakku E (2019) Farmers and their data: an examination of farmers' reluctance to share their data through the lens of the laws impacting smart farming. NJAS-Wagening J Life Sci 90:100301. https://doi.org/10.1016/j.njas.2019.04.007
Wolfert S, Ge L, Verdouw C, Bogaardt MJ (2017) Big data in smart farming–a review. Agric Syst 153:69–80. https://doi.org/10.1016/j.agsy.2017.01.023

Xiong Y, Ge Y, Grimstad L, From PJ (2020) An autonomous strawberry-harvesting robot: design, development, integration, and field evaluation. J Field Robot 37(2):202–224

Zhai Z, Martínez JF, Beltran V, Martínez NL (2020) Decision support systems for agriculture 4.0: survey and challenges. Comput Electron Agric 170:105256. https://doi.org/10.1016/j.compag.2020.105256

Zhang N, Wang M, Wang N (2002) Precision agriculture—a worldwide overview. Comput Electron Agric 36(2–3):113–132. https://doi.org/10.1016/S0168-1699(02)00096-0

Applications of AI in Future Farming

12

Aparna G. Kumar, B. Gayathri,
and Radhakrishnan Edayileveetil Krishnankutty

Abstract

Global food production faces serious threats as the world population increase significantly by 2050. At the same time, approximately 24 billion tons of fertile soil are lost annually and water scarcity affects nearly 40 % of the global population. Here comes the relevance of artificial intelligence (AI) offers, which offers transformative solutions to these challenges by enabling smarter and data-driven farming practices. This chapter comprehensively explores how AI technologies are being applied across all stages of modern agriculture. The role of sensor networks, satellite imagery and drone surveillance to collect real-time field data, and utilisation of machine learning algorithms to process and guide precision soil management, irrigation scheduling and nutrient application are discussed in detail. Advanced computer vision systems to detect crop diseases and pest infections at early stages, for timely and targeted remedies to reduce chemical usage has also been included. AI-powered yield prediction models provide accurate harvest forecasts, while automated robots perform for planting, weeding, praying and harvesting open-up novel opportunities with minimal human labour. Environmental stability strengthening through AI-assisted water conservation, carbon footprint monitoring and biodiversity management has also been included. The chapter also addresses major adoption barriers such as poor rural connectivity and high implementation costs. Evidence from global case studies confirms that AI could increase the yield of 10-20 % water saving of 20-40 %labour reduction of 30-50 % and pesticide reduction to 40-6- %, establishing AI as an essential pillar for future sustainable agriculture.

A. G. Kumar · B. Gayathri · R. E. Krishnankutty (✉)
School of Biosciences, Mahatma Gandhi University, Kottayam, Kerala, India
e-mail: radhakrishnanek@mgu.ac.in

R. E. Krishnankutty et al. (eds.), *Smart Agriculture*,
https://doi.org/10.1007/978-981-95-8785-8_12

1 Introduction

The global community currently faces an unparalleled agricultural challenge as the global population estimate reaches 10 billion by 2050, a significant increase from current numbers (Le Mouël and Forslund, 2017). This requires a profound transformation in the food production systems to meet this rising demand because the traditional farming methods alone will not serve the purpose where climate change has drastically affected growing conditions through rising temperatures, altered rainfall patterns, and more frequent extreme weather conditions (Hillier et al. 2018). These shifts have shortened the growing seasons and reduced crop yields; drought now affect around 1.5 billion people while floods destroy millions of hectares of agricultural land each year. Even a narrow temperature increase of 1–2 °C can reduce the yields of major staple crops by 10–25%, placing billions at the edge of food insecurity (Hu et al. 2024; Mirzabaev et al. 2023).

Thus, Agriculture is facing serious challenges. In the last 40 years, about one-third of the world's farmland has been lost due to erosion, pollution, city growth, and damage to the soil (Borrelli et al. 2017). Every year, around 24 billion tons of fertile soil are lost. Water shortages affect nearly 40% of people worldwide, and finding enough workers for farming is becoming harder. Many people are moving from villages to cities, and older farmers are retiring (Chen et al. 2025). At the same time, fewer young people are choosing farming as a career. Because of this, it's difficult to expand farmland in a way that's good for the environment and affordable. To keep farming going with fewer workers, many countries are turning to machines and new technologies based approaches (Sharma and Shivandu, 2024). Tools like soil sensors, weather trackers, satellites, drones, and smart farm equipment help farmers to understand their fields better. These tools allow farmers to care for each plant more precisely, use water and fertilizers wisely, and treat different areas of soil based on specific needs. The focus of the chapter is to explore automation and robotics as part of the mechanization shift worldwide and consider ways to integrate AI across supply chains evaluating contributions to environmental sustainability and the economic impacts of adopting AI (Dabic-Miletic 2024).

2 The Architecture of AI-Driven Agriculture

The AI-powered farms utilize extensive data-collection networks to monitor various aspects of agricultural operations in a modern agricultural farm allowing optimization of farm management and enhancing productivity (Wei et al. 2023). The use of environmental sensors, which include weather stations that continuously measure physical parameters such as temperature, humidity, precipitation, wind speed, solar radiation, and atmospheric pressure at regular intervals allows better farm management (Benos et al. 2021). The soil monitoring networks also deploy sensors at regular intervals to assess the soil moisture, temperature, pH, electrical conductivity, and nutrient levels for meeting the changing soil requirements (Islam et al. 2023). Also, certain specialized plant health monitors also adds to modern agriculture by monitoring chlorophyll content, the temperature of the leaves, and a set of other indicators of plant health (Sishodia et al. 2020). Remote sensing system, such as drones, and satellite can provide some useful imagery across the electromagnetic spectrum, and establish a measure of utility or value for the assessment of crop growth in this multispectral imaging process (Tsouros et al. 2019). Furthermore, remote monitoring of machinery can also positively impact productive efficiency of farm practices, as mechanization such as GPS-enabled farm equipment collects valuable information such as location, positioning, fuel usage, and performance (Shrivastava et al., 2023).

Using sensor data, together with cloud-based analytics, have taken this further down through digitization processes using computational methods to provide analysis on very large datasets of information to identify patterns in many cases, and build predictive models (Khanna and Kaur 2019). Furthermore, machine learning pipelines can automate the training and fine-tuning of artificial intelligence (AI) models, and allow for human involvement as needed to ensure a model is still fit-for-purpose on a regular basis over time (Liakos et al. 2018). An important characteristic is that the farmers are presented with the complex data in the common field, using visual representation or dashboards that are readily available, to make informed decisions concerning various aspects of producing crops (Farooq et al. 2019). This supervised learning, examples of which are disease classification systems, that uses computer vision to elicit information on the presence of plant pathogens and nutrient deficiencies, and the impact of these on the whole field (Jiang et al. 2019). In addition, the quality grading system compares crop quality against market standards, both visual and chemical, and price forecasting models predict prices based on supply and demand fundamentals (Saleem et al. 2019).

The following unsupervised learning approaches could advance the agricultural analyses by finding patterns in data, without outcomes defined by a user. For a best example, the use of clustering algorithms used to identify the management zones within fields based on various factors (Benos et al. 2021; Gavioli et al. 2019), while anomaly detection systems pinpoint unusual patterns in crop growth or equipment performance, signaling potential issues in the field. In addition, pattern recognition algorithms are leveraged to identify the relationships and associations between environmental aspects and crop responses, that had not previously been established on the farm level (Liakos et al. 2018; Gautron et al. 2022).

The advanced uses of AI-based Reinforcement Learning, optimizes decision-making based on the principles of trial and error (Ahmad et al. 2022). For example, irrigation optimization systems use trial and error to learn and determine an effective irrigation strategy by implementing different strategies into an intelligent-learning framework to measure the allocation through the crop (Saikai et al. 2023). Likewise, the applications of AI in pesticide management identify the most effective time to treat, the dosage required for treatment, and guides the timing and the systems for harvest adoption, ensuring maximum yield, quality, and profitability (Logeshwaran et al. 2024).

The agriculture sector has also been transformed through the merger of computer-based vision and imaging techniques such as multispectral and hyperspectral imaging, allowing sophisticated identification of crop health, that is, spectral traits which are not visible to the human eye (Lowe et al. 2017; Ram et al. 2024). Visibility spectrum examination and analysis allow for the examination of plant health with respect to color analysis, growth stages, and detection of weeds. The capacity of near-infrared analysis has additionally enabled the computation of vegetation indices and biomass estimates, as well as evaluation of water stress and protein concentration while thermal infrared provides information on plant temperature and potential disease markers (Behmann et al. 2015; Dolatabadian et al. 2025). The ability to use complex algorithms, including convolutional neural networks (CNNs) primarily to identify plant disease has created AI image analysis (Nagasubramanian et al. 2019; Ahmad et al. 2022). Advanced object detection methods, such as YOLO and the R-CNN series can identify and locate many objects in a single image (Badgujar et al. 2024; Chen et al. 2019) and newly standard image segmentation methods were able to classify imagery at the pixel level facilitating recognition of plant components, soil types, weeds, and diseases (Sapkota and Karkee 2025). Local processing of this data is enhanced through edge computing that relieves bandwidth burdens and can respond to critical conditions faster (or more automatically) without cloud data connection while also ensuring that the data is filtered and transformed prior to its transmission, facilitating autonomous operation if a connection is lost (Sathya et al. 2024; O'Grady et al. 2019). By exploiting the Cloud Integration System to secure data transmission and storage, a real-time synchronization across devices, historical data analysis, and the integration of weather services is maintained (Dozono et al. 2022; Chergui and Kechadi 2022).

In a nutshell, the architecture of AI-driven agriculture is characterized by a sophisticated interplay of the data collection (Dommetti et al. 2025), processing (Rugji et al. 2024), and analytical technologies enhancing operational efficiency empowering farmers to make data-informed decisions and driving productivity and sustainability in the agricultural sector (Weraikat et al. 2024).

3 Precision Agriculture—Smart Farming in Soil Management and Optimization

The most crucial milestone of precision agriculture is the AI-powered soil management (Fig. 12.1), which enable farmers to analyze and optimize soil conditions with best precision (Wadoux et al. 2023). Soil analysis systems are being applied to produce large-scale, deep maps characterizing vertical variation in soil properties in agricultural fields (Khosla et al. 2019). Maps are site-specific analyses such as pH profiling and nutrient analysis, and have value in identifying areas for lime application to correct pH and ameliorate nutrient availability (Nawar et al. 2024).

In addition to mapping acidity, we can determine the concentrations of major nutrients (nitrogen, phosphorus, potassium, and key micronutrients) as well as organic matter (Hengl et al. 2017). This helps farmers to assess soil health,

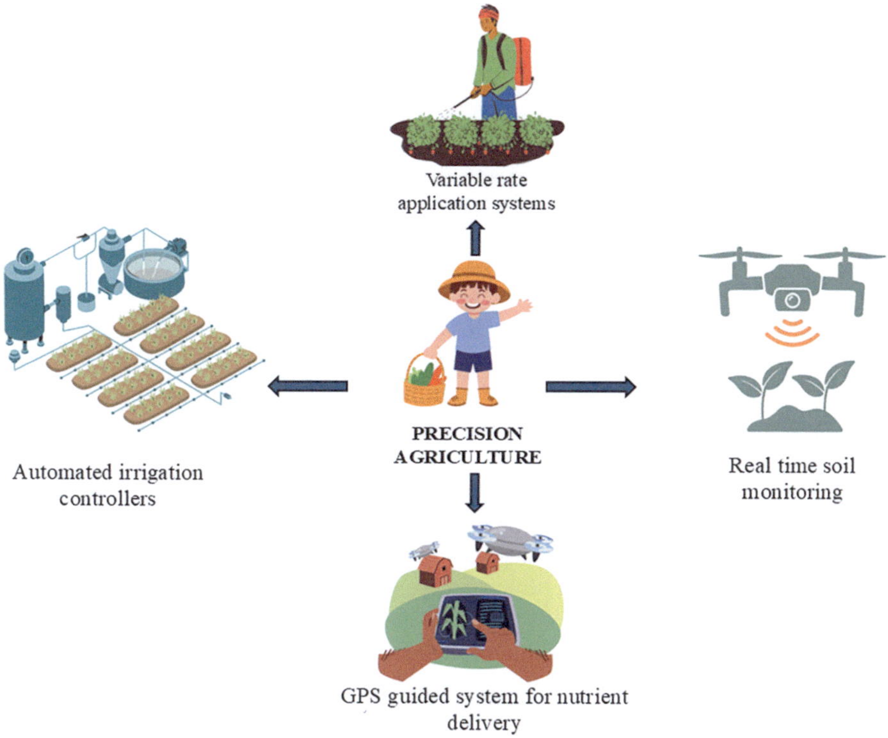

Fig. 12.1 Application of Precision Agriculture Technologies for Enhanced Crop Management and Sustainability where advanced technologies and data-driven systems are used to optimize farming practices at micro levels. The approach aims to increase productivity, reduce input waste, and promote environmental stewardship

extrapolate carbon sequestration potential, and evaluate water-holding capacity and drainage characteristics (Franzluebbers 2020). This information promotes on-farm, site-specific management to support increased productivity while maintaining soil fertility over the long-term (Kumar et al. 2021a, b).

Real-time soil systems tend to be oriented toward continuously observing changes in soil conditions, collecting dynamic processes and data, and supporting advisories to facilitate decisions (Sharma et al. 2024). For example, moisture management is an integral aspect, and multi-depth sensors are used to measure soil moisture at delineated intervals. These values allow predictive models to be used that predict moisture changes in accordance with prevailing weather patterns, enabling interventions at the right time (Zhang et al. 2023). Automatic irrigation systems work on moisture thresholds; making sure they do not allow the moisture levels to fall below the threshold while understanding water logging and drought stress specific for that crop (Kumar et al. 2023). In addition to these moisture conditions, real-time monitors continue to monitor nutrient cycling. It tracks nitrogen's mineralization rate from organic matter, phosphorous availability based on soil pH and moisture conditions, and is aware of potassium leaching through subsurface layers (Brady and Weil 2016). They use this to maintain that the appropriate levels of nutrients are being given to the crop, in a timely manner, and achieve an elevated point of growth while not allowing nutrients to be wasted by flushing through leaching excess moisture (Liu et al. 2020).

Variable rate application systems are able to adjust input levels prescribed by soil conditions based on a real-time context. For example, instead of simply using GPS to contextually alter fertilizer use, they may provide few meter-changing rates to utilize prescription maps that have been developed by using soil test and crop nutrient uptake information (Mulla 2013). The same principle would apply to liming for a pH correction. Organic matter may be added at levels determined, based on soil health assessments, while gypsum are only applied on sodium-impacted soils, or at a time when micronutrients are likely to be deficient in plant growth. Collectively, these systems help the farm to maintain soil fertility that will enable maximum crop yields (Schepers et al. 2004).

Irrigation systems in water-scarce regions maximize water utilization while ending with productivity. Improved evapotranspiration models utilize a series of environmental parameters to inform crop water needs such as temperature, humidity, wind speed, and solar radiation (Allen et al. 1998). The irrigation schedule can vary depending on the growth stage of the plant and how much water is available in the soil which is affected by the texture and organic matter present (Pereira et al. 2015). Selecting irrigation systems to maximize efficiency based on water loss through transpiration as measured by wire frame development and leaf coverage, can provide greater irrigation reliability (Evans et al. 2013).

Greenhouse settings and systems manage temperature by heating and cooling versus ventilation. Limiting high humidity is also required to limit disease outbreaks (Hoseinzadeh and Garcia 2024; Hemming et al. 2019). CO_2 enrichment within greenhouse enhances photosynthesis, however some adaptations and light-management periods are required for low-light periods (Hoseinzadeh and Garcia 2024). Open field environments use strategies such as windbreaks and mulch to limit evapotranspiration and temperature and moisture management. Shade management strategies are also available during hot periods and microsprinkler system also have frost protection in crops (Cao et al. 2022). Nutrient management is significantly enhanced through comprehensive nutrient planning powered by AI (Acharya et al. 2024). Utilizing advanced models leads developers to calculate more accurately the specific nutrient requirements of plants based on crop removal calculations including a nutrient uptake template for each species with crop yield goals (Sharma et al. 2024). For example, assessing estimated growth stage-specific uptake rates of nutrients as part of an evaluation of harvest indices (Wu and Ma, 2015), as well as considering economically viable situations of plants grown together and spatial variability of nutrients due to poor nutrient allocation, for example, in effluent zones, leads to clarity and provides immediate value the farm system. These considerations are represented through estimates of soil nutrient supply where nutrient supply is tested in soil-available nutrients, calculations of organic matter mineralization, and understand supports for previous crop residues, manure, and compost

(Senapaty and Ray, 2023; Soy and Prakash 2025). The purpose of Variable Rate Fertilizer in open field, is further supported through GPS-guided applications where systems can deliver nutrients only to the spatial areas required as opposed to blanket application to the entire open field (Hemming et al. 2019; Hoseinzadeh and Garcia 2024). Prescription maps are generated to maximize the effectiveness of the work derived from grid sampling of soil and then used with real-time sensor information that supports zone-based management of nutrients to be applied to similar soil areas. Genetic optimization algorithms have improved the efficiency in which fertilizers can be allocated (Mishra and Mishra 2024). The timing of application of nutrients/fertilizers are strategically planned around weather conditions, crop growth stages, and soil temperature and moisture levels to minimize potential losses (Dommetti et al. 2025; Wu and Ma, 2015).

The use of organic and biological fertilizers in the crop field could also be optimized through AI-powered systems. Composting processes are refined and maintained by monitoring temperature and moisture levels, assessing microbial activity, and predicting nutrient content, which informs the optimal application timings. The management of biological fertilizers relies on the assessment of the growth phase of soil microbiomes to establish the best time to apply beneficial bacteria and fungi (Balaska et al. 2023; Javaid et al. 2022).

In conclusion, precision agriculture relies on advanced technologies to manage the use of soil, water, and nutrients in a field. This full-farm approach offers benefits to yield alongside alleviating and promoting sustainable farming to meet the challenges of future population.

4 Monitoring and Advanced Plant Health Assessment for Crops

Advanced crop monitoring systems are integrating a range of technologies that not only facilitate a comprehensive evaluation of the plants' health but also help to detect problems that may not be visible to the naked eye. Advanced crop monitoring is supplying a multi-modal health-assessment toolbox that has included spectral analysis and computations of various vegetation index in order to assess plant health and crop yield (Xue and Su 2017). One of the most common vegetation index was termed the Normalized Difference Vegetation Index (NDVI) that assesses the reflected red and near-infrared energy. It is typically stated from 0.2 to 0.9 indicating a progression from unhealthy vegetation to healthy vegetation. From the perspective of advanced yield prediction and the management of nitrogen, NDVI helps farmers optimize fertilization strategies for optimal plant growth (Rouse et al. 1974). Enhanced Vegetation Index (EVI) specifically would be of value in relation to the coverage of vegetation accessing any changes, while also supporting the evaluation of the rate of change of plant health (Huete et al. 2002). The Green Normalized Difference Vegetation Index (GNDVI) assesses chlorophyll content of vegetation based on measures of vegetation reflectance, using green light rather than red light, providing a much more valuable index for estimating chlorophyll content as it relates to the tracking of nitrogen deficiency as well as assessing the stages of growth (Gitelson et al. 1996). The Photochemical Reflectance Index (PRI) is the closest to a plant stress indicator due to light use efficiency and is an indicator of photosynthetic efficiency, which can also indicate water stress (Gamon et al. 1992). Together, these indices offer complete insight on the physiological status of a plant and allow for proper management decisions (Sudu et al. 2022).

Thermal imaging technology has increased the measurements to additional metrics such as using temperature differences in plant measurement to quantify plant stress condition based on temperature. An example of this is Crop Water Stress Index (CWSI) to determine potential water stress using canopy temperature measurements to help to schedule irrigation (Cohen et al. 2021). Other means of thermal imaging methods can also measure temperature differences related to diseases, including allowing for early detection of pathogens before visible symptoms are found on the plant. The thermal patterns can also measure heat stress, and pest damage, when it is indicated for possible action to be taken on the crop (DeJonge et al. 2015).

Chlorophyll fluorescence, and as well light-adaptation analyses, is a highly technical analysis, monitoring photosynthesis efficiency in field agriculture settings. Chlorophyll fluorescence consists of methods to measure the efficiency of Photosystem II through quantum yield measures, also photochemical and nonphotochemical quenching is measured (Murchie and Lawson 2013). The measures resulting from analyses can be converted to quantify stress responses (Kalaji et al. 2016). Recent advances in portable fluorometers have enabled field-based assessments of photosynthetic performance under varying environmental conditions (Baker 2008).

4.1 Disease Detection and Management

The application of computer vision-based disease identification systems greatly boosts the capability for accurately detecting plant diseases to a level which would far exceed human capability. The use of deep learning techniques of convolutional neural networks enables the construction of very large datasets of plant disease images which can be used to identify all types of common crop diseases, including fungal diseases such as rust and blight, bacterial diseases such as crown gall and fire blight, viral diseases such as mosaic viruses, and concurrently identify nutritional disorders associated with nitrogen, phosphorus, and potassium deficiencies (Too et al. 2021). There have been recent breakthroughs in deep learning models that accurately detect plant diseases with many models providing better than 99% accuracy on relatively large datasets with thousands of diseased and healthy plant images (Mohanty et al. 2016).

This ability to monitor the progression of diseases over time and through multiple stages of development is very valuable for good crop management. Automated monitoring systems used in the field can continually assess the severity of infection, track the extent of its movement in space and time, identify the primary foci of infection, and describe features of the disease trajectory pathophysiology (Bock et al. 2021). By enabling continual surveillance of disease, timely and targeted intervention can improve crop health and crop yield (Mahlein 2016). Today, automated detection systems can take this type of surveillance further using machine learning and deep learning approaches that monitor disease in real time and allow for more rapid, accurate interventions through continual surveillance (Nawaz et al. 2025).

Drone-based surveillance of disease monitoring is also a valuable element of real-time disease surveillance, in which regular flights high-resolution images can be taken every day or week of the entire field, which can then be processed through automated systems. Integrating drone data with sensor networks, including plant-based sensors that detect microclimate conditions and predict disease risk is the next logical step (Rahman et al. 2025). The enrich application of Leaf Wetness Sensors and spore traps to provide important information for management plan development takes that even further. The predictive disease modeling utilizes the potential of AI systems to forecast infectious disease outbreaks before they tend to occur by assessing the environmental factors such as temperature and humidity (Sahoo et al. 2024). Such models can also predict the risk of diseases, particularly fungal infections influenced by rainfall patterns and spore dispersion. The epidemiological modeling further deepens the understanding of plant diseases by calculating disease spread rates and assessing the dynamics of pathogen populations, which is crucial for determining economic thresholds for intervention (Akter et al. 2024; Mohyuddin et al. 2024).

4.2 Pest Monitoring and Management

The automated pest detection systems harness the AI-powered technology to provide early warning signals for insect attacks, significantly improving pest management strategies (Kariyanna and Sowjanya 2024; Venkateswara and Padmanabhan 2025). The advanced imaging systems facilitate the identification and counting of pest insects through use of computer vision techniques by performing automated analysis of pheromone trap images for species identification and population trend analysis (Hong et al. 2020, 2021) along field-based detection methods that recognize crop damage patterns and feeding signs of insects.

Machine learning models have also assisted in pest modeling predictions via the analysis of phenology, in the case of pest predictions predominantly tracking degree-day accumulation to enable staging of insect development, predicting generation timing, peak activity periods, and eventual overwintering survival rates (Grünig et al. 2021). The population dynamics modeling can be used to examine the relationship between predators and pests on a crop land spatially defining how pest populations are influenced through management, immigration and emigration pattern, reproduction and environmental factors which could inhibit or reduce pest populations (Duffy et al. 2022).

The strategy to embed Integrated Pest Management (IPM) strategies are very much enhanced in the future and optimized with AI systems that include both biological and chemical control methods to maximize the level of effectiveness. Artificial intelligence (AI)-based modeling can notify farmers the best time to release beneficial insects and notify habitat manipulations to support the introduction on natural enemies to the pest, effectively helping in pest management (Kogan et al. 2019). Chemical control measures that are extensive and optimized use of AI modeling by optimizing pesticide resistance management, pesticide application timing, and pesticide drift mitigation that produces an effective, synergistic, and sustainable pest management program (Angon et al. 2023). Recent advances in machine learning algorithms have enabled more accurate prediction of pest appearance using environmental sensors, taking into account air temperature and relative humidity for daily pest occurrence forecasting (Rustia et al. 2021). Artificial intelligence (AI)-enhanced IPM systems now provide comprehensive approaches that combine biological, cultural, and chemical control strategies while maintaining ecosystem health and reducing reliance on synthetic pesticides (Singh et al. 2024).

5 Harvest Planning and Yield Forecasting

Modern yield prediction systems are delivering accurate harvest predictions using advanced data-driven, data-fusion technologies. Many of these systems employ multi-model frameworks to improve crop yield estimates, leveraging different types of machine learning (Van Klompenburg et al. 2020). Below are the most practical applications of these types of application models. For example, ensemble models such as Random Forests are extremely successful as they deploy several decision trees to summarize how complex, nonlinear associations exist in agricultural datasets (Jeong et al. 2016). Similar to these application models emphasizing which factors matter for yield, actionable insights are provided. These insights guide farm managers and growers to make informed and timely decisions that support the decision-making process (Shahhosseini et al. 2021). All of these application models are built on robust models that typically exhibit accuracy levels between 85 and 90% for major crop types and growing conditions, across growing seasons (Liu et al. 2024). Recent advances in multi-modal data-fusion techniques have further enhanced prediction accuracy by integrating diverse data sources including satellite imagery, weather data, and soil information (Chen et al. 2025).

Time series forecasting and seasonal decomposition methods complement all farm management systems and are useful methods to see yield trends incorporated into adjoining growing seasons and help to map crop yield trends against seasonal variation. Time series allows for a decomposition of yield trends over the long-term to understand the contributing cycles to productivity (Roberts et al. 2017). Seasonally tracking a succession of trends can help forecasting which make sense of the mapping for both farm managers and growers to equip the management practices to accommodate yield variations, ranging from planning farm inputs to informing farm actions (Van Klompenburg et al. 2020). The updates of crop yields made available in real time through forecasting processes allow for weekly updates to yield

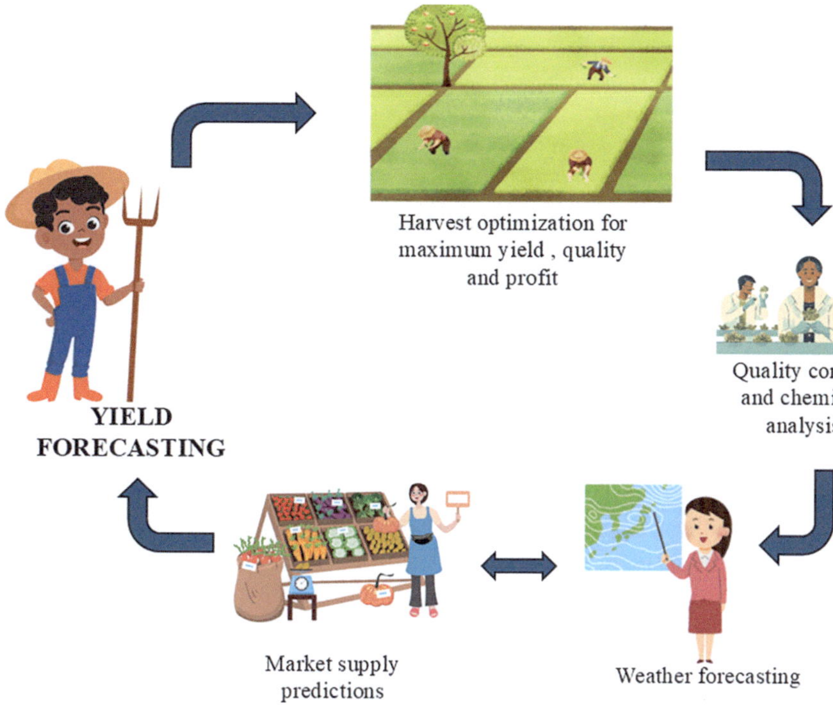

Fig. 12.2 Practical Applications of AI-Enhanced Yield Forecasting Systems in Agriculture where the integration of artificial intelligence into yield forecasting systems, showcasing how these technologies transform agricultural decision-making and productivity. Artificial intelligence (AI)-powered forecasting tools leverage real-time data and predictive modeling to deliver actionable insights across the crop lifecycle

forecasts, which incorporate prediction approaches by growth stages of the plant and weather influences to deliver site-specific management practices to update practices on the field according to the latest data (Fig. 12.2) (Elbasi et al. 2023).

For all of the possible yield prediction systems, it operate over various spatial and temporal resolutions and can generate a yield mapping for various fields at a site. This granularity allows for the identification of within-field variations and management zone-specific forecasts, which are crucial for efficient harvest logistics planning (Nikhil et al. 2024). On a broader scale, regional and national forecasting aggregates production estimates, aiding in market supply predictions, export potential assessments, and supporting policy-planning efforts (Van Klompenburg et al. 2020).

5.1 Optimization of Harvest Based on AI Systems

The regular optimization of harvest timing is critical for ensuring maximum yield with excellent quality and profitability. The AI systems are proficient at determining the precise moment for harvesting the crops by using various technologies to assess crop maturity and quality. The visual assessment and the automation employed color analysis are important for evaluating the ripeness of fruit, while the size and shape measurements ensure optimal development before the harvest followed by detection of defects and quality grading (Wang et al. 2019). Chemical maturity analysis through near-infrared (NIR) spectroscopy enables a sophisticated measurement of the sugar content, moisture levels, oil content, and the levels of protein in crops postharvest (Porep et al. 2015). The assessment of physical properties (firmness, density, texture, and size distribution) is an important component, as one needs to ensure that the crops meet market grading standards and are acceptable for processing (Baiano 2017). Advanced NIR applications have demonstrated exceptional accuracy in grading cereals according to protein content at harvest time, enabling premium pricing for higher quality grains (Jiang et al. 2022).

Weather factors can influence decisions made regarding the timing of harvesting; the modern AI-based technology application is facilitating real-time weather forecasts for better management of harvest conditions, mainly regarding timing of harvest (Sharma et al. 2022). For example, the applications provide an immediate weather forecast that assists growers to find extended rain-free windows at harvest, condition of soil moisture, and saturated weight, so machinery can get off and on wet areas, and field drying potential in real-time; all of which assist with planning and maintain quality (Elbeltagi et al. 2021). While the weather risks cannot be avoided by the farmer, with the appropriate application of real-time demand forecasts, the farmer can change their plans and work with field conditions to avoid a harvest disaster and establish suitable alternative harvesting plans based on market demand and their field conditions.

This proactive strategy includes planning for storage, figuring out optimal resource use in hurdles of harvesting at once, and timing the ultimate market opportunity for latest returns. Another beneficial function of AI application is that it can automate harvest coordination which manages complex steps during the busiest time of a farm's year. There is so much to schedule with harvest, the logistics of equipment, labor, transport, and storagethat applying AI based systems analysis would prove valuable to the characters who want to employ the most efficient combinations of available resources, as harvesting must occur in a timely and well managed manner (Marinello et al. 2017). Secondly, quality control is paramount to maintain product quality standards and therefore quality control, or inspection, is essential during the harvest process. There are two advantages to quality control, or inspection, during the harvest/quality control phase. The first and foremost benefit of harvesting quality control or inspection is real-time inspection for product specifications (Verdouw et al. 2016). Automated grade segregation and quality as well as prevention of contamination hazards, in order to enhance quality standards, can be done by automated separation of quality grades as well as preventing the risks associated with contamination. Next, the value of traceability systems for accountability through the supply chain and enabling professionals to properly manage and organize the postharvest produce (Charlebois et al. 2014). Together predictive practice and appropriate harvest planning reflect quality yield forecasting to improve agricultural productivity (Bacco et al. 2019a, b). Recent stored data and IoT-enabled coordinated supply chain collaborative achievements benefitted from advances in harvest management coordination with downstream processing and distribution networks (Balaji and Arshinder 2016). In addition, automated harvest systems that utilize robotics and AI technologies can identify and harvest crops at their optimum maturity while continuing to make harvest decisions to maximize operations by reducing postharvest losses (Sishodia et al. 2020). The establishment of blockchain-based traceability systems provides new ways of ensuring postharvest quality management, with its immutable records of crop handling, from hands on the farm to the consumers (Zhang et al. 2022).

6 Autonomous Field Operations Using Automated Robotics in Agriculture

The core of advancements in robotics are the autonomous tractors and equipment designed to navigate complex agricultural environments with remarkable precision (Fountas et al. 2020a, b). Global Positioning System (GPS) localization combined with Real-Time Kinematic (RTK) positioning, or RTK-GPS, is an essential technology for modern field operations with centimeter-level accuracy (Keicher and Seufert 2012). Centimeter-level precision will facilitate consistency of field practices, automated boundary detection, and reliable avoidance of collisions ensuring safe machinery operations (Emmi et al. 2014). Like GPS-based systems, computer vision is also important for navigation (Kamilaris and Prenafeta-Boldú 2018). Whether on different parts of the field, or in specific zones, farmers can specify individual seed varieties and even control how deep a seed is planted based on the type of soil. When combined with precision planting and cultivation technology, the benefits to modern-day agriculture are immeasurable (Lowenberg-DeBoer et al. 2020).

With advanced planting, cultivating, and transport technologies, there are also improved precision in seed placement, using differentiation of how many seeds are planted through specific areas across the field, where the soil has different conditions, enabling more rapid production (Bacco et al. 2019a, b). Among the advantages mentioned above, SLAM also fits into such systems, enabling features such as simultaneous construction or mapping of field locations, dynamic obstacle recognition and avoidance, and ideal path planning or adjustments, and is advantageous if coordinating multi-machine autonomy in the field and enables improved operational efficiency (Cheein and Carelli 2013). While planting, the planting technician can account for moisture level in the soil to reduce compaction, manage leftover crop material, and delineate seed depth or placement (Koller and Lanini 2005). These conditions, along with appropriate quality checks throughout the procedures can ensure planting is accomplished to the highest level. Likewise, automated and semi-automated cultivation and weeding processes are enhancing the precision and accuracy with which farmers can perform field work (Fennimore and Cutulle 2019). Machines will facilitate navigation between rows of crops, while calculating the optimum depth based on constituents in the soil. Machines can also account for speed of travel, so as not to damage adjacent crops. For weeds within the rows, dedicated systems can identify and remove each weed individually, using a mechanical tool that quickly detaches weeds by the root (Young et al. 2014). This process can also be extended to spraying herbicides, with less chance of damage to crops and less chemical usage than a blanket application (Gerhards and Oebel 2006).

6.1 Robotic Harvesting Systems

Robotic technologies are contributing to the harvesting of special crops by improving efficiency and discontinuing labor (Zhang et al. 2020). In apples, specialized equipment with several three-dimensional vision systems locate and assess the quality of the fruit by way of color. These vision systems are designed to pick apples without excessive force, and consequently, no risk exists for physical injuries (Silwal et al. 2017). Sort, pack, and complete, including postharvest sorting and packing to meet market requirement, all the processes can be incorporated.

Strawberry harvesting technologies have several robotic arms which are synchronized to harvest berries according to handling sequences designed to prevent high quality losses (Xiong et al. 2020). The machines operate continuously and can achieve high yields during peak harvest season. Similar technologies exist for tomato harvesting machines, which are being used in greenhouse and field productions. The tomato machines use a detection mechanism for ripeness and delicately cut the stem for high quality harvest.

Modern combine harvesters for grain production are modern advanced machines (Spekken et al. 2013). These complex finely tuned machines can vary many operational requirements (fan speed, sieve configurations, rotor speed, etc.) based on the crop. The effort to adapt equipment for the best harvest will also improve grain losses as well as consumers' and producers' quality expectations. Combine harvesters have field measurements (e.g., grain flow to tank, yield mapping, moisture, protein, etc.) which helps to inform growers on total productivity (Leroux et al. 2018).

6.2 Robotic Milking Systems

Automation has advanced very quickly in order to enhance efficiencies in dairy operations and animal management systems (Berckmans 2014). Automated milking system that can identify the cow, provide health checks of udder health, and monitor quality aspects of the milk during milking (de Koning and van der Vorst 2017). This allows the dairy producer to have improved efficiency and productivity, while managing to protect or maintain the health of their livestock and dairy quality.

There are many different information systems that can provide an overview of integrated observations of the activity monitoring, chewing activity, and reproductive activity (Dolecheck et al. 2015). The information system will allow for the detection of changes in those behaviors that may provide the earliest warning signs of things like lameness, so that the farmer can take prompt and careful interventions to assist with maintaining health and performance in the herd (Rutten et al. 2013).

6.3 Precision Application Robotics

Robotic systems are benefitting in improving the accuracy and management of pesticide and fertilizer application (Giles and Billing 2015). As it pertains to the new delivery systems of pesticides, they can identify individual weeds by species and apply pesticides with the appropriate dosage, instead of a blanket application for the treatment of multiple species independently (Dammer and Wartenberg 2007). This has the ability to improve chemical dispersal, and the effectiveness of treatment, while minimizing environmental impact and damage to beneficial organisms involved in pest management (Miller and Butler Ellis 2000). The same approach applies for fungicide applications, in that the precision of application can be restricted to only areas where disease symptoms are expressing themselves, and the application itself would take into consideration disease resistance, and the increase in areas called thresholds that can be minimized to provide inputs (Oerke 2006).

The same thinking applies with fertilizer applications, as newer pieces of equipment are designed to apply nutrients to the specific areas and not in general, but plant-specific (Mulla 2013). In many instances, the same equipment is used more stringently to combine and apply fertilizer mixtures directly to the plant root zone, therebyenhancing the plant growth and nutrient uptake. With the benefit of soil sensors, if the management put in an additional component in providing information assistance, it can fit back in to management decisions. (Corwin and Lesch 2005).

In many instances, input efficiencies and the overall efficiencies of a system have improved due to best management practices on agricultural systems, in agriculture-supporting robotic systems (Pedersen et al. 2006). It is being undertaken at the industry level and with nearly all types of agriculture. These continued developments in automation and robotics will enhance efficiency not only today, but also in future (Bechar and Vigneault 2016). Autonomous machinery, precision harvesting tools, and targeted input systems are all contributing to a transition that can facilitate more organized and sustainable agriculture (Stafford 2000). Additional innovations in robotic navigation and control systems continue to enhance field operations (Reid et al. 2000). Advanced sensor technologies are enabling real-time monitoring and adjustment of application parameters (Adamchuk et al. 2004). Machine learning algorithms are being integrated to optimize treatment decisions based on field variability (Kerry and Oliver 2004).

7 Environmental Monitoring and Sustainability

Environmental monitoring systems are well-established as decision-support systems to accompany information management on the conditions of the field that supports plant growth, yield, and development (Leroux et al. 2018). Environmental monitoring systems collect field-related information based on a sensor network design, using sensor technologies with field-related measuring accuracy and spatial and temporal resolution that allow for "local" weather forecasts on a finer scale (Ojha et al. 2015). This is compared to local area weather forecasting in agriculture. The benefit of being able to collect and capture real-time (or near real-time) data means that with the specific implementation of decision-support, there is expected to be increased realism in making decisions to improve productivity based on a physical weather system and resource management, and specifically climate. Understanding first local, rather than general weather conditions, is a relatively major advantage of these types of systems.

Environmental monitoring systems provide information on where microclimates exist within the field (Kalma et al. 2008). Microclimates refer to small-scale atmospheric variability within a field that can affect a crop's development. The types of microclimates measured include, but are not limited to, the air and canopy temperatures, soil temperatures at various depths, and agro-climatic zones defined by heat accumulation thresholds, with forecasts of early frost being issued (Snyder et al. 2005). Monitoring humidity throughout the height of the crop also enables Vapor Pressure Deficit (VPD) calculations to estimate plant water needs and is used as a threshold for understanding the length of time leaf surfaces are wet in terms of the increased likelihood of a fungal infection (Sentelhas et al. 2008).

Evaluation of wind patterns is another valuable component (Aylor 2003). Sensors measure wind speed, direction, and turbulence and use them to help guide the sprays, recognize the benefits of wind as a source of pollination for wind-pollinated crops, and assess airborne potential for infecting crops with disease. Air-quality measurements also assist in this aspect by measuring dust, ozone, and carbon dioxide levels (Ainsworth et al. 2012). The monitoring of CO_2 is particularly important in controlled environments to maximize photosynthesis, whereas the modification of air pressure may indicate plant responses to stress (Hatfield and Prueger 2015). All of these measurements provide an overall view of the condition of the crop as well as the surrounding environment (Steenwerth et al. 2014). Advanced wireless sensor networks enable continuous data collection across large agricultural areas (Morais et al. 2008). Integration of multiple sensor types provides comprehensive environmental characterization for precision agriculture applications (Bramley and Hamilton 2004).

7.1 Water Quality and Conservation

Water management is a vital part of sustainable agriculture, with emerging technologies helping to monitor and manage water (Xia et al. 2020). The quality of irrigation water remains of primary importance as water usability can be verified by monitoring the salinity and pH levels of water to assure crop usability without plant damage, protecting plants, and preventing long-lasting negative impact to the soil (Rhoades et al. 1992). The systems will reveal and monitor nutrient charge in the water and transportation of water in and out of the land, as well as serve as an early warning device for chemical or biological contaminants for crops and the natural surrounding environment. By monitoring runoff and drainage, the farmer can measure the amount of quality water that enters his/her field, if they are losing significant nutrients, for example, nitrogen or phosphorus, the soil particles that are leaving the land, and the potential for erosion of their land (Li et al. 2011). This information will play a key role in forming a plan to manage the pollution of local rivers and lakes (Zhang et al. 2014; Harmel et al. 2009).

7.2 Soil Health Monitoring and Carbon Sequestration

Sustainable agriculture relies on long-term observations of soil conditions (Suman et al. 2022). One important aspect is changes in organic matter and the levels of carbon held in the soil. Knowledge of how organic materials decompose, and how much carbon the soil can hold, is valuable in judging what level of sustainability farming practices are able to maintain over time with respect to the health of the soil and long-term productivity (Bai et al. 2023). Also, monitoring biological activity reflects the functional aspects of soil. Biological activity is comprised of the microbial diversity, the diversity of population, the diversity of enzyme activity (to inform functions), and the functional efficiency of nutrient recycling (García-Orenes et al. 2013; Nannipieri et al. 2017). Together, biological activity reflects the overall condition of the soil to help informed decisions, maintain health, and stability of crop yield

7.3 Sustainable Management of Resources in AI-Powered Field

A. *Water Use Efficiency Optimization*

Water Use Efficiency (WUE) is a measure of the total quantity of plant material produced compared to the quantity of water used, and farmers can use the tracking of WUE to compare different types of crops and how they affect water usage under different farming methods (Hatfield et al. 2019). Those differences may also provide quite a bit of education, particularly when farmers select crops and are consciously working with the limited water that is available to promote varieties of crop development (Sinclair et al. 2005).

Another key practice is called deficit irrigation where potential crop water uptake is intentionally reduced during developmental stages of crop growth that are not as critical to plant growth, while water availability is ensured during very critical stages of developmental conditions, such as flowering of the crop (Fereres et al. 2007). Again, this can conserve water, often enhance crop quality, and ensures that water usage is maximized (Du et al. 2015). All of these practices emphasize responsible water use, and not only promote a more efficient use of water in agriculture, but also, improved agriculture's resiliency to future climate and water impacts (Tardieu et al. 2014).

B. *Energy Efficiency and Carbon Footprint Reduction*

The AI system will carry out full carbon footprint analyses to best illustrate the environmental impact on a farm (Wang et al. 2024; Mana et al. 2024; Kumari et al. 2025). This includes calculating emissions derived from producing and applying inputs such as fertilizer, assessing the impacts associated with transport logistics, examining the factor of energy needed to process the crops, and providing potential avenues for carbon offsetting based on changes in practices in the field (Ali et al. 2025).

C. *Biodiversity Conservation*

Artificial intelligence (AI) supports the adoption of agricultural practices that actively enhance, endure, and conserve biodiversity (Shivaprakash et al. 2022). Through habitat management, these systems can help to identify and protect wildlife corridors, optimize landscapes to support pollinator habitats, conserve natural enemies of pests, and protect native species (Meshram et al. 2025). This expands into integrated landscape management, where AI can optimize crop rotations and cover crop selection not just for agronomic benefits but also for the ecosystem services they provide, such as pollination and control of pests (Pandey and Mishra 2024). It can also aid in the design of agroforestry systems and the integration of farming practices with broader watershed management goals.

D. *Precision Conservation Agriculture*

I. *Soil Conservation Optimization*

The precision conservation utilizes AI to design and implement highly effective and site-specific soil conservation strategies (Awais et al. 2023). It begins with sophisticated erosion risk assessment that combines slope analysis, calculations of rainfall erosivity, and assessments of crop cover for creating detailed risk maps. This simultaneously allows for the targeted implementation of conservation practices and the evaluation of their effectiveness (Pace et al. 2025). Artificial intelligence (AI) also informs advanced cover crop management by helping to select the optimal species mix, determining the best planting and termination times, and analyzing the economic benefits of these practices. The tillage optimization is another key area, where AI aids in the development of minimal tillage strategies, assesses the risk of soil compaction, plans for residue management, and provides guidance on selecting the most appropriate equipment for the focused conservation of tillage (Zhang et al. 2024).

II. *Nutrient Loss Prevention*

The critical environmental challenge is preventing the loss of nutrients from fields into surrounding ecosystems. Artificial intelligence (AI)-powered systems address this through the precision application timing, which synchronizes fertilizer applications with crop uptake patterns and avoids periods where weather would lead to runoff or leaching, thereby achieving both economic and environmental optimization (Balan et al. 2025). Furthermore, AI enhances the management of buffer strips by using data models to determine their optimal width and placement and selecting the most effective plant species for nutrient filtration, scheduling necessary maintenance for continuous monitoring performance in protecting water quality (Talaviya et al. 2020).

8 Supply Chain Integration and Postharvest Management

The modern agricultural landscape demands seamless and intelligent integration of the entire supply chain, from the field to the fork (Kashyap et al. 2025). This integration is powered by advanced technologies, for minimizing waste, maximizing profitability, and ensuring the delivery of high-quality, safe food to consumers. The effective postharvest management forms the backbone of this integrated system, preserving the value of agricultural produce through sophisticated handling, storage, and transportation processes (Fig. 12.3) (Fadiji et al. 2023). By leveraging the artificial intelligence, robotics, and data analytics, the stakeholders can now optimize every step of the journey wherein transforming traditional linear chains into dynamic, responsive, and efficient networks.

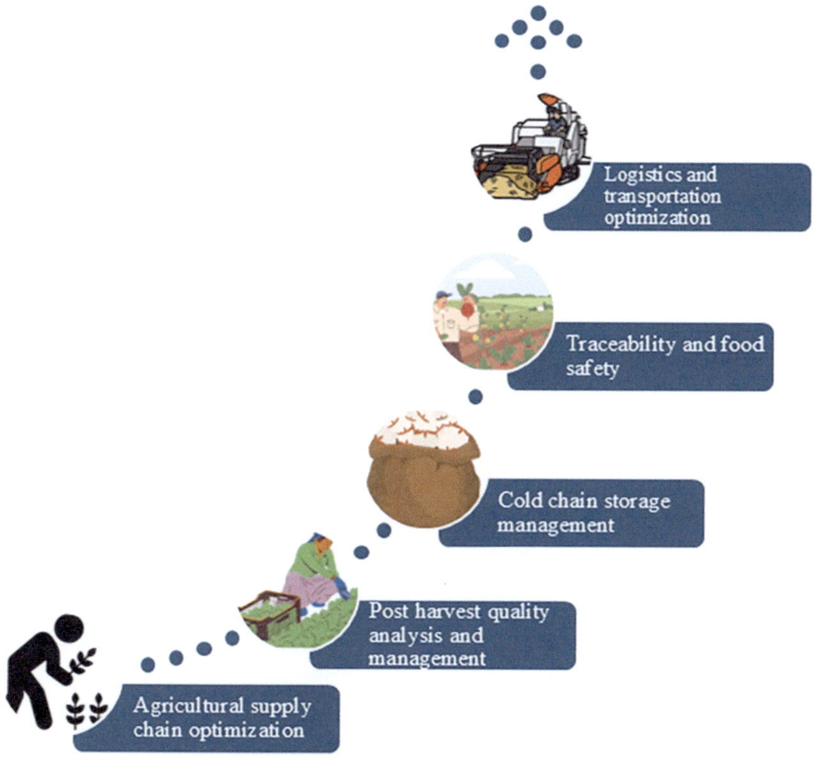

Fig. 12.3 AI-Powered Postharvest Optimization Strategies illustrates a suite of AI-driven postharvest strategies designed to maximize the commercial value of agricultural produce. The framework integrates advanced technologies across multiple stages of the postharvest pipeline for Enhancing Market Value of Harvested Crops

8.1 Agricultural Supply Chain Optimization

The foundation of an efficient supply chain lies in the accurate anticipation of market needs and strategic decision-making (Dhal and Kar 2024). The sophisticated machine learning models predicts consumer demand by analyzing the social media sentiment to identify emerging food trends, integrating economic indicators such as disposable income and demographic shifts, and modeling historical seasonal consumption patterns alongside with weather data (Elufioye et al. 2024). Furthermore, these systems conduct a comprehensive export market analysis for assessing international trade flows and foreign market demands to provide a global perspective.

Complementing demand forecasting, an advanced price forecasting systems utilizes sophisticated algorithms to predict the commodity price movements (Halder et al. 2025; Olufemi-Phillips et al. 2024). This system performs real-time analysis of futures contracts and options markets, integrate the global production estimates with the consumption forecasts, and quantify the potential impact of weather events on worldwide production and prices (Garcia et al. 2015). They also model the effects of changing government policies on market dynamics. This intelligence empowers the farmer with market timing optimization, allowing them to make informed selling decisions (Isengildina-Massa et al. 2011). The AI models can also analyze the complex relationships between harvest timing, quality premiums, and storage costs; calculate optimal storage periods based on projected prices; compare the merits of forward contracts against potential spot market prices; and recommend hedging strategies to mitigate financial risks associated with price volatility (Adjemian et al. 2013).

8.2 Postharvest Technology and Quality Management

The most crucial task upon harvest is maintaining and verifying product quality which is yet a noncomprisable task (Cubero et al. 2011). The automated sorting and the grading systems which are driven by computer vision ensure consistent and objective quality control of the harvest (Blasco et al. 2016). The advanced computational imaging systems perform precise size and shape measurements to ensure quality standards that are met with market specifications, conduct spectral analysis for accurate color and ripeness assessment to identify and classify the surface blemishes or foreign contaminants in the harvested crop. Beyond this initial grading, AI-powered quality prediction models forecast the product's future shelf life and condition by predicting the progression of ripeness, estimate longevity under various storage scenarios, and finally model the rate of quality deterioration during transit to ensure the product's attributes which are matched and maintained to the specific quality requirements of different market channels (Pereira et al. 2018).

Also, the packaging process is simultaneously enhanced by AI automation (Folinas et al. 2019). The Robotic systems can also optimize the packaging operations by selecting the most efficient package sizes for different retail channels by applying accurate labels regarding variety, origin, and quality grade, and embedding crucial traceability information for food safety compliance (Bosona and Gebresenbet 2013). This automation also extends to sustainability, with systems capable of selecting the packaging materials to minimize impact on the environment (Silvenius et al. 2014).

8.3 Cold Chain and Storage Management After Harvest

The efficient storage systems play a vital role in maintaining the quality and extending the shelf life of perishable goods (James et al. 2016). Advanced control mechanisms help to regulate temperature and humidity with precision, ensuring that each product is kept in ideal conditions. This careful balance prevents moisture loss and avoids condensation, both of which are crucial for preserving freshness and reducing spoilage during storage.

Certain products are stored under the controlled-environmental conditions to further prolongs freshness (Palou et al. 2011). These systems also manage the air circulation to ensure uniform conditions throughout the facility of storage. Integral to this operation is predictive maintenance, where AI monitors the health of critical equipment like compressors, manages sensor calibration, optimizes energy consumption to reduce costs, and activates emergency response protocols to prevent spoilage in case of equipment failure (Rong et al. 2011).

8.4 Traceability and Food Safety

The confidence of a customer and the regulatory compliance are built on complete transparency and proven food safety (Galvez et al. 2018). The blockchain technology, often integrated with AI, creates an immutable and transparent record from the farm to the consumer. This provides verifiable proof of a product's origin and the production methods and a complete history of all processing and handling steps, GPS-monitored transportation tracking, and this grants consumers access to the product's full history. Concurrently, the AI systems actively monitors for food safety issues, enabling real-time detection of biological or chemical contaminants and validating the integrity of the cold chain storage throughout distribution along with managing rapid and targeted product recalls if necessary (Tsang et al. 2019), and automating the documentation required for regulatory compliance (Kamilaris et al. 2019).

8.5 Logistics and Transportation Optimization

The final physical link in the supply chain is the efficient movement of goods (Mancini 2013). Artificial intelligence (AI) algorithms optimize the transportation through sophisticated route planning systems that can calculate the most efficient multi-stop delivery routes by maximizing truck load capacity without compromising the product quality and integrate real-time traffic data for dynamic route adjustments with plans for minimal fuel consumption (Defryn et al. 2016). The fleet management systems utilize AI to schedule predictive vehicle maintenance, monitor and optimize driver performance for safety and efficiency, and track the cargo conditions including temperature, humidity, and vibration in real-time during transit, reduces storage time by allowing goods to be transferred directly from incoming trucks to outgoing vehicles (Mercier et al. 2017).

9 Economic Impact and Financial Management Using AI-Based Approaches

The modern farming is shifting toward a more strategic way of handling money and it's no longer just about tracking expenses or reacting to what went wrong last season (Castle et al. 2016). Today's farmers have access to the tools that help them to think ahead, plan smarter, reduce risks, and make decisions with confidence. With insights coming from weather data, soil sensors, market trends, and more, they're able to fine-tune everything from daily input costs to long-term investments. This kind of adaptability means they can adapt quickly when conditions change, sidestep financial trouble, and build farms that are not only productive but also financially stable and ready for the future (Pivoto et al. 2018).

9.1 Precision Economics and Cost Optimization

Financial planning in agriculture is becoming sharper and more practical, especially attempting to reduce costs without affecting productivity (Schimmelpfennig 2016). Farms are not just tracking their inputs anymore, but also exploring every step in their production processes to determine where there can be savings. Think of fertilizers; these were usually spread on the same schedule each year without regards to market prices, appropriate timing, and nutrient absorption by the crops (Colaço and Bramley 2018). Seed selection has evolved too. It is not just deciding which seed is the cheapest-farmers are now weighing yield potential, disease resistance to local pathogens, and market value before selecting a variety (Finger et al. 2019). Pesticide usage is done in roughly the same way, where the cost base of treatment is balanced against the expected damage of the pest (Lamichhane et al. 2015). Even labor is being planned in a more methodological way. Farmers evaluate whether machinery is needed or if letting people do the work is simply more productive for the job at hand, deciding on a case-by-case basis, which helps to save time and money (Peng et al. 2022).

Pesticide application is similarly calculated, with the cost of the treatment being compared to the damage from pests. Even labor is planned with more thought. Farmers consider when it makes more sense to use machinery as opposed to labor when making their decisions depending on which route saves time and money (Fleming et al. 2000).

9.2 Financial Planning and Budgeting

The seasonal cash flow modeling provides a dynamic forecast of financial inflows and outflows throughout the production cycle (Barry et al. 2012). This allows farm managers to optimize the timing of input purchases based on price forecasts and cash availability, coordinate the grain marketing strategies with upcoming financial obligations, and optimize the credit utilization to minimize interest expenses. This financial foresight also extends to the tax planning, enabling the strategic timing of income and expenses for an improved tax efficiency (Boehlje and Lins 2015).

For the long-term capital allocation, AI-powered investment analysis can provide a rigorous framework for major decisions such as calculating detailed ROI, payback periods, and internal rates of return for the investments in new technologies (Tey and Brindal 2012). It also provides an analytical support for critical land decisions, whether involving the evaluation of the purchase of additional acreage or the profitability of rental agreements, based on precise productivity data and appreciation potential (Bigelow et al. 2016). This same analytical rigor can be applied to the infrastructure management for storage or irrigation, even to the sustainability investments goals, both quantifying the financial returns alongside environmental benefits (Higgins et al. 2017).

10 Challenges and Barriers to AI Adoption in Agriculture

The integration of advanced technologies into agriculture presents a range of challenges that extend beyond the technical domain (Rose et al. 2016). One of the primary difficulties lies in data acquisition and management. Agriculture can often be tough as extreme temperature, moisture, and vibration creates bias in sensors (Bramley and Ouzman 2019). This means sensors are constantly needing to be recalibrated without knowing if it was used correctly, and a lack of standardized protocols exists. The process of collecting and integrating diverse datasets into a single platform is complicated by inconsistencies in format, as well as spatial and temporal resolution. In addition, the volume of imagery to be analyzed at high resolutions (huge number and continuous influx of imaging data), as well as continuous incoming sensor data increases demands on storage capacities, processing infrastructure, and internet bandwidth, particularly in rural areas with limited domains for internet service providers (Wolfert et al. 2017).

Algorithmic modeling introduces the above issues. Predictive models must be able to address the wider variability in the environmental (geographic location, season, annual variations in weather, as well as the number of tested genotypes in species of crops) (Liakos et al. 2018). Additionally, models must also be interpretable by end users to promote decision-making. Further, costs of implementation are prohibitive. They include initial costs of infrastructure, hardware and software, plus ongoing costs of subscriptions and maintenance. Small- and medium-sized farms must do the scrutinizing costs since return on investment may not be realistic

(Vecchio et al. 2020). Eventually due to payback periods and uncertainty with some indirect benefits such as risk mitigation, adoption tends to be narrow.

There are often social and cultural reasons for the limited uptake of advanced agri-technology. Many farmers, particularly from older generations, have been apprehensive about moving from conventional practices to digital (Michels et al. 2020). There is a general reluctance of complicated tools, in particular this unease is enhanced when one cannot see how the tool produces decisions. Other concerns include data protection, accountability for what happens if the system fails, as well as reliance on third-party providers (Fleming et al. 2018).

These factors are complicating in rural areas as well. Inadequate internet connectivity often renders large data sets impractical (Saiz-Rubio and Rovira-Más 2020). Power supply, frequently out of reach, also make modern systems impossible to run. Moreover, the regulatory environment around the use of this technology remains fluid (Bronson 2018). There is a lack of clear guidelines for data ownership, privacy protection, and liability for autonomous transactions. Consequently, farmers and technology developers must also deal with an ambiguous legal landscape (Carolan 2017).

11 Future Horizons and Emerging Technologies

Agriculture is currently transforming at an unprecedented rate, which rivals the impact of the Green Revolution (Pingali 2012). This transition is a result of the increased food production demanded by a growing population, while needing to conserve natural resources, reduce environmental degradation, and adapt to changing climatic conditions (Godfray et al. 2010). The use of new technologies is allowing farms to be much more precise and responsive to cultivation needs (Zhang and Kovacs 2012). These systems automatically adjust farming practices based on changing conditions in a field, thus allowing for better management of crops and resources. Through the integration of multiple innovations, agriculture is progressing toward a model where decisions are informed by real-time data, operations are more sustainable, and food systems are better equipped to ensure long-term productivity and resilience (Gebbers and Adamchuk 2010).

11.1 Next-Generation AI Applications for Climate-Resilient Agriculture

The new forecasting systems give farmers early warnings about floods, droughts, or storms, so they can take action before damage happens (Hansen et al. 2019). Another key focus is improving the soil's ability to trap carbon from the air, which helps to fight climate change while also making the land healthier. To make crops stronger, researchers are finding the exact traits that help plants to survive with less water or in very hot conditions (Cattivelli et al. 2008). They're also working on ways to cool plants during heatwaves and grow crops in salty or damaged soils. In places where diseases are changing quickly, new crop varieties are being developed to resist these threats (Singh et al. 2023).

12 Cognitive Agriculture Systems and Implementation Strategies

The smart AI systems can plan work, use resources wisely, check for risks, and keep improving how the farm runs (Walter et al. 2017). They act like the brain of the farm, collecting information from different places to make better decisions that save time, money, and protect nature. Farmers will be able to talk to these systems using simple voice commands. They can ask questions, get advice, and receive reports about how their farm is doing. These tools will also help farmers to learn from each other by sharing good ideas and farming methods across different regions and climates (Eastwood et al. 2017).

First,the basics like putting sensors in the soil, setting up weather stations, improving internet in villages, training people, and testing the tools in small farms should be ensured. Next, useful tools like machines that adjust how much fertilizer or water is used, systems that watch crops closely, and methods to guess how much harvest will come should be implemented (Vasconez et al. 2019). In the final step, advanced tools like self-driving tractors, systems that help manage risks, track the supply chain, and make farming eco-friendlier and more sustainable shall be incorporated(Boursianis et al. 2022).

13 Global Perspectives and Case Studies

Artificial intelligence (AI)-powered machines and models are increasingly appearing in agriculture, providing compensated forecasting and management possibilities, with astonishing evidence of great results around the world, with different challenges based on the region (Sharma et al. 2023). In developed countries, options can be seen by leading-edge initiatives, such as climate-controlled greenhouse management in the Netherlands, with AI-based feasibility driven by extreme resource efficiencies. Israel uses AI to pioneer precision irrigation and drought-tolerant crops in arid environments, while Japan developed robotic solutions to support its aging farmer population (Bacco et al. 2019a, b).

In developing countries, AI is being deployed through appropriate and accessible technology. In India, farmers are getting help through mobile phones that give them advice about crops, weather updates, and market prices. This is especially useful for small farmers who may not have access to big resources (Kumar et al. 2021a, b). In Africa, satellite images are used to watch for droughts and to guess how much crop will grow in large areas (Li et al. 2024). In Latin America, new farming methods are being used to grow more food while also protecting the rainforests (de Oliveira et al. 2020). Big farming companies have also shown how technology can help. (Shepherd et al. 2018).

14 Conclusion: The Transformative Era of Agriculture Through AI

The present-day status of farming after incorporating the technological advancements has a yield increase of 10–25%, water-use reductions of 20–40%, labor savings of 30–50%, and pesticide use cut by 40–60%. These statistics demonstrate improvements in product quality and efficiency without sacrificing environmental safety. The change includes improving accuracy at the individual plant level at a farm. This change in our agricultural approach can also allow us to confront our climate change challenges and improve sustainability, while building economic resiliency with a source of reduced input cost and risk. Farmers can also expect a continued focus and investment in education, infrastructure, financing, and the long-term plan and vision is needed to enable government policies for self-sufficiency to be beneficial for farms that are adaptable to change, farms that improve soil health, respect ecosystems, sequester carbon, and increase biodiversity. It is clear that the potential of AI may be able to create an adaptability transition within the global food system which may allow climate-resilient supply-chain systems to be established and reduce waste and increase sustainability for generations.

References

Acharya S, Sankhyan NK, Baghla K, Kashyap M (2024) Artificial intelligence equipped nutrient management in agriculture. Vigyan Varta 5(6):98–101. https://vigyanvarta.in/adminpanel/upload_doc/VV_0624_21.pdf

Adamchuk VI, Hummel JW, Morgan MT, Upadhyaya SK (2004) On-the-go soil sensors for precision agriculture. Comput Electron Agric 44(1):71–91. https://doi.org/10.1016/j.compag.2004.03.002

Adjemian MK, Garcia P, Irwin SH, Smith A (2013) Non-convergence in domestic commodity futures markets: causes, consequences, and remedies. Economic Research Report, (ERS-115). USDA Economic Research Service. https://doi.org/10.2139/ssrn.2238783

Ahmad I, Yang Y, Yue Y, Ye C, Hassan M (2022) Deep learning based detector YOLOv5 for identifying insect pests. Appl Sci. Retrieved 1 Sept 2025, from https://www.mdpi.com/2076-3417/12/19/10167

Ainsworth EA, Yendrek CR, Sitch S, Collins WJ, Emberson LD (2012) The effects of tropospheric ozone on net primary productivity and implications for climate change. Annu Rev Plant Biol 63:637–661. https://doi.org/10.1146/annurev-arplant-042110-103829

Akter J, Kamruzzaman M et al (2024) Artificial intelligence in American agriculture: a comprehensive review of spatial analysis and precision farming for sustainability. IEEE. Retrieved 1 Sept 2025, from https://ieeexplore.ieee.org/abstract/document/10796075/

Ali Z, Muhammad A, Lee N, Waqar M, Lee SW (2025) Artificial Intelligence for sustainable agriculture: a comprehensive review of AI-driven technologies in crop production. Sustainability. Retrieved 2 Sept 2025, from https://www.mdpi.com/2071-1050/17/5/2281

Allen RG, Pereira LS, Raes D, Smith M (1998) Crop evapotranspiration-guidelines for computing crop water requirements. FAO Irrigation and drainage paper 56. Food and Agriculture Organization of the United Nations

Angon PB, Das A, Anjum N, Mondal S, Datta A, An R et al (2023) Integrated pest management (IPM) in agriculture and its role in maintaining ecological balance and biodiversity. Adv Agric 2023:5546373. https://doi.org/10.1155/2023/5546373

Awais M, Naqvi SMZA, Zhang H, Li L, Zhang W, Awwad FA, Ismail EAA, Khan MI, Raghavan V, Hu J (2023) AI and machine learning for soil analysis: an assessment of sustainable agricultural practices. Bioresour Bioprocess 10(1). https://doi.org/10.1186/S40643-023-00710-Y. Springer

Aylor DE (2003) Spread of plant disease on a continental scale: role of aerial dispersal of pathogens. Ecology 84(8):1989–1997. https://doi.org/10.1890/01-0619

Bacco M, Barsocchi P, Ferro E, Gotta A, Ruggeri M (2019a) The digitisation of agriculture: a survey of research activities on smart farming. Array 3-4:100009. https://doi.org/10.1016/j.array.2019.100009

Bacco M, Berton A, Gotta A, Caviglione L (2019b) IEEE 802.11 WiFi for vehicular networking: overview, experimental evaluation, and future challenges. Mob Inf Syst 2019:8718315. https://doi.org/10.1155/2019/8718315

Badgujar C, Poulose A, Gan H (2024) Agricultural object detection with You Only Look Once (YOLO) algorithm: a bibliometric and systematic literature review. Comput Electron Agric. Elsevier. Retrieved 1 Sept 2025, from https://www.sciencedirect.com/science/article/pii/S0168169924004812

Bai X, Huang Y, Ren W, Coyne M, Jacinthe PA, Tao B, Hui D, Yang J, Matocha C (2023) Responses of soil carbon sequestration to climate-smart agriculture practices: a meta-analysis. Glob Chang Biol 29(20):5194–5209. https://doi.org/10.1111/gcb.16817

Baiano A (2017) Applications of hyperspectral imaging for quality assessment of liquid foods: a review. J Food Eng 214:10–15. https://doi.org/10.1016/j.jfoodeng.2017.06.012

Baker NR (2008) Chlorophyll fluorescence: a probe of photosynthesis in vivo. Annu Rev Plant Biol 59:89–113. https://doi.org/10.1146/annurev.arplant.59.032607.092759

Balaji M, Arshinder K (2016) Modeling the causes of food wastage in Indian perishable food supply chain. Resour Conserv Recycl 114:153–167. https://doi.org/10.1016/j.resconrec.2016.07.016

Balan G, Kumar V, Sakthivel AR (2025) Machine learning and artificial intelligence methods and applications for post-crisis supply chain resiliency and recovery. Supply Chain Analytics 10(11):100121. Elsevier. Retrieved 2 Sept 2025, from https://www.sciencedirect.com/science/article/pii/S2949863525000214

Balaska V, Adamidou Z, Vryzas Z, Gasteratos A (2023) Sustainable crop protection via robotics and artificial intelligence solutions. Machines. Retrieved 1 Sept 2025, from https://www.mdpi.com/2075-1702/11/8/774

Barry PJ, Sonka ST, Lajili K (2012) Vertical coordination, financial structure, and the changing theory of the firm. Am J Agric Econ 74(5):1219–1229. https://doi.org/10.2307/1242778

Bechar A, Vigneault C (2016) Agricultural robots for field operations: concepts and components. Biosyst Eng 149:94–111. https://doi.org/10.1016/j.biosystemseng.2016.06.014

Behmann J, Mahlein AK, Rumpf T, Römer C, Plümer L (2015) A review of advanced machine learning methods for the detection of biotic stress in precision crop protection. Precision Agric 16(3):239–260. https://doi.org/10.1007/S11119-014-9372-7. Springer

Benos L, Tagarakis AC, Dolias G, Berruto R, Kateris D, Bochtis D (2021) Machine learning in agriculture: a comprehensive updated review. Sensors. https://doi.org/10.3390/s21113758

Berckmans D (2014) Precision livestock farming technologies for welfare management in intensive livestock systems. Rev Sci Tech 33(1):189–196. https://doi.org/10.20506/rst.33.1.2273

Bigelow D, Borchers A, Hubbs T (2016) US farmland ownership, tenure, and transfer. USDA-ERS Econ Inform Bull 161. https://doi.org/10.22004/ag.econ.249444

Blasco J, Munera S, Aleixos N, Cubero S, Molto E (2016) Machine vision-based measurement of quality of cereal grains. J Food Eng 185:42–50. https://doi.org/10.1016/j.jfoodeng.2016.04.004

Bock CH, Barbedo JG, Del Ponte EM, Bohnenkamp D, Mahlein AK (2021) From visual estimates to fully automated sensor-based measurements of plant disease severity: status and challenges for improving accuracy. Phytopathology 112(2):337–371. https://doi.org/10.1094/PHYTO-03-21-0111-RVW

Boehlje M, Lins D (2015) Risks and risk management in an industrialized agriculture. Agric Finance Rev 58(1):1–16. https://doi.org/10.1108/00214669810006010

Borrelli P, Robinson DA, Fleischer LR, Lugato E, Ballabio C, Alewell C, Meusburger K, Modugno S, Schütt B, Ferro V, Bagarello V, Van Oost K, Montanarella L, Panagos P (2017) An assessment of the global impact of 21st century land use change on soil erosion. Nat Commun 8(1):1–13. https://doi.org/10.1038/s41467-017-02142-7

Bosona T, Gebresenbet G (2013) Food traceability as an integral part of logistics management in food and agricultural supply chain. Food Control 33(1):32–48. https://doi.org/10.1016/j.foodcont.2013.02.004

Boursianis AD, Papadopoulou MS, Diamantoulakis P, Liopa-Tsakalidi A, Barouchas P, Salahas G et al (2022) Internet of things (IoT) and agricultural unmanned aerial vehicles (UAVs) in smart farming: a comprehensive review. Internet of Things 18:100187. https://doi.org/10.1016/j.iot.2020.100187

Brady NC, Weil RR (2016) The nature and properties of soils, 15th edn. Pearson

Bramley RGV, Hamilton RP (2004) Understanding variability in winegrape production systems 1. Within vineyard variation in yield over several vintages. Aust J Grape Wine Res 10(1):32–45. https://doi.org/10.1111/j.1755-0238.2004.tb00006.x

Bramley RGV, Ouzman J (2019) Farmer attitudes to the use of sensors and automation in fertilizer decision-making: nitrogen fertilization in the Australian grains industry. Precis Agric 20(1):157–175. https://doi.org/10.1007/s11119-018-9589-y

Bronson K (2018) Smart farming: including rights holders for responsible agricultural innovation. Technol Innov Manag Rev 8(2):7–14. https://doi.org/10.22215/timreview/1135

Cao X, Yao Y, Li L, Zhang W, An Z, Zhang Z, Xiao L, Guo S, Cao X, Wu M, Luo D (2022) Igrow: a smart agriculture solution to autonomous greenhouse control. Proc AAAI Conf Artif Intell. https://ojs.aaai.org/index.php/AAAI/article/view/21440 36:11837

Carolan M (2017) Publicising food: big data, precision agriculture, and co-experimental techniques of addition. Sociol Rural 57(2):135–154. https://doi.org/10.1111/soru.12120

Castle MH, Lubben BD, Luck JD (2016) Factors influencing the adoption of precision agriculture technologies by Nebraska producers. In: Proceedings of the 13th International Conference on Precision Agriculture. International Society of Precision Agriculture

Cattivelli L, Rizza F, Badeck FW, Mazzucotelli E, Mastrangelo AM, Francia E et al (2008) Drought tolerance improvement in crop plants: an integrated view from breeding to genomics. Field Crop Res 105(1–2):1–14. https://doi.org/10.1016/j.fcr.2007.07.004

Charlebois S, Sterling B, Haratifar S, Naing SK (2014) Comparison of global food traceability regulations and requirements. Compr Rev Food Sci Food Saf 13(5):1104–1123. https://doi.org/10.1111/1541-4337.12101

Cheein FA, Carelli R (2013) Agricultural robotics: unmanned robotic service units in agricultural tasks. IEEE Ind Electron Mag 7(3):48–58. https://doi.org/10.1109/MIE.2013.2252957

Chen Y, Lee W, Gan H, Peres N, Fraisse C, Sensing YZ-R (2019) Strawberry yield prediction based on a deep neural network using high-resolution aerial orthoimages. Remote Sens. Retrieved 1 Sept 2025, from https://www.mdpi.com/2072-4292/11/13/1584

Chen Y, Zhang L, Wang H, Liu X (2025) Multi-modal data fusion and deep ensemble learning for accurate crop yield prediction. arXiv preprint, arXiv:2502.06062. https://doi.org/10.48550/arXiv.2502.06062

Chergui N, Kechadi MT (2022) Data analytics for crop management: a big data view. J Big Data 9(1). Springer. https://doi.org/10.1186/S40537-022-00668-2

Cohen Y, Alchanatis V, Meron M, Saranga Y, Tsipris J (2021) Assessment for crop water stress with infrared thermal imagery in precision agriculture: a review and future prospects for deep learning applications. Comput Electron Agric 182:106020. https://doi.org/10.1016/j.compag.2021.106020

Colaço AF, Bramley RGV (2018) Do crop sensors promote improved nitrogen management in grain crops? Field Crop Res 218:126–140. https://doi.org/10.1016/j.fcr.2018.01.007

Corwin DL, Lesch SM (2005) Apparent soil electrical conductivity measurements in agriculture. Comput Electron Agric 46(1–3):11–43. https://doi.org/10.1016/j.compag.2004.10.005

Cubero S, Aleixos N, Molto E, Gomez-Sanchis J, Blasco J (2011) Advances in machine vision applications for automatic inspection and quality evaluation of fruits and vegetables. Food Bioprocess Technol 4(4):487–504. https://doi.org/10.1007/s11947-010-0411-8

Dammer KH, Wartenberg G (2007) Sensor-based weed detection and application of variable herbicide rates in real time. Crop Prot 26(3):270–277. https://doi.org/10.1016/j.cropro.2005.08.018

Defryn C, Sörensen K, Cornelissens T (2016) The selective vehicle routing problem in a collaborative environment. Eur J Oper Res 250(2):400–411. https://doi.org/10.1016/j.ejor.2015.09.059

DeJonge KC, Taghvaeian S, Trout TJ, Comas LH (2015) Comparison of canopy temperature-based water stress indices for maize. Agric Water Manag 156:51–62. https://doi.org/10.1016/j.agwat.2015.03.023

Dhal S, Kar D (2024) Transforming agricultural productivity with AI-driven forecasting: innovations in food security and supply chain optimization. Forecasting. Retrieved 2 Sept 2025, from https://www.mdpi.com/2571-9394/6/4/46

Dabic-Miletic S (2024) The challenges of integrating AI and robotics in sustainable WMS to improve supply chain economic resilience. J Ind Intell, 2(2):119–131

Dolatabadian A, Neik TX, Danilevicz MF, Upadhyaya SR, Batley J, Edwards D (2025) Image-based crop disease detection using machine learning. Plant Pathol 74(1):18–38. https://doi.org/10.1111/PPA.14006. Wiley Online Library

Dolecheck KA, Silvia WJ, Heersche G Jr, Chang YM, Ray DL, Stone AE et al (2015) Behavioral and physiological changes around estrus events identified using multiple automated monitoring technologies. J Dairy Sci 98(12):8723–8731. https://doi.org/10.3168/jds.2015-9645

Dommetti B, Reddy ES, Saravanan TR (2025) Revolutionizing agricultural practices: a comprehensive exploration of farm era's farm management application with advanced data analysis, geospatial mapping, and real-time monitoring modules. AIP Conf Proc 3257(1). https://doi.org/10.1063/5.0279836/20578195/020139_1_5.0279836.PDF

Dozono K, Amalathas S, Saravanan R (2022) The impact of cloud computing and artificial intelligence in digital agriculture. In: Proceedings of Sixth International Congress on Information and Communication Technology, vol 235, pp 557–569. https://doi.org/10.1007/978-981-16-2377-6_52

Du T, Kang S, Zhang J, Li F, Yan B (2015) Water use efficiency and yield responses of cotton to alternate partial root-zone drip irrigation in the arid area of Northwest China. Irrig Sci 33(1):19–30. https://doi.org/10.1007/s00271-014-0445-7

Duffy C, Fealy R, Fealy RM (2022) Machine learning for detection and prediction of crop diseases and pests: a comprehensive survey. Agriculture 12(9):1350. https://doi.org/10.3390/agriculture12091350

Eastwood C, Klerkx L, Nettle R (2017) Dynamics and distribution of public and private research and extension roles for technological innovation and diffusion: case studies of the implementation and adaptation of precision farming technologies. Glob Food Sec 12:1–12. https://doi.org/10.1016/j.gfs.2016.12.001

Elbasi E, Zaki C, Topcu AE, Abdelbaki W, Zreikat AI, Cina E, Shdefat A, Saker L (2023) Crop prediction model using machine learning algorithms. Appl Sci. https://doi.org/10.3390/app13169288

Elbeltagi A, Kumari N, Dharpure JK, Mokhtar A, Alsafadi K, Kumar M et al (2021) Prediction of combined terrestrial evapotranspiration index (CTEI) over large river basin based on machine learning approaches. Water 13(4):547. https://doi.org/10.3390/w13040547

Elufioye O, Ike C, Odeyemi O, Usman FO, Mhlongo NZ (2024) Ai-Driven predictive analytics in agricultural supply chains: a review: assessing the benefits and challenges of ai in forecasting demand and optimizing supply in agriculture. Comput Sci IT Res J 5(2):473–497. Retrieved 2 Sept 2025, from https://www.researchgate.net/profile/Olubusola-Odeyemi/publication/378287475_AI-DRIVEN_PREDICTIVE_ANALYTICS_IN_AGRICULTURAL_SUPPLY_CHAINS_A_REVIEW_ASSESSING_THE_BENEFITS_AND_CHALLENGES_OF_AI_IN_FORECASTING_DEMAND_AND_OPTIMIZING_SUPPLY_IN_AGRICULTURE/links/65d0dece476dd15fb343ffb4/AI-DRIVEN-PREDICTIVE-ANALYTICS-IN-AGRICULTURAL-SUPPLY-CHAINS-A-

REVIEW-ASSESSING-THE-BENEFITS-AND-CHALLENGES-OF-AI-IN-FORECASTING-DEMAND-AND-OPTIMIZING-SUPPLY-IN-AGRICULTURE.pdf

Emmi L, Gonzalez-de-Soto M, Pajares G, Gonzalez-de-Santos P (2014) New trends in robotics for agriculture: integration and assessment of a real fleet of robots. Sci World J 2014:404059. https://doi.org/10.1155/2014/404059

Evans RG, LaRue J, Stone KC, King BA (2013) Adoption of site-specific variable rate sprinkler irrigation systems. Irrig Sci 31(4):871–887. https://doi.org/10.1007/s00271-012-0365-x

Fadiji T, Bokaba T, Fawole OA, Twinomurinzi H (2023) Artificial intelligence in postharvest agriculture: mapping a research agenda. Front Sustain Food Syst 7. https://doi.org/10.3389/FSUFS.2023.1226583/FULL

Farooq A, Hu J, Jia X (2019) Analysis of spectral bands and spatial resolutions for weed classification via deep convolutional neural network. GIScience Remote Sens 56(2):207–227. https://doi.org/10.1080/15481603.2018.1513491

Fennimore SA, Cutulle M (2019) Robotic weed control in specialty crops. Annu Rev Phytopathol 57:309–328. https://doi.org/10.1146/annurev-phyto-082718-100022

Fereres E, Soriano MA, Bacelar EA, Cachorro P, Costantini E, Domínguez A, Drescher A, Girona J, Jiménez L, Jiménez M, Martínez-Cob A (2007) Deficit irrigation for reducing agricultural water use. J Exp Bot 58(2):147–159. https://doi.org/10.1093/jxb/erl165

Finger R, Swinton SM, El Benni N, Walter A (2019) Precision farming at the nexus of agricultural production and the environment. Ann Rev Resour Econ 11:313–335. https://doi.org/10.1146/annurev-resource-100518-093929

Fleming RA, Marsh TL, Marsh SR (2000) Precision agriculture: economics of nitrogen management in corn using site-specific crop response estimates. Am J Agric Econ 82(2):251–263. https://doi.org/10.1111/0002-9092.00024

Fleming A, Jakku E, Lim-Camacho L, Taylor B, Thorburn P (2018) Is big data for big farming or for everyone? Perceptions in the Australian grains industry. Agron Sustain Dev 38(3):1–10. https://doi.org/10.1007/s13593-018-0501-y

Folinas D, Manikas I, Manos B (2019) Traceability data management for food chains. Br Food J 108(8):622–633. https://doi.org/10.1108/00070700610682337

Fountas S, Mylonas N, Malounas I, Rodias E, Hellmann Santos C, Pekkeriet E (2020a) Agricultural robotics for field operations. Sensors. Retrieved 1 Sept 2025, from https://www.mdpi.com/1424-8220/20/9/2672

Fountas S, Mylonas N, Malounas I, Rodias E, Santos CH, Pekkeriet E (2020b) Agricultural robotics for field operations. Sensors 20(9):2672. https://doi.org/10.3390/s20092672

Franzluebbers AJ (2020) Soil health and carbon management. Food Energy Secur 5(4):231–244. https://doi.org/10.1002/fes3.96

Galvez JF, Mejuto JC, Simal-Gandara J (2018) Future challenges on the use of blockchain for food traceability analysis. TrAC Trends Anal Chem 107:222–232. https://doi.org/10.1016/j.trac.2018.08.011

Gamon JA, Peñuelas J, Field CB (1992) A narrow-waveband spectral index that tracks diurnal changes in photosynthetic efficiency. Remote Sens Environ 41(1):35–44. https://doi.org/10.1016/0034-4257(92)90059-S

Garcia P, Irwin SH, Smith A (2015) Futures market failure? Am J Agric Econ 97(1):40–64. https://doi.org/10.1093/ajae/aau067

García-Orenes F, Morugán-Coronado A, Zornoza R, Scow K (2013) Changes in soil microbial community structure influenced by agricultural management practices in a Mediterranean agro-ecosystem. PLoS One 8(11):1–8. https://doi.org/10.1371/journal.pone.0080522

Gautron R, Maillard O, Preux P, Corbeels M, Sabbadin R (2022) Reinforcement learning for crop management support: review, prospects and challenges. Comput Electron Agric. Retrieved 1 Sept 2025, from https://www.sciencedirect.com/science/article/pii/S0168169922004999

Gavioli A, de Souza E, Bazzi C, Schenatto K, Betzek NM (2019) Identification of management zones in precision agriculture: an evaluation of alternative cluster analysis methods. Biosyst Eng 181:86–102. Elsevier. Retrieved 1 Sept 2025, from https://www.sciencedirect.com/science/article/pii/S1537511018303295

Gebbers R, Adamchuk VI (2010) Precision agriculture and food security. Science 327(5967):828–831. https://doi.org/10.1126/science.1183899
Gerhards R, Oebel H (2006) Practical experiences with a system for site-specific weed control in arable crops using real-time image analysis and GPS-controlled patch spraying. Weed Res 46(3):185–193. https://doi.org/10.1111/j.1365-3180.2006.00504.x
Giles DK, Billing RC (2015) Deployment and performance of a UAV for crop spraying. Chem Eng Trans 44:307–312. https://doi.org/10.3303/CET1544052
Gitelson AA, Kaufman YJ, Merzlyak MN (1996) Use of a green channel in remote sensing of global vegetation from EOS-MODIS. Remote Sens Environ 58(3):289–298. https://doi.org/10.1016/S0034-4257(96)00072-7
Godfray HCJ, Beddington JR, Crute IR, Haddad L, Lawrence D, Muir JF et al (2010) Food security: the challenge of feeding 9 billion people. Science 327(5967):812–818. https://doi.org/10.1126/science.1185383
Grünig M, Razavi NM, Calanca P, Mazzi D, Wegner JD, Pellissier L (2021) Applying deep neural networks to predict incidence and phenology of plant pests and diseases. Ecosphere 12(10):e03791. https://doi.org/10.1002/ecs2.3791
Halder S, Islam M, Mamun Q, Mahboubi A, Walsh P, Islam MZ (2025) A comprehensive survey on AI-enabled secure social industrial internet of things in the agri-food supply chain. Smart Agric Technol 11:100902. Elsevier. Retrieved 2 Sept 2025, from https://www.sciencedirect.com/science/article/pii/S2772375525001352
Hansen J, Hellin J, Rosenstock T, Fisher E, Cairns J, Stirling C et al (2019) Climate risk management and rural poverty reduction. Agric Syst 172:28–46. https://doi.org/10.1016/j.agsy.2018.01.019
Harmel RD, Smith DR, Haney RL, Dozier M (2009) Nitrogen and phosphorus runoff from cropland and pasture fields fertilized with poultry litter. J Soil Water Conserv 64(6):400–412. https://doi.org/10.2489/jswc.64.6.400
Hatfield JL, Prueger JH (2015) Temperature extremes: effect on plant growth and development. Weather Clim Extremes 10:4–10. https://doi.org/10.1016/j.wace.2015.08.001
Hatfield JL, Dold C, Dold C (2019) Water-use efficiency: advances and challenges in a changing climate. Front Plant Sci 10:1–14. https://doi.org/10.3389/fpls.2019.00103
Hemming S, de Zwart F, Elings A, Righini I, Petropoulou A (2019) Remote control of greenhouse vegetable production with artificial intelligence—greenhouse climate, irrigation, and crop production. Sensors 19:1807. Retrieved 1 Sept 2025, from https://www.mdpi.com/1424-8220/19/8/1807
Hengl T, Leenaars JG, Shepherd KD, Walsh MG, Heuvelink GB, Mamo T et al (2017) Soil nutrient maps of Sub-Saharan Africa: assessment of soil nutrient content at 250 m spatial resolution using machine learning. Nutr Cycl Agroecosyst 109(1):77–102. https://doi.org/10.1007/s10705-017-9870-x
Higgins A, Hajkowicz S, Bui E (2017) A multi-objective model for environmental investment decision making. Comput Oper Res 35(1):253–266. https://doi.org/10.1016/j.cor.2006.02.027
Hillier J, Bayala J, Centre A, Rusinamhodzi KL, Vermeulen SJ, Dinesh D, Howden SM, Cramer L, Thornton PK (2018) Transformation in practice: a review of empirical cases of transformational adaptation in agriculture under climate change. Front Sustain Food Syst 2:65. https://doi.org/10.3389/FSUFS.2018.00065/FULL
Hong S, Kim S, Kim E, Lee C, Lee J, Agriculture DL (2020) Moth detection from pheromone trap images using deep learning object detectors. Agriculture 10(5):170. Retrieved 2 Sept 2025, from https://www.mdpi.com/2077-0472/10/5/170
Hong S, Nam I, Kim S, Kim E, Lee C, Ahn S, Park IK, Kim G (2021) Automatic pest counting from pheromone trap images using deep learning object detectors for Matsucoccus thunbergianae monitoring. Insects 12:342. Retrieved 2 Sept 2025, from https://www.mdpi.com/2075-4450/12/4/342
Hoseinzadeh S, Garcia DA (2024) Can AI predict the impact of its implementation in greenhouse farming? Renew Sustain Energy Rev 197:114423. Elsevier. Retrieved 1 Sept 2025, from https://www.sciencedirect.com/science/article/pii/S1364032124001461

Hu T, Zhang X, Khanal S, Wilson R, Leng G, Toman EM et al (2024) Climate change impacts on crop yields: a review of empirical findings, statistical crop models, and machine learning methods. Environ Model Softw 179:106119. Elsevier. Retrieved 1 Sept 2025, from https://www.sciencedirect.com/science/article/pii/S1364815224001804
Huete A, Didan K, Miura T, Rodriguez EP, Gao X, Ferreira LG (2002) Overview of the radiometric and biophysical performance of the MODIS vegetation indices. Remote Sens Environ 83(1–2):195–213. https://doi.org/10.1016/S0034-4257(02)00096-2
Isengildina-Massa O, Irwin SH, Good DL, Gomez JK (2011) The impact of situation and outlook information in corn and soybean futures markets: evidence from WASDE reports. J Agric Appl Econ 43(4):565–581. https://doi.org/10.1017/S1074070800000043
Islam M, Oliullah K, Kabir M, Alom M, Mridha M (2023) Machine learning enabled IoT system for soil nutrients monitoring and crop recommendation. J Agric Food Res 14:100880. Retrieved 1 Sept 2025, from https://www.sciencedirect.com/science/article/pii/S2666154323003873
James SJ, James C, Evans JA (2016) Modelling of food transportation systems–a review. Int J Refrig 60:309–319. https://doi.org/10.1016/j.ijrefrig.2015.08.018
Javaid M, Haleema A, Khanb I, Suman R (2022) Advanced Agrochem. https://www.academia.edu/download/113039859/Published_paper_on_AI_in_Agriculture.pdf
Jeong JH, Resop JP, Mueller ND, Fleisher DH, Yun K, Butler EE et al (2016) Random forests for global and regional crop yield predictions. PLoS One 11(6):e0156571. https://doi.org/10.1371/journal.pone.0156571
Jiang P, Chen Y, Liu B, He D, Liang C (2019) Real-time detection of apple leaf diseases using deep learning approach based on improved convolutional neural networks. IEEE Access 7:59069–59080. https://doi.org/10.1109/ACCESS.2019.2914929
Jiang H, Xu W, Chen Q (2022) Evaluating the potential of hyperspectral imaging and multivariate analysis for seed variety identification. Biosyst Eng 219:68–76. https://doi.org/10.1016/j.biosystemseng.2022.04.022
Kalaji HM, Jajoo A, Oukarroum A, Brestic M, Zivcak M, Samborska IA et al (2016) Chlorophyll a fluorescence as a tool to monitor physiological status of plants under abiotic stress conditions. Acta Physiol Plant 38(4):102. https://doi.org/10.1007/s11738-016-2113-y
Kalma JD, McVicar TR, McCabe MF (2008) Estimating land surface evaporation: a review of methods using remotely sensed surface temperature data. Surv Geophys 29(4–5):421–469. https://doi.org/10.1007/s10712-008-9037-z
Kamilaris A, Prenafeta-Boldú FX (2018) Deep learning in agriculture: a survey. Comput Electron Agric 147:70–90. https://doi.org/10.1016/j.compag.2018.02.016
Kamilaris A, Fonts A, Prenafeta-Boldŏ FX (2019) The rise of blockchain technology in agriculture and food supply chains. Trends Food Sci Technol 91:640–652. https://doi.org/10.1016/j.tifs.2019.07.034
Kariyanna B, Sowjanya M (2024) Unravelling the use of artificial intelligence in management of insect pests. Smart Agric Technol 8:100517. Elsevier. Retrieved 2 Sept 2025, from https://www.sciencedirect.com/science/article/pii/S2772375524001229
Kashyap A, Shukla O, Prakash S, Kumar R (2025) Artificial intelligence in agriculture: unveiling trends in supply chain advancements. Sustain Futures 9:100709. Elsevier. Retrieved 2 Sept 2025, from https://www.sciencedirect.com/science/article/pii/S2666188825002758
Keicher R, Seufert H (2012) Automatic guidance for agricultural vehicles in Europe. Comput Electron Agric 25(1–2):169–194. https://doi.org/10.1016/S0168-1699(99)00037-5
Kerry R, Oliver MA (2004) Average variograms to guide soil sampling. Int J Appl Earth Obs Geoinf 5(4):307–325. https://doi.org/10.1016/j.jag.2004.07.005
Khanna A, Kaur S (2019) Evolution of Internet of Things (IoT) and its significant impact in the field of precision agriculture. Comput Electron Agric 157:218–231. https://doi.org/10.1016/j.compag.2018.12.039
Khosla R, Fleming K, Delgado JA, Shaver TM, Westfall DG (2019) Use of site-specific management zones to improve nitrogen management for precision agriculture. J Soil Water Conserv 57(6):513–518. https://doi.org/10.2489/jswc.57.6.513

Kogan M, Crist E, Jepson PC (2019) New integrated pest management paradigm for the modern age. J Integrated Pest Manage 10(1):12. https://doi.org/10.1093/jipm/pmz010
Koller M, Lanini WT (2005) Site-specific herbicide applications based on weed maps provide effective control. Calif Agric 59(3):182–187. https://doi.org/10.3733/ca.v059n03p182
de Koning K, van der Vorst Y (2017) Automatic milking-common practice on dairy farms. In: The First North American Conference on Precision Dairy Management, vol 2010, pp 52–67
Kumar A, Paritosh K, Pareek N, Vivekanand V (2021a) Smart farming for improving agricultural management. Egypt J Remote Sens Space Sci 24(3):971–981. https://doi.org/10.1016/j.ejrs.2021.08.007
Kumar A, Paritosh K, Pareek N, Vivekanand V (2021b) Conservation agriculture effects on soil water holding capacity and water-saving varied with management practices and agroecological conditions: a review. Agronomy 11(9):1681. https://doi.org/10.3390/agronomy11091681
Kumar A, Singh R, Patel S, Sharma K (2023) IoT-based precision irrigation system with real-time soil moisture monitoring for sustainable agriculture. J Agric Eng 54(2):1234–1247. https://doi.org/10.4081/jae.2023.1456
Kumari K, Nafchi AM, Mirzaee S, Abdalla A (2025) AI-driven future farming: achieving climate-smart and sustainable agriculture. Agri Eng 7:89. Retrieved 2 Sept 2025, from https://www.mdpi.com/2624-7402/7/3/89
Lamichhane JR, Dachbrodt-Saaydeh S, Kudsk P, Messéan A (2015) Toward a reduced reliance on conventional pesticides in European agriculture. Plant Dis 100(1):10–24. https://doi.org/10.1094/PDIS-05-15-0574-FE
Le Mouël C, Forslund A (2017) How can we feed the world in 2050? A review of the responses from global scenario studies. Eur Rev Agric Econ 44(4):541–591. https://doi.org/10.1093/erae/jbx006
Leroux C, Jones H, Pichon L, Guillaume S, Lamour J, Taylor J et al (2018) GeoFIS: an open source, decision-support tool for precision agriculture data. Agriculture 8(6):73. https://doi.org/10.3390/agriculture8060073
Li S, Elliott JA, Tiessen KH, Yarotski J, Lobb DA, Flaten DN (2011) The effects of multiple beneficial management practices on hydrology and nutrient losses in a small watershed in the Canadian Prairies. J Environ Qual 40(5):1627–1642. https://doi.org/10.2134/jeq2011.0054
Li S, Wang J, Zhang L, Chen Y, Liu X (2024) Integration of remote sensing and machine learning for precision agriculture: a comprehensive perspective on applications. Agronomy 14(9):1975. https://doi.org/10.3390/agronomy14091975
Liakos K, Busato P, Moshou D, Pearson S, Bochtis D (2018) Machine learning in agriculture: a review. Sensors 18(8):2674. https://doi.org/10.3390/s18082674. Retrieved 1 Sept 2025, from https://www.mdpi.com/1424-8220/18/8/2674
Liu X, Wang Y, Chen L, Zhang M (2020) Real-time nutrient cycling monitoring in agricultural soils using wireless sensor networks. Precis Agric 21(4):789–805. https://doi.org/10.1007/s11119-019-09698-1
Liu Y, Wang S, Chen J, Zhang M (2024) Crop yield prediction using machine learning: an extensive and systematic literature review. Expert Syst Appl 249:123456. https://doi.org/10.1016/j.eswa.2024.123456
Logeshwaran J, Srivastava D, Kumar KS, Rex MJ, Al-Rasheed A, Getahun M, Soufiene BO (2024) Improving crop production using an agro-deep learning framework in precision agriculture. BMC Bioinformatics 25(1). https://doi.org/10.1186/S12859-024-05970-9
Lowe A, Harrison N, French AP (2017) Hyperspectral image analysis techniques for the detection and classification of the early onset of plant disease and stress. Plant Methods 13(1):80. Springer. https://doi.org/10.1186/S13007-017-0233-Z
Lowenberg-DeBoer J, Huang IY, Grigoriadis V, Blackmore S (2020) Economics of robots and automation in field crop production. Precis Agric 21(2):278–299. https://doi.org/10.1007/s11119-019-09667-5
Mahlein AK (2016) Plant disease detection by imaging sensors–parallels and specific demands for precision agriculture and plant phenotyping. Plant Dis 100(2):241–251. https://doi.org/10.1094/PDIS-03-15-0340-FE

Mana A, Allouhi A, Hamrani A, Rehman S, El Jamaoui I, Jayachandran K (2024) Sustainable AI-based production agriculture: exploring AI applications and implications in agricultural practices. Smart Agric Technol 7:100416. Elsevier. Retrieved 2 Sept 2025, from https://www.sciencedirect.com/science/article/pii/S2772375524000212

Mancini S (2013) A real-life multi depot multi period vehicle routing problem with a heterogeneous fleet: formulation and adaptive large neighborhood search based matheuristic. Transp Res Part C: Emerg Technol 70:100–112. https://doi.org/10.1016/j.trc.2015.06.016

Marinello F, Pezzuolo A, Cillis D, Sartori L (2017) Application of ISO 11783 standard for automated operations in agriculture: a simulation approach for tractor-implement interaction. Biosyst Eng 158:1–9. https://doi.org/10.1016/j.biosystemseng.2017.03.005

Mercier S, Villeneuve S, Mondor M, Uysal I (2017) Time–temperature management along the food cold chain: a review of recent developments. Compr Rev Food Sci Food Saf 16(4):647–667. https://doi.org/10.1111/1541-4337.12269

Meshram K, Mishra U, Array UR (2025) Application of artificial intelligence in agri-tech, environmental and biodiversity conservation. Array 26:100412. Elsevier. Retrieved 2 Sept 2025, from https://www.sciencedirect.com/science/article/pii/S2590005625000396

Michels M, Fecke W, Feil JH, Musshoff O, Pigisch J, Krone S (2020) Smartphone adoption and use in agriculture: empirical evidence from Germany. Precis Agric 21(2):403–425. https://doi.org/10.1007/s11119-019-09675-5

Miller PCH, Butler Ellis MC (2000) Effects of formulation on spray nozzle performance for applications from ground-based boom sprayers. Crop Prot 19(8–10):609–615. https://doi.org/10.1016/S0261-2194(00)00080-6

Mirzabaev A, Kerr R, Hasegawa T et al (2023) Severe climate change risks to food security and nutrition. Clim Risk Manag 39:100473. Elsevier. Retrieved 1 Sept 2025, from https://www.sciencedirect.com/science/article/pii/S2212096322000808

Mishra H, Mishra D (2024) AI for data-driven decision-making in smart agriculture: from field to farm management. In: Artificial intelligence techniques in smart agriculture. Springer, pp 173–193. https://doi.org/10.1007/978-981-97-5878-4_11

Mohanty SP, Hughes DP, Salathé M (2016) Using deep learning for image-based plant disease detection. Front Plant Sci 7:1419. https://doi.org/10.3389/fpls.2016.01419

Mohyuddin G, Khan M, Haseeb A et al (2024) Evaluation of machine learning approaches for precision farming in smart agriculture system: a comprehensive review. IEEE Access 12:60155. Retrieved 1 Sept 2025, from https://ieeexplore.ieee.org/abstract/document/10504121/

Morais R, Fernandes MA, Matos SG, Serôdio C, Ferreira PJSG, Reis MJCS (2008) A ZigBee multi-powered wireless acquisition device for remote sensing applications in precision viticulture. Comput Electron Agric 62(2):94–106. https://doi.org/10.1016/j.compag.2007.12.004

Mulla DJ (2013) Twenty five years of remote sensing in precision agriculture: key advances and remaining knowledge gaps. Biosyst Eng 114(4):358–371. https://doi.org/10.1016/j.biosystemseng.2012.08.009

Murchie EH, Lawson T (2013) Chlorophyll fluorescence analysis: a guide to good practice and understanding some new applications. J Exp Bot 64(13):3983–3998. https://doi.org/10.1093/jxb/ert208

Nagasubramanian K, Jones S, Singh AK, Sarkar S, Singh A, Ganapathysubramanian B (2019) Plant disease identification using explainable 3D deep learning on hyperspectral images. Plant Methods 15(1). https://doi.org/10.1186/S13007-019-0479-8

Nannipieri P, Ascher J, Ceccherini MT, Landi L, Pietramellara G, Renella G (2017) Microbial diversity and soil functions. Eur J Soil Sci 68(1):12–26. https://doi.org/10.1111/ejss.4_12398

Nawar S, Corstanje R, Halcro G, Mulla D, Mouazen AM (2024) Rapid in-field soil analysis of plant-available nutrients and pH for precision agriculture—a review. Precis Agric 25(4):1789–1820. https://doi.org/10.1007/s11119-024-10181-6

Nawaz M, Nazir T, Javed A, Tariq U, Yong HS, Khan MA, Cha J (2025) A review on automated plant disease detection: motivation, limitations, challenges, and recent advancements for future research. J King Saud Univ-Comput Inf Sci 37(1):102009. https://doi.org/10.1016/j.jksuci.2024.102009

Nikhil U, Pandiyan A, Raja S, Stamenkovic Z (2024) Machine learning-based crop yield prediction in south india: performance analysis of various models. Computers 13:137. Retrieved 2 Sept 2025, from https://www.mdpi.com/2073-431X/13/6/137

O'Grady M, Langton D, O'Hare GMP (2019) Edge computing: a tractable model for smart agriculture? Artif Intell Agric 3:42–51. Elsevier. Retrieved 1 Sept 2025, from https://www.sciencedirect.com/science/article/pii/S2589721719300339

Oerke EC (2006) Crop losses to pests. J Agric Sci 144(1):31–43. https://doi.org/10.1017/S0021859605005708

Ojha T, Misra S, Raghuwanshi NS (2015) Wireless sensor networks for agriculture: the state-of-the-art in practice and future challenges. Comput Electron Agric 118:66–84. https://doi.org/10.1016/j.compag.2015.08.011

de Oliveira LFP, Moreira AP, Silva MF (2020) Implementation of artificial intelligence in agriculture for optimisation of irrigation and application of pesticides and herbicides. Artif Intell Agric 4:58–73. https://doi.org/10.1016/j.aiia.2020.03.002

Olufemi-Phillips A, Ofodile OC, Toromade AS, Abbey Ngochindo Igwe N, Eyo-Udo L (2024) Utilizing predictive analytics to manage food supply and demand in adaptive supply chains. Int J Business Manage 21(1):90–112. Retrieved 2 Sept 2025, from https://www.researchgate.net/profile/Amarachi-Olufemi-Phillips/publication/387312396_Utilizing_Predictive_Analytics_to_Manage_Food_Supply_and_Demand_in_Adaptive_Supply_Chains/links/6767fe3bfb9aff6eaae6ebdc/Utilizing-Predictive-Analytics-to-Manage-Food-Supply-and-Demand-in-Adaptive-Supply-Chains.pdf

Pace R, Schiano Di Cola V, Monti MM, Affinito A, Cuomo S, Loreto F, Ruocco M (2025) Artificial intelligence in soil microbiome analysis: a potential application in predicting and enhancing soil health—a review. Discov Appl Sci 7(2):85. https://doi.org/10.1007/S42452-024-06381-4

Palou E, López-Malo A, Barbosa-Cánovas GV, Welti-Chanes J, Swanson BG (2011) Polyphenoloxidase activity and color of blanched and high hydrostatic pressure treated banana puree. J Food Sci 64(1):42–45. https://doi.org/10.1111/j.1365-2621.1999.tb09857.x

Pandey D, Mishra R (2024) Towards sustainable agriculture: harnessing AI for global food security. Artif Intell Agric. Elsevier. Retrieved 2 Sept 2025, from https://www.sciencedirect.com/science/article/pii/S2589721724000151

Pedersen SM, Fountas S, Have H, Blackmore BS (2006) Agricultural robots—system analysis and economic feasibility. Precis Agric 7(4):295–308. https://doi.org/10.1007/s11119-006-9014-9

Peng J, Zhao Z, Liu D (2022) Impact of agricultural mechanization on agricultural production, income, and mechanism: evidence from Hubei Province, China. Front Environ Sci 10:1–17. https://doi.org/10.3389/fenvs.2022.838686

Pereira LS, Allen RG, Smith M, Raes D (2015) Crop evapotranspiration estimation with FAO56: past and future. Agric Water Manag 147:4–20. https://doi.org/10.1016/j.agwat.2014.07.031

Pereira LFS, Janzen T, Campos AD, Chalco-Cano Y (2018) Fuzzy modeling of the effects of irregular samples on batch fermentation of cheese whey. J Food Eng 162:29–37. https://doi.org/10.1016/j.jfoodeng.2015.04.005

Pingali PL (2012) Green revolution: impacts, limits, and the path ahead. Proc Natl Acad Sci 109(31):12302–12308. https://doi.org/10.1073/pnas.0912953109

Pivoto D, Waquil PD, Talamini E, Finocchio CPS, Dalla Corte VF, de Vargas Mores G (2018) Scientific development of smart farming technologies and their application in Brazil. Inf Process Agric 5(1):21–32. https://doi.org/10.1016/j.inpa.2017.12.002

Porep JU, Kammerer DR, Carle R (2015) On-line application of near infrared (NIR) spectroscopy in food production. Trends Food Sci Technol 46(2):211–230. https://doi.org/10.1016/j.tifs.2015.10.002

Rahman A, Syed HH, Khan MS, Ahmad K, Ishaq A, Khan AR et al (2025) Integrating IoT sensors and machine learning for sustainable precision agroecology: enhancing crop resilience and resource efficiency through data-driven strategies, challenges, and future prospects. Discov Agric 3(1):24. https://doi.org/10.1007/s44279-025-00247-y

Ram B, Oduor P, Igathinathane C, Howatt K, Sun X (2024) A systematic review of hyperspectral imaging in precision agriculture: analysis of its current state and future prospects. Comput

Electron Agric. Elsevier. Retrieved 1 Sept 2025, from https://www.sciencedirect.com/science/article/pii/S0168169924004289

Reid JF, Zhang Q, Noguchi N, Dickson M (2000) Agricultural automatic guidance research in North America. Comput Electron Agric 25(1–2):155–167. https://doi.org/10.1016/S0168-1699(99)00036-3

Rhoades JD, Kandiah A, Mashali AM (1992) Water quality for agriculture (irrigation and drainage paper 29). Food and Agriculture Organization of the United Nations, Rome. Available at: https://www.fao.org/4/t0234e/t0234e00.htm

Roberts MJ, Braun NO, Sinclair TR, Lobell DB, Schlenker W (2017) Comparing and combining process-based crop models and statistical models with some implications for climate change. Environ Res Lett 12(9):095010. https://doi.org/10.1088/1748-9326/aa7f33

Rong A, Akkerman R, Grunow M (2011) An optimization approach for managing fresh food quality throughout the supply chain. Int J Prod Econ 131(1):421–429. https://doi.org/10.1016/j.ijpe.2009.11.026

Rose DC, Sutherland WJ, Parker C, Lobley M, Winter M, Morris C et al (2016) Decision support tools for agriculture: towards effective design and delivery. Agric Syst 149:165–174. https://doi.org/10.1016/j.agsy.2016.09.009

Rouse JW, Haas RH, Schell JA, Deering DW (1974) Monitoring vegetation systems in the Great Plains with ERTS. In: Third ERTS Symposium, NASA SP-351, pp 309–317

Rugji J, Erol Z, Taşçı F, Musa L, Hamadani A, Gündemir MG, Karalliu E, Siddiqui SA (2024) Utilization of AI–reshaping the future of food safety, agriculture and food security–a critical review. Crit Rev Food Sci Nutr 65(26):5136–5180. Taylor & Francis. https://doi.org/10.1080/10408398.2024.2430749

Rustia DJ, Chao JJ, Chiu LY, Wu YF, Chung JY, Hsu JC, Lin TT (2021) Prediction of pest insect appearance using sensors and machine learning. Sensors 21(14):4846. https://doi.org/10.3390/s21144846

Rutten CJ, Velthuis AG, Steeneveld W, Hogeveen H (2013) Invited review: sensors to support health management on dairy farms. J Dairy Sci 96(4):1928–1952. https://doi.org/10.3168/jds.2012-6107

Sahoo L, Mohapatra D, Raghuvanshi HR, Kumar S, Kaur R, Chawla R, Afreen N (2024) Transforming agriculture through artificial intelligence: advancements in plant disease detection, applications, and challenges. J Adv Biol Biotechnol 27:381–388. Retrieved 1 Sept 2025, from https://www.researchgate.net/profile/Sonal-Kumar-3/publication/379837071_Transforming_Agriculture_through_Artificial_Intelligence_Advancements_in_Plant_Disease_Detection_Applications_and_Challenges/links/661e105066ba7e2359dd300b/Transforming-Agriculture-through-Artificial-Intelligence-Advancements-in-Plant-Disease-Detection-Applications-and-Challenges.pdf

Saikai Y, Peake A, Chenu K (2023) Deep reinforcement learning for irrigation scheduling using high-dimensional sensor feedback. PLoS Water 2(9):e0000169. https://doi.org/10.1371/JOURNAL.PWAT.0000169

Saiz-Rubio V, Rovira-Más F (2020) From smart farming towards agriculture 5.0: a review on crop data management. Agronomy 10(2):207. https://doi.org/10.3390/agronomy10020207

Saleem MH, Potgieter J, Arif KM (2019) Plant disease detection and classification by deep learning. Plants 8(11):468. https://doi.org/10.3390/plants8110468

Sapkota R, Karkee M (2025) Comparing YOLOv11 and YOLOv8 for instance segmentation of occluded and non-occluded immature green fruits in complex orchard environment. http://arxiv.org/abs/2410.19869

Sathya D, Thangamani R, Balaji BS (2024) The revolution of edge computing in smart farming. In: Intelligent robots and drones for precision agriculture. Springer. Part F2461, pp 351–389. https://doi.org/10.1007/978-3-031-51195-0_17

Schepers JS, Blackmer TM, Wilhelm WW, Resende M (2004) Transmittance and reflectance measurements of corn leaves from plants with different nitrogen and water supply. J Plant Physiol 161(2):185–195. https://doi.org/10.1078/0176-1617-01098

Schimmelpfennig D (2016) Farm profits and adoption of precision agriculture. USDA-ERS Econ Res Rep 217. https://doi.org/10.22004/ag.econ.249766

Senapaty M, Ray A (2023) IoT-enabled soil nutrient analysis and crop recommendation model for precision agriculture. Computers 12:61. Retrieved 1 Sept 2025, from https://www.mdpi.com/2073-431X/12/3/61

Sentelhas PC, Dalla Marta A, Orlandini S, Santos EA, Gillespie TJ, Gleason ML (2008) Suitability of relative humidity as an estimator of leaf wetness duration. Agric For Meteorol 148(3):392–400. https://doi.org/10.1016/j.agrformet.2007.09.011

Shahhosseini M, Hu G, Archontoulis SV (2021) Forecasting corn yield with machine learning ensembles. Front Plant Sci 11:1120. https://doi.org/10.3389/fpls.2020.01120

Sharma K, Shivandu SK (2024) Integrating artificial intelligence and Internet of Things (IoT) for enhanced crop monitoring and management in precision agriculture. Sens Int 5:100292. Elsevier. Retrieved 1 Sept 2025, from https://www.sciencedirect.com/science/article/pii/S2666351124000147

Sharma A, Kumar R, Singh P (2022) Smart weather data management based on artificial intelligence and big data analytics for precision agriculture. Agriculture 13(1):95. https://doi.org/10.3390/agriculture13010095

Sharma A, Jain A, Gupta P, Chowdary V (2023) Artificial intelligence in aSgriculture: benefits, challenges, and trends. Appl Sci 13(13):7405. https://doi.org/10.3390/app13137405

Sharma P, Kumar V, Singh A, Patel R (2024) Advanced technologies of soil moisture monitoring in precision agriculture: a comprehensive review. Comput Electron Agric 218:108674. https://doi.org/10.1016/j.compag.2024.108674

Shepherd M, Turner JA, Small B, Wheeler D (2018) Priorities for science to overcome hurdles thwarting the full promise of the 'digital agriculture' revolution. J Sci Food Agric 100(14):5083–5092. https://doi.org/10.1002/jsfa.9346

Shivaprakash KN, Swami N, Mysorekar S, Arora R, Gangadharan A, Vohra K, Jadeyegowda M, Kiesecker JM (2022) Potential for artificial intelligence (AI) and machine learning (ML) applications in biodiversity conservation, managing forests, and related services in India. Sustainability 14:7154. https://doi.org/10.3390/su14127154

Shrivastava P, Tewari VK, Gupta C, Singh G (2023) IoT and radio telemetry based wireless engine control and real-time position tracking system for an agricultural tractor. Discover Internet of Things, 3(1):6

Silvenius F, Gronman K, Katajajuuri JM, Soukka R, Koivupuro HK, Virtanen Y (2014) The role of household food waste in comparing environmental impacts of packaging alternatives. Packag Technol Sci 27(4):277–292. https://doi.org/10.1002/pts.203

Silwal A, Davidson JR, Karkee M, Mo C, Zhang Q, Lewis K (2017) Design, integration, and field evaluation of a robotic apple harvester. J Field Robot 34(6):1140–1159. https://doi.org/10.1002/rob.21715

Sinclair TR, Purcell LC, Sneller CH (2005) Crop transformation and the challenge to increase yield potential. Trends Plant Sci 9(2):70–75. https://doi.org/10.1016/j.tplants.2003.12.008

Singh RP, Hodson DP, Huerta-Espino J, Jin Y, Bhavani S, Njau P et al (2011) The emergence of Ug99 races of the stem rust fungus is a threat to world wheat production. Annu Rev Phytopathol 49:465–481. https://doi.org/10.1146/annurev-phyto-072910-095423

Singh V, Sharma N, Singh S (2023) Drones in plant disease assessment, efficient monitoring, and detection: a way forward to smart agriculture. Agronomy 13(6):1524. https://doi.org/10.3390/agronomy13061524

Singh K, Sharma R, Kumar A (2024) A deep learning model for predicting risks of crop pests and diseases from sequential environmental data. Plant Methods 20(1):15. https://doi.org/10.1186/s13007-023-01122-x

Sishodia R, Ray R, Singh SK (2020) Applications of remote sensing in precision agriculture: a review. Remote Sens 12(19):3136. https://doi.org/10.3390/rs12193136. Retrieved 1 Sept 2025, from https://www.mdpi.com/2072-4292/12/19/3136

Snyder RL, Spano D, Duce P, Baldocchi D (2005) An introduction to environmental biophysics. Springer Science & Business Media. https://doi.org/10.1007/b100273

Soy A, Prakash VV (2025) AI-Powered soil nutrient mapping for optimized crop rotation strategies with advanced multispectral analysis and precision variable rate technology. In: SHS Web of Conferences. EDP Sciences. https://doi.org/10.1051/iciaites/2025042701062

Spekken M, Anselmi AA, Molin JP (2013) A simple method for filtering spatial data and determining the width of the boundary strip in on-the-go experiments. Biosyst Eng 115(4):470–480. https://doi.org/10.1016/j.biosystemseng.2013.04.005

Stafford JV (2000) Implementing precision agriculture in the 21st century. J Agric Eng Res 76(3):267–275. https://doi.org/10.1006/jaer.2000.0577

Steenwerth KL, Hodson AK, Bloom AJ, Carter MR, Cattaneo A, Chartres CJ et al (2014) Climate-smart agriculture global research agenda: scientific basis for action. Agric Food Secur 3(1):11. https://doi.org/10.1186/2048-7010-3-11

Sudu B, Rong G, Guga S, Li K, Zhi F, Guo Y et al (2022) Retrieving SPAD values of summer maize using UAV hyperspectral data based on multiple machine learning algorithm. Remote Sens 14(21):5407. https://doi.org/10.3390/rs14215407

Suman J, Rakshit A, Ogireddy SD, Singh S, Gupta C, Chandrakala J (2022) Microbiome as a key player in sustainable agriculture and human health. Front Soil Sci 2:1–25. https://doi.org/10.3389/fsoil.2022.821589

Talaviya T, Shah D, Patel N, Yagnik H, Shah M (2020) Implementation of artificial intelligence in agriculture for optimisation of irrigation and application of pesticides and herbicides. Artif Intell Agric 4:58–73. Elsevier. Retrieved 2 Sept 2025, from https://www.sciencedirect.com/science/article/pii/S258972172030012X

Tardieu F, Simonneau T, Muller B (2014) The physiological basis of drought tolerance in crop plants: a scenario-dependent probabilistic approach. Annu Rev Plant Biol 65:733–759. https://doi.org/10.1146/annurev-arplant-050213-040000

Tey YS, Brindal M (2012) Factors influencing the adoption of precision agricultural technologies: a review for policy implications. Precis Agric 13(6):713–730. https://doi.org/10.1007/s11119-012-9273-6

Too EC, Yujian L, Njuki S, Yingchun L (2021) A comparative study of fine-tuning deep learning models for plant disease identification. Comput Electron Agric 161:272–279. https://doi.org/10.1016/j.compag.2018.03.032

Tsang YP, Choy KL, Wu CH, Ho GTS, Lam CHY, Koo PS (2019) An Internet of Things (IoT)-based risk monitoring system for managing cold supply chain risks. Ind Manag Data Syst 118(7):1432–1462. https://doi.org/10.1108/IMDS-09-2017-0384

Tsouros D, Bibi S, Sarigiannidis PG (2019) A review on UAV-based applications for precision agriculture. Information 10:349. Retrieved 1 Sept 2025, from https://www.mdpi.com/2078-2489/10/11/349

Van Klompenburg T, Kassahun A, Catal C (2020) Crop yield prediction using machine learning: a systematic literature review. Comput Electron Agric 177:105709. https://doi.org/10.1016/j.compag.2020.105709. Elsevier. Retrieved 2 Sept 2025, from https://www.sciencedirect.com/science/article/pii/S0168169920302301

Vasconez JP, Kantor GA, Auat Cheein FA (2019) Human–robot interaction in agriculture: a survey and current challenges. Biosyst Eng 179:35–48. https://doi.org/10.1016/j.biosystemseng.2018.12.005

Vecchio Y, Agnusdei GP, Miglietta PP, Capitanio F (2020) Adoption of precision farming tools: the case of Italian farmers. Int J Environ Res Public Health 17(3):869. https://doi.org/10.3390/ijerph17030869

Venkateswara S, Padmanabhan J (2025) Deep learning based agricultural pest monitoring and classification. Sci Rep 15:8684. https://doi.org/10.1038/s41598-025-92659-5

Verdouw CN, Wolfert J, Beulens AJ, Rialland A (2016) Virtualization of food supply chains with the internet of things. J Food Eng 176:128–136. https://doi.org/10.1016/j.jfoodeng.2015.11.009

Wadoux AMJC, Minasny B, McBratney AB (2023) AI and machine learning for soil analysis: an assessment of sustainable agricultural practices. Bioresour Bioprocess 10(1):93. https://doi.org/10.1186/s40643-023-00710-y

Walter A, Finger R, Huber R, Buchmann N (2017) Opinion: smart farming is key to developing sustainable agriculture. Proc Natl Acad Sci 114(24):6148–6150. https://doi.org/10.1073/pnas.1707462114

Wang C, Liu B, Liu L, Zhu Y, Hou J, Liu P, Li X (2019) A review of deep learning used in the hyperspectral image analysis for agriculture. Artif Intell Rev 54(7):5205–5253. https://doi.org/10.1007/s10462-021-10018-y

Wang Q, Li Y, Li R (2024) Ecological footprints, carbon emissions, and energy transitions: the impact of artificial intelligence (AI). Humanit Soc Sci Commun 11:1–18. https://doi.org/10.1057/s41599-024-03520-5

Wei Y, Han C, Yu Z (2023) An environment safety monitoring system for agricultural production based on artificial intelligence, cloud computing and big data networks. J Cloud Comput 12(1):83. Springer. https://doi.org/10.1186/S13677-023-00463-1

Weraikat D, Šorič K, Žagar M, Sokač M (2024) Data analytics in agriculture: enhancing decision-making for crop yield optimization and sustainable practices. Sustainability 16(17):7331. Retrieved 1 Sept 2025, from https://www.mdpi.com/2071-1050/16/17/7331

Wolfert S, Ge L, Verdouw C, Bogaardt MJ (2017) Big data in smart farming–a review. Agric Syst 153:69–80. https://doi.org/10.1016/j.agsy.2017.01.023

Wu W, Ma B (2015) Integrated nutrient management (INM) for sustaining crop productivity and reducing environmental impact: a review. Sci Total Environ 512:415–427. Elsevier. Retrieved 1 Sept 2025, from https://www.sciencedirect.com/science/article/pii/S0048969715000030

Xia Y, Zhang M, Tsang DCW, Li L, Du B, Li Q, Poon CS, Ok YS (2020) Recent advances in control technologies for non-point source pollution with nitrogen and phosphorous from agricultural runoff: current practices and future prospects. Appl Biol Chem 63(1):1–25. https://doi.org/10.1186/s13765-020-0493-6

Xiong Y, Ge Y, Grimstad L, From PJ (2020) An autonomous strawberry-harvesting robot: design, development, integration, and field evaluation. J Field Robot 37(2):202–224. https://doi.org/10.1002/rob.21889

Xue J, Su B (2017) Significant remote sensing vegetation indices: a review of developments and applications. J Sens 2017:1353691. https://doi.org/10.1155/2017/1353691

Young SL, Meyer GE, Woldt WE (2014) Future directions for automated weed management in precision agriculture. In: Automation: the future of weed control in cropping systems. Springer, pp 249–259. https://doi.org/10.1007/978-94-017-9454-4_12

Zhang Q, Kovacs JM (2012) The application of small unmanned aerial systems for precision agriculture: a review. Precis Agric 13(6):693–712. https://doi.org/10.1007/s11119-012-9274-5

Zhang Y, Collins AL, Murdoch N, Lee D, Naden PS (2014) Cross sector contributions to river pollution in England and Wales: updating waterbody-scale information to support policy delivery for the water framework directive. Environ Sci Pol 42:16–32. https://doi.org/10.1016/j.envsci.2014.04.010

Zhang Z, Kayacan E, Thompson B, Chowdhary G (2020) High precision control and deep-learning-based corn stand counting algorithms for agricultural robot. Auton Robot 44(7):1289–1302. https://doi.org/10.1007/s10514-020-09915-y

Zhang X, Li Y, Lv Z, Wang J, Chen X, Lv H et al (2022) Research on cold chain logistics traceability system of fresh agricultural products based on blockchain. Comput Intell Neurosci 2022:1957957. https://doi.org/10.1155/2022/1957957

Zhang H, Li J, Chen W, Wang S (2023) Machine learning enabled IoT system for soil nutrients monitoring and crop recommendation. Smart Agric Technol 4:100187. https://doi.org/10.1016/j.atech.2023.100187

Zhang W, Tang P, Zhao L, Huang Q (2024) Integration of remote sensing and machine learning for precision agriculture: a comprehensive perspective on applications. Agronomy 14(9):1–28. https://doi.org/10.3390/agronomy14091975

Genotyping Advancements for Mitigating Biotic Stress-Induced Damage in Wheat

13

Chandranandani Negi, Mansi Singh,
Neeraj Kumar Vasistha, Dharmendra Singh,
Krishan Kumar, and Pritesh Vyas

Abstract

Wheat being a staple food crop around the globe, is severely threatened by the diseases which results in huge crop losses. These diseases include wheat rust, spot blotch, fusarium head blight, powdery mildew, and barley yellow head virus. The pathogens of these diseases keep evolving and make them robust to cause pathogenesis in a wide range of temperatures and elevation. It is this resilience of these pathogens that helps them to spread widely in various geographical regions and cause epidemic in extreme cases. To prevent these losses, crop breeders keep on working on strategies for disease-resistant wheat varieties around the globe. To develop disease-resistant varieties in wheat, it is necessary to identify genomic regions (QTLs, Genes, Alleles) associated with disease resistance. Wild wheat relatives are a potential genomic resource of such genomic regions and have resistance against many present-day pathogens. These wild wheat relatives are used to introgress disease-resistant genes into elite wheat cultivars and the breeding material so developed is used for advanced molecular studies. Various genotyping techniques such as GBS, GWAS, DArT assay, and the next-generation sequencing techniques along with advanced bioinformatics

C. Negi · K. Kumar
Department of Science & Technology, Jayoti Vidyapeeth Women's University, Jaipur, Rajasthan, India

M. Singh · P. Vyas (✉)
School of Applied and Life Sciences, Uttaranchal University, Dehradun, Uttarakhand, India

N. K. Vasistha
Department of Genetics & Plant Breeding, Rajiv Gandhi University, Itanagar, Arunachal Pradesh, India

D. Singh
Government Model College, Jhabua, Madhya Pradesh, India

Department of Biotechnology, Central University of Kashmir, Ganderbal, Jammu and Kashmir, India

R. E. Krishnankutty et al. (eds.), *Smart Agriculture*,
https://doi.org/10.1007/978-981-95-8785-8_13

tools utilize this material and identify the significant disease-resistant genomic regions. The RNA-Seq provides transcriptome analysis in addition to the sequencing data. This review presents a comprehensive account about wheat breeding for disease resistance, with utilization of the breeding material for genomic analysis using various genotyping and sequencing techniques to identify disease-resistant genomic regions.

Keywords

Wheat · *Aegilops* · Genotyping · Rust · Powdery Mildew · Spot Blotch

1 Introduction

Wheat (*Triticum aestivum* L., $2n = 6x = 42$, AABBDD), with three inter-related diploid genomes of ~17 Gb, out of which more than 80% is made up of repetitive sequences. Wheat is a staple food of 35% people in the world supplying around 20% of the protein consumption in human diet (Arumuganathan and Earle 1991; Shiferaw et al. 2013). The bread wheat comprises of around 95% of wheat cultivated across the globe while the remaining percentage is of durum wheat (Shewry 2009). Wheat being one of the largest cultivated cereals around the globe with over 215 M ha of cultivated land (Babu et al. 2020). Domestication, selection, and cultivation of wheat by people have led to tremendous improvement of its agronomical traits. According to the statistics, nearly 734 million tonnes of wheat production was achieved in the year 2018 (http://www.fao.org/faostat/en/#data/QC). However, losses incurred due to pathogens are drastic where stripe and stem rust could lead to 100% while leaf rust can cause 50% yield loss during severe outbreak (Bhardwaj et al. 2016). In Asia, nearly 46% of yield loss is caused by stripe rust outbreak (Singh et al. 2004).

Among numerous fungal wheat diseases, stem rust, powdery mildew (PM), spot blotch (SB) and yellow rust (YR) are most detrimental. There are three rust diseases of wheat, namely stem rust, leaf rust, and stripe rust, all infested through the pathogens of the *Basidiomycete* family, namely leaf rust by *Puccinia triticina* Eriks (Pt), stem rust by *P. graminis* Pers. f. sp. *tritici* Eriks. & Henn. (Pgt), and stripe rust by *P. striiformis* Westend. f. sp. *tritici* (Pst) (McIntosh et al. 1995).

Genomic studies of diverse populations provide the estimation of genetic determinants which control traits in plant, with their role in abiotic or biotic stress resistance. Analysis through precise phenotyping and genotyping during early generations recombinant inbred lines (RILs) gives us the genomic regions bearing disease resistance in the selected lines and panels (Bansal and Bariana 2017). In the recent years, Genome-wide association studies (GWAS), Genotyping-by-Sequencing (GBS), Whole genome resequencing (WGRS), and RNA sequencing (RNA-Seq) are being widely used for the identification of QTLs, MTAs, and genes related to development and growth activities and their relation with environmental stresses in different tissues and stages in crop plant life cycle (Yu et al. 2006).

Next-generation sequencing (NGS) paves the way at whole genome level for the determination of nucleotides which are responsible for phenotypic changes. Systems like Heliscope™ Single Molecule Sequencer, Single-Molecule Real-Time (SMRT™) Sequencer by Pacific Biosciences (PacBio), AB SOLiD System (Sequencing by Oligo Ligation Detection), Roche 454 sequencing system (pyrosequencing), Illumina Sequencers (sequencing-by-synthesis), Oxford Nanopore Technologies (ONT) sequencing, and the Ion Personal Genome Machine have the capability to generate multiple long sequence reads in a short time (Barba et al. 2014; Bräutigam and Gowik 2010). During the last decade, these genotyping platforms have led to the emergence of various NGS-based molecular markers, which has equipped scientists with elaborate knowledge of the genomes facilitating precise mapping of targeted loci.

Application of NGS-based 90K SNP arrays with bulk segregant analysis (BSA) results for quick identification of markers linked to a particular genomic area or quantitative trait loci (QTL). In this, bulk of individuals with phenotypes of two extremities with respect to a targeted trait along with contrasting parents are sampled and genotyped. Further, statistical analysis of markers provides the putative regions responsible for the trait (Michelmore et al. 1991). Next-generation sequencing (NGS) data deciphers a vast information regarding gene expressions which is an important tool to understand the complexities of biological functions. Utilizing GWAS for the identification of QTLs in the selected population can be additionally accompanied with RNA-Seq data for further analysis of the gene expression. Employing of Diversity Array Technology (DArT) for hybridization-based genotyping (that exploits microarray technology) is done for identification of DNA variations at numerous genomic simultaneously without depending on sequence information (Wenzl et al. 2004). Technological advancements have combined DArT with NGS, which offers more flexibility for characterizing the germplasm. In DArTSeq method of complexity reduction, only the genomic coding regions are targeted. Complexity reduction is achieved using restriction enzymes which is combined with NGS technology for rapid SNP discovery. Diversity arrays technology (DArT) markers have been widely utilized for investigating disease resistance in wheat recombinant lines having gene transfer from wild relatives (Crossa et al. 2007; Millet et al. 2014). Presently, Illumina sequencer is the commonly used system utilizing sequencing-by-synthesis principle and bridge amplification with Illumina HiSeq 1500, HiSeq 2000, HiSeq 2500, HiSeq 3000 or HiSeq 4000 alternatives (Ari and Arikan 2016). Illumina sequencers however produce short reads while SMRT sequencer being able to generate longer reads (Shaheenuzzamn et al. 2020).

Initially the genome assembly for bread wheat (Chinese Spring), *Triticum urartu* and *Aegilops tauschii* was done using Roche 454 and Illumina sequencing approaches (Brenchley et al. 2012; Jia et al. 2013; Ling et al. 2013). Utilization of orthologous gene pairs was done for searching the draft genome sequences of *Ae. tauschii* accession AL8/78 and the hexaploid wheat cv. Chinese Spring 454 contigs for identifying homologous contig or scaffold sequences developing polymorphic markers. Utilization of such contig and scaffold sequences was done for screening of simple sequence repeat (SSR) motifs by employing BatchPrimer3 program (You et al. 2008). Although these assemblies were highly fragmented and contained numerous unordered scaffolds, these works gave the initial picture of wheat genome.

Wheat diseases harshly impact the quality and yield of grain around the globe, with new races evolved to become more harmful and widely infecting the wheat growing areas. Utilizing genetic control strategies is the most impactful, economical, and sustainable way for controlling the diseases and minimizing yield losses. This is done by identifying and using the novel genetic sources. There is need to uncover novel QTLs/genes and pyramid them for achieving high levels of long-term resistance. In the past few years, discovery through accurate mapping of several disease-resistance loci/genes have led to huge change in wheat breeding approach. Over 220 rust-resistance genes including 83 for stripe rust (SR), 79 LR, and 60 stem rust resistance genes have been officially named and catalogued utilizing biparental populations (Li et al. 2020a, b, c; McIntosh et al. 2014). Rust fungi, being obligate parasites needs a living host for completing its lifecycle. The present review gives a comprehensive insight of wheat diseases, their impact on wheat production, and application of advanced and new genomic approaches for identification of novel MTAs/genes and their utilization for sustained crop production.

2 Diseases Severely Affecting Wheat Production and Associated Resistant Genes

Several diseases are responsible for causing huge losses in crop production in wheat. The causal agent of most of these diseases is fungi which forms a hyphal network and spreads in the mesophyll tissue (Fig. 13.1). The morphological appearance is different but the manifestation of the hyphal network is similar. The major diseases are the rust diseases, powdery mildew, spot blotch, fusarium head blight (FHB), barley yellow dwarf virus. The FHB is concerning for wheat production and studies regarding to the disease have led to identification of around 50 unique QTLs along with *Fhb1*, which provide resistance to the diverse wheat population (Bai et al. 2018), although genomic studies are prominently promising for discovering the genomic regions and resistance genes. Similarly, another problem related to this cereal is Barley yellow dwarf virus (BYDV). Since the primary genetic pool of wheat lacks resistance for the same, the genes imparting resistance originates from wheat wild relatives as three genes *bdv2*, *bdv3,* and *bdv4* from *Thinopyrum intermedium* (Zhang et al. 2009) were introgressed into wheat cultivars.

In case of yellow rust, at present, globally 88% of the wheat production is at risk to it, resulting for a loss of more than five million tons, having an approximate market value of USD 1 billion annually (Schwessinger 2017). 84 permanently designated, 100 temporarily designated stripe rust-resistance genes, and 363 QTLs found in wheat (Haider et al. 2023; Klymiuk et al. 2022; McIntosh et al. 2010) whereas the research on finding LR resistance in wheat have been done by wheat researchers throughout the world led to characterization of *Lr* genes from wild wheat relatives such as *Agropyron elongatum* (*Lr24*) (Schachermayr et al. 1995), *Aegilops tauschii*

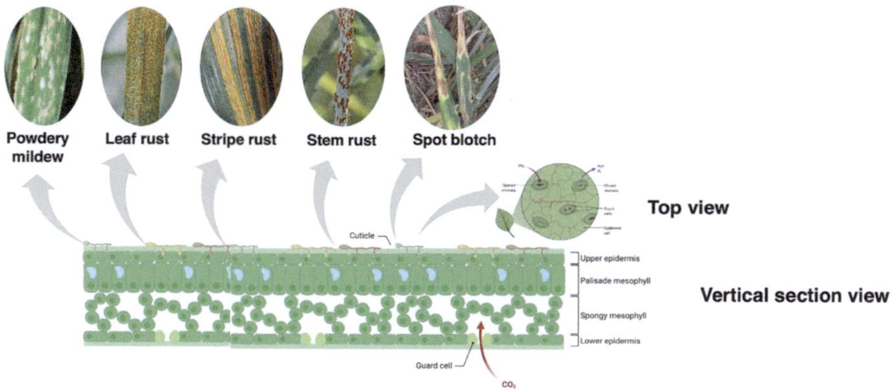

Fig. 13.1 Disease-infection process begins as pathogen spore lands on the leaf surface started to germinate under the favorable conditions and form germ tube. The germ tube before entering the cell through stomatal opening form appressorium. More advanced infection structures developed from the appressorium are infection peg, sub-stomatal vesicles, and primary infection hyphae. These further transformed to haustorial mother cell (HMC) which extracts the nutrition required for pathogen growth inside the cell

(*Lr21*) [Huang and Gill 2001), *Aegilops umbellulata* (*Lr9*) (Gupta et al. 2005) and *Secale cereale* (*Lr26*) (Zeller 1973) and *Aegilops longissima* (*Lr85*) (Sharma et al. 2024).

In SR, out of many alien genes derived from wheat wild relatives, *Sr32* and *Sr39*, introgressed from the short arm of Chr 2 utilizing separate accessions of *Ae. Speltoides,* which maintains resistivity against various pathogen strains of the SR. Diversity arrays technology (DArT) markers associated with *Lr34*/*Yr18*/*Pm38*, *Lr46*/*Yr29*/*Pm39*, *Sr2*, *Sr6*, *Sr25* are present for marker-assisted selection (Bhardwaj et al. 2019). For powdery mildew, the identification of 88 PM genes or alleles have been identified at 66 loci (*Pm1* to *Pm66*) (Li et al. 2020a, b, c; Zhu et al. 2020). In addition to this, above 30 temporarily designated *Pm* genes have also been described and designated to their corresponding wheat chromosome (Li et al. 2020a, b, c). In wheat, mapping of several (powdery mildew) PM resistance genes have been achieved, with cloning of *Pm2*, *Pm3*, *Pm8* and *Pm21* (Cao et al. 2011; Hurni et al. 2013; Sánchez-Martín et al. 2016; Yahiaoui et al. 2004). In SB, 3 formally designated spot blotch (*Sb*) resistance genes (*Sb1* to *Sb3*) have been identified (Kumar et al. 2015; Lillemo et al. 2013; Lu et al. 2016).

These fungal attacks trigger several immune responses (Fig. 13.2) which results in various physiological changes in the plant and brings further responses for their defense mechanism. The extreme prevalence of any of the above-mentioned diseases with continuous evolution of new races and changes in their infection pattern may lead to severe food crises. Therefore, a huge attention is required for identifying, characterizing, and utilizing durable disease-resistant genes from different sources. Progress and expansion in several molecular approaches have been done through which various biological processes and phenomenon like RNA expression, genomic variation, DNA-protein interaction can be tracked and high-throughput genome sequencing is possible.

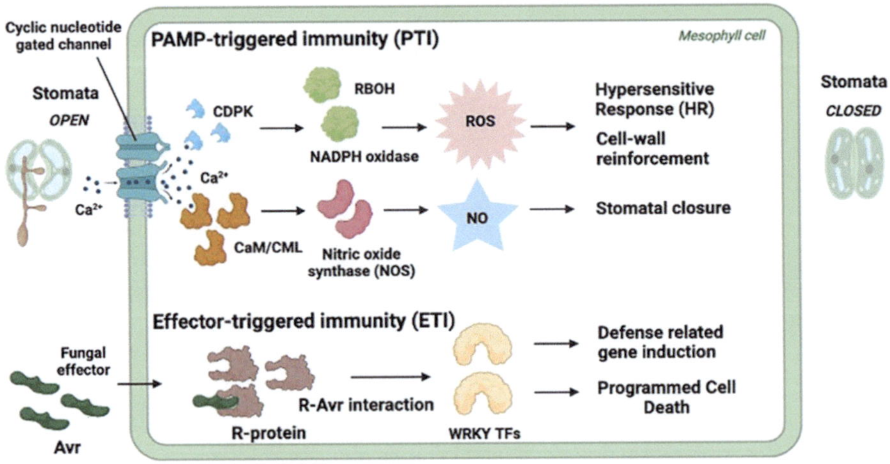

Fig. 13.2 Wheat–fungal pathogen interaction at molecular level occurred via PAMP (pathogen associated molecular pattern)-triggered immunity (PTI) where pathogen spores are received mechanically through cyclic nucleotide gated channels such as mechanosensitive channels and voltage gated channels lead to change in ion-flux (Deller et al. 2011). The influx of Ca^{2+} lead to activation of defense signaling via calcium-dependent protein kinase (CDPK) and calmodulin (CaM)/calmodulin-like (CML) proteins. These proteins further activate respiratory burst oxidase homolog (RBOH) leading to generation of reactive oxygen species (ROS) such as peroxide, superoxide, hydroxyl radical, singlet oxygen and alpha oxygen resulted in hypersensitive response (HR) and cell wall reinforcement. Also, defense signaling activates nitric oxide synthase (NOS) leading to release of nitric oxide responsible for stomatal closure. The next level of plant immunity is effector triggered immunity (ETI) when fungal effector avirulence (Avr) protein interact with resistance (R) protein within cell lead to activation of WRKY transcription factors resulting in induction of other defense-related genes, causing programmed cell death (PCD) of infected cells (Yuan et al. 2021). The response is visualized on leaf surface as necrotrophic spots

3 Recent Advances in Genomics and Next-Generation Sequencing (NGS) Approaches

Improvements in NGS technology with associated bioinformatics tools have revolutionized the wheat genomic studies. Various software models and tools are being utilized for analyzing the NGS data, after breeding wheat for resistance to diseases, which have brought rapid progress in the breeding programs. These tools can be utilized depending on the need of the users covering genome browsing and annotation, aligning the sequence reads with reference genome sequence, base-calling, and detecting the polymorphism. Several alignment tools have emerged for fast alignment of wide sets of short reads with identifying gaps and mismatches to wheat genome sequence, being a landmark for the exploration of marker trait association (Table 13.1). Huge applicability of NGS includes construction of genetic maps, exploring the structural rearrangements, precise study of polymorphism or mutations, transcriptome analysis, estimation of genomic diversity, and facilitation of appropriate markers to show genome variation. Application of NGS for detection of multiple polymorphic DNA sequence markers in a short span of time has made this

Table 13.1 Various methods used in the genotyping of breeding material

S. No.	Genotyping methods	Platform	Chemistry	References
1.	Genotyping by sequencing	Roche 454 GS FLX; Illumina	Genome complexity reduction	He et al. (2014)
2.	Infinium assay	Illumina	Hybridization and extension on BeadChip	Adler et al. (2013)
3.	GoldenGate assay	Illumina	Hybridization of Allele-specific oligos or locus-specific oligos on Array Matrix or BeadChip	González-Neira (2013)
4.	DArT assay	Diversity Array technology	Microarray hybridization	Akbari et al. (2006), Pascual et al. (2020)
5.	KASPar assay	KBiosciences	Competitive allele-specific PCR amplification on Fluidigm's dynamic array	Smith and Maughan (2015)
6.	BeadXpress	Illumina	Hybridization on VeraCode microbeads in 96 well plate or 8 well strip	Thomson et al. (2012)
7.	SSR	Generic designed manually	Hybridization and PCR amplification	Jaiswal et al. (2017)

technology an important tool in breeding (Shendure and Ji 2008). Various high throughput and precise approaches consisting of SNPs (single nucleotide polymorphisms), DArT-Seq (Diversity Array Technology Sequencing), GBS, and population-specific targeted genotyping-by-sequencing (tGBS) have provided researchers with helpful information about genomic regions conferring disease resistance in wheat (Nsabiyera et al. 2020).

Genotyping-by-sequencing (GBS) displays numerous single nucleotide polymorphism (SNPs) in plant chromosomes. This method utilizes enzymatic activities for eliminating most of the repeated regions, resulting in the reduction of genome representation for increment in the efficiency of sequencing (Elshire et al. 2011). Major NGS platforms include Solexa (Illumina), SOLID (ABI), and 454 (Roche). Being high multiplexing system, this technique was designed in the Buckler lab for Illumina NGS platform, which allows study of large genomic species consisting of huge diversity. This approach exploits exceptionally high throughput of NGS, through which simultaneous analysis of multiple genomic regions of interest is done with precisely designed probes being amplified and later sequenced. Prior information of genome is not a prerequisite for the use of this technology since it simultaneously performs SNP detection and genotyping. Hence, it surpasses the drawbacks of the conventional PCR or array-based SNP genotyping techniques. When complemented with the phenotypic data, GBS approach became an efficient approach to rapidly map and identify the genes associated with disease resistance in wheat (Edwards et al. 2013). Wheat populations are genotyped using various genotyping arrays including 9K iSelect Illumina, Axiom Wheat HD genotyping array, 90K iSelect, and with technological advancements these platforms are being more convenient and efficient (Neupane 2022; Subramaniam et al. 2024; Sun et al. 2020; Wang et al. 2014; Winfield et al. 2016).

Poland et al. 2012, used GBS technique which resulted in incorporation of 1000s of markers in the bread wheat map. The application of GBS on wheat lines highlighted GBS as an economical and powerful platform for genomic-assisted breeding. Cheng et al. 2020, studied, 140 wheat lines in China for resistance to PM and revealed thirty-six varieties exhibiting resistance in all four environments. Many species within the genus *Aegilops* have been efficiently utilized as a promising genomic resource for imparting resistance to various diseases in wheat improvement programs. In a study conducted by (Edae et al. 2016), stem rust resistance loci on chromosome 2U in *Aegilops umbellulata* biparental mapping population for stem rust races including Ug99 was done utilizing GBS SNP markers. This work significantly reported the first QTL identified for stem rust from *Ae. umbellulata*. This major finding can further be beneficial for introgression of novel stem rust resistance gene into commercial varieties. In addition Edae and Rouse 2019, confirmed the results of previous study on a new biparental population and performed BSA-Seq (Bulk Segregant Analysis RNA-Sequencing) on both the populations. The BSR-Seq study provided the candidate resistance genes and the transcriptome assembly of the resistant and susceptible bulks provided expression of different genes associated with stem rust introgressed from *Aegilops umbellulate.*

To accomplish an advance breeding work, it is prerequisite to have information regarding genetic diversity among parental varieties for the targeted traits. Diversity arrays technology (DArT) is a method for determining genetically diversified marker (Jaccoud et al. 2001). This technology provides high throughput, low cost per data, and sequence-independent genotyping with simultaneous information of several hundred to thousands of polymorphic loci spreading on the genome. Diversity arrays technology (DArT) arrays in wheat were developed by Akbari et al. 2006. Diversity arrays technology (DArT) process begins with assembling DNA samples representing the germplasm anticipated, which further gets analyzed with technology. Sample group usually belongs to primary gene pool of the species being investigated with expanding to secondary or even tertiary gene pool. Diversity arrays technology (DArT) is hybridization-based strategy utilizing microarray platform for the detection of restriction site polymorphism with enzymes which are methylation sensitive. Gaál et al. 2024, in their recent work, utilized wild wheat relative *Aegilops biuncialis* accessions for introgression in wheat to obtain early flowering, drought tolerance, and rust resistance. They applied DArTseq technology for genotyping backcrossed populations obtained from the crosses, with identification of 8998 wheat-specific and 11,952 *Aegilops*-specific Silico-DArT markers. Diversity arrays technology (DArT) array has also been utilized for polyploid species consisting of complex genome in an order to determine the level of introgression of diverse parent in the interspecific hybrids or evaluating progenies obtained from interspecific crosses (Eberhard et al. 2010; Millet et al. 2014; Szőke-Pázsi et al. 2024). Diversity arrays technology (DArT) along with single nucleotide polymorphism (SNP) markers are referred as third-generation DNA marker system. Single nucleotide polymorphism (SNP) arises due to various types of substitutions, which results in the polymorphism in base composition of genome. This polymorphism is utilized as markers for figuring out desired trait. When executing the genomic selection, utilization of genomic data is done for predicting the efficiency of wheat genotypes for targeted traits by involvement of high-density molecular markers, like SNPs, for capturing the genetic variations in the genome (Li et al. 2024; Van Tassell et al. 2008). After the advancement in SNP markers and high throughput computational system, concrete findings over genomic data are achievable. In a work by Daetwyler et al. 2013, screening of 250 morphologically different landrace accessions was achieved for the identifying stem rust, leaf rust, and stripe rust resistance with utilization of an Illumina Infinium 9K wheat SNP chip technology.

The genotyping platform GBS achieves marker discovery and genotyping simultaneously, which results in in minimal ascertainment bias. This bias occurs with fixed-array genotyping, when marker data is not acquired from a random sample of the polymorphisms in the population of interest and as such there is no representations of the reference set in assaying polymorphism for the new genotype sets. Moreover, GBS-derived markers are uniform in distribution over the genome as compared to DArT markers, which clusters and exhibit low density on D genome.

Genome-wide association studies (GWAS) is a statistical approach to map QTLs which utilizes linkage disequilibrium (LD) to detect associations between genotype and phenotype. It is a statistical method that successfully identifies candidate alleles which controls the targeted traits. Genome-wide association studies (GWAS) also has its application in the selected population that can be additionally accompanied with RNA-Seq data for further analysis of the expressivity of gene (Liu et al. 2018; Xu et al. 2022).

Genome-wide association studies (GWAS) mainly focuses on the rapid screening of germplasm collection to find the genes of interest with additionally finding out the recombination in them for achieving high resolution mapping. It is advantageous over other approaches since it utilizes existing natural populations with huge diversity with a purpose to record more recombination events occurring throughout the evolutionary process contributing to a higher resolution. Therefore, it becomes an useful tool in the identification of rare alleles (Jin-Long et al. 2012). However, this technique gives type I error or false positives because of its population structure. Recently, Alemu et al. 2021, conducted GWAS for yellow rust resistance on 300 durum accessions, including 39 cultivars and 261 landraces, to identify genetic-resistance loci. Eleven SNPs related to *Yr* resistance were identified on four chromosomes, that is, 1A, 1B, 2B, and 5A. Gardiner et al. 2020, used 3 approaches for mapping the yellow rust-resistance gene *Yr7* and for identifying associated SNPs in wheat. They used QTL mapping, fine mapping, and GWAS of resistant and susceptible double-haploid bulks for identifying the candidate gene. In an extensive work, Kumar et al. 2020, strived for potential source of resistance in a panel of 483 wheat genotypes for stem, stripe, and leaf rust, utilizing 35 K SNP array. Seven genomic regions were associated with all the three rusts on chromosome groups 1 and 6 along with 2B. Moreover, identification of 16, 18, and 27 QTL for YR, LR, and SR, was done respectively. In a study by El Hanafi et al. 2021, GWAS was utilized for 196 genotypes on the basis of 10,477 SNP markers. In this, 85 genotypes were found to be resistant while 22 genotypes were moderately resistant. Genome-wide association studies (GWAS) using mixed linear model discovered 23 markers on chromosomes 2A, 2B, 2D, and 7B associated with adult plant resistance. Other similar work utilizing GWAS has been also done to achieve stripe and stem rust resistance in wheat (Megerssa et al. 2020; Mourad et al. 2018; Mu et al. 2020; Vikram et al. 2021; Marone et al. 2024). Genome-wide association studies (GWAS) also led to the detection of significant MTAs for FHB which have been identified (Arruda et al. 2015; Verges et al. 2021). In a study by Silva et al. 2022, evaluation for the assessing the disease in wheat with identifying genomic regions imparting resistance to the BYDV was done.

One of the recent approaches, RNA sequencing, provide information about gene expression while giving an insight on the complex processes occurring within the plant. This method highlights the variations in the expression of genes, thus represent plasticity of genome toward stress through spatiotemporal expression signatures of genes. This method strategies direct sequencing of cDNA obtained from RNA transcriptome sequences. RNA-Seq analysis provide the RNA transcript levels (Wang et al. 2009). Even though coding sequences account for only 1–2% of the genome, such regions have high functionality (Li et al. 2012; Pennisi 2001). Ramirez-Gonzalez et al. 2015, identified markers for *Yr15* using RNA-Seq of bulked pools for identification of SNPs associated with the gene. Over 27,000 genes with SNPs were investigated between the parents and later categorized by the results from the bulks sequenced. In a recent work by He et al. 2021, durum wheat TRI 1796 was found to confer high resistance to 22 isolates of *Blumeria graminis* f. sp. *tritici* (*Bgt*). Using the bulked segregant RNA-Seq (BSR-Seq) analysis together with molecular analysis, a new powdery mildew resistance gene *Pm68* was mapped to the terminal part of the short arm of chromosome 2B. Dobon et al. 2016, conducted the *Puccinia striiformis* f. sp. *tritici* pathogen and wheat interaction study by utilizing RNA-Seq approach. Since, RNA-Seq does not require prior genome-sequence knowledge, it analyses both host and pathogen transcriptomes, allowing the study of host-hijacking by the pathogens for disease development and impact analysis to the host plant's mechanism during the pathogen infection. Hiebert et al. 2020, identified a gene *SuSr-D1* (Suppressor of Stem rust resistance1, D-genome), on chromosome arm 7DL in cultivar "Canthatch." The time-course RNA-Seq analysis of the *SuSr-D1* or Med15 mutants showed resistance to several races of stem rust. In another study Jia et al. 2020, utilized Chinese wheat cultivar Jimai 23 for BSR-Seq to develop series of markers for mapping *PmJM23* gene. *PmJM23* was found on *Pm2* locus at short arm of chromosome 5D (5DS). *PmJM23* provides a broad resistance which is a better than *Pm2* alleles. (Li et al. 2020a, b, c), worked on RNA-Seq of Chinese Spring (CS)- *Ae. longissima* $4S^{l}$(4B) disomic substitution line (TA3465) to develop $4S^{l}$—specific markers for transferring *Bgt* resistance gene from $4S^{l}$ locus through induced homoeologous recombination. In a study Ma et al. 2019, developed a set of 199 recombinant inbred lines from a cross between wheat line 20,828 and a susceptible cultivar Chuannong 16. These RILs were evaluated for resistance to stripe rust races followed by genotyping using Wheat 55K SNP array and SSR markers. The association analysis revealed stripe rust resistance QTL *QYr.sicau-1B.1* for all-stage resistance on chromosome arm 1BL. In another study, Narang et al. 2020, reported introgression of a major dominant gene *LrAp* from wild wheat species *Aegilops peregrina* to the long arm of wheat chromosome 6B. The genetic mapping of *LrAp* was done using BSA together with resistance gene-enrichment sequencing and the markers so developed would further facilitate marker-assisted gene pyramiding of *LrAp*. Identification of a NLR cluster was achieved through integrating long-term DNA reads and RNA sequencing. Further, location of *Pm69* was confirmed in this cluster and was introgressed and pyramided with *Yr15* (Li et al. 2024).

In order to achieve high throughput mutation mapping, various approaches are being utilized, few of them include MutMap (Abe et al. 2012), MutChromSeq (Sánchez-Martín et al. 2016), AgRenSeq (Arora et al. 2019), Targeted Chromosome-based Cloning via Long-range Assembly (TACCA) (Thind et al. 2017), Simultaneous Identification of Multiple Causal Mutations (SIMM) (Steuernagel et al. 2017).

MutChromSeq (Mutant Chromosome Sequencing) technique was discovered by Sanchez-Martin in wheat for Cloning *Pm2* gene (Sánchez-Martín et al. 2016). This approach is independent of recombination-based mapping and identifying the mutation, instead targeting all DNA sequences. Another approach, namely, AgRenSeq (Association Genetics with R-Gene Enrichment Sequencing) works for cloning of resistance (R) gene from variable germplasm including wild relatives of the crop. Need for reference genome is removed in this approach since it directly identifies the nucleotide-binding/leucine-rich regions conferring resistance. As such wild diversity can be explored and sequenced for identifying genes imparting resistance. In wheat, identification of four stem rust resistance genes; *Sr33*, *Sr45*, *Sr46*, and *SrTA1662* was achieved using this method (Arora et al. 2019). An advanced version for R-gene-enrichment sequencing (RenSeq) is MutRenSeq Approach. This technique uses RenSeq data of wild parent and EMS-derived loss-of-resistance mutants to find out the R-gene complements, for identifying R-gene/causal mutations (Steuernagel et al. 2017). Utilization of this method was done to clone *Yr7, Yr5* and *YrSP* in wheat (Marchal et al. 2018).

4 Conclusion

It is inevitable that climate change is happening very rapidly and is very difficult to deal with. The other factor which is of great concern is the evolution of new pathogens due to this climate change. The invasion of these pathogens in new geographical areas is an alarming situation. The researchers need to find new genes or alleles to counter these pathogens and introgress them into the elite cultivars, such that large-scale crop losses may be prevented. Rapid improvements in genotyping methods and NGS technology with decreased cost have made sequence-based approach a very practical and suitable way to study the genome as well as populations. In the past few years, the potential to look within the genome has transformed the wheat breeding with more possibilities to apply NGS for mitigating the increased threat of these diseases for food quantity and quality.

Acknowledgments The authors acknowledge their respective organization for providing them necessary facilities for carrying out this work.

Competing Interests The authors declare that they have no conflict of interest.

FundingThis research did not receive any funding.

References

Abe A, Kosugi S, Yoshida K, Natsume S, Takagi H, Kanzaki H, Matsumura H, Yoshida K, Mitsuoka C, Tamiru M, Innan H (2012) Genome sequencing reveals agronomically important loci in rice using MutMap. Nat Biotechnol 30(2):174–178

Adler AJ, Wiley GB, Gaffney PM (2013) Infinium assay for large-scale SNP genotyping applications. J Vis Exp 81:e50683

Akbari M, Wenzl P, Caig V, Carling J, Xia L, Yang S, Uszynski G, Mohler V, Lehmensiek A, Kuchel H, Hayden MJ (2006) Diversity arrays technology (DArT) for high-throughput profiling of the hexaploid wheat genome. Theor Appl Genet 113(8):1409–1420

Alemu SK, Huluka AB, Tesfaye K, Haileselassie T, Uauy C (2021) Genome-wide association mapping identifies yellow rust resistance loci in Ethiopian durum wheat germplasm. PLoS One 16(5):e0243675

Ari Ş, Arikan M (2016) Next-generation sequencing: advantages, disadvantages, and future. In: Ari Ş, Muzaffer A (eds) Plant omics: trends and applications. Springer, pp 109–135

Arora S, Steuernagel B, Gaurav K, Chandramohan S, Long Y, Matny O, Johnson R, Enk J, Periyannan S, Singh N, Asyraf Md Hatta M (2019) Resistance gene cloning from a wild crop relative by sequence capture and association genetics. Nat Biotechnol 37(2):139–143

Arruda MP, Brown PJ, Lipka AE, Krill AM, Thurber C, Kolb FL (2015) Genomic selection for predicting Fusarium head blight resistance in a wheat breeding program. T P G 8(3):01

Arumuganathan K, Earle E (1991) Nuclear DNA content of some important plant species. Plant Mol Biol Rep 99(3):208–218

Babu P, Baranwal DK, Harikrishna DP, Bharti H, Joshi P, Thiyagarajan B, Gaikwad KB, Bhardwaj SC, Singh GP, Singh A (2020) Application of genomics tools in wheat breeding to attain durable rust resistance. Front Plant Sci 11:567147

Bai G, Su Z, Cai J (2018) Wheat resistance to Fusarium head blight. Can J Plant Pathol 40(3):336–346

Bansal U, Bariana H (2017) Advances in identification and mapping of rust resistance genes in wheat. In: Bansal U, Bariana H (eds) Wheat rust diseases: methods and protocols. Springer, New York, pp 151–162

Barba M, Czosnek H, Hadidi A (2014) Historical perspective, development and applications of next-generation sequencing in plant virology. Viruses 6(1):106–136

Bhardwaj SC, Prasad P, Gangwar OP, Khan H, Kumar S (2016) Wheat rust research-then and now. Indian J Agric Sci 86(10):1231–1244

Bhardwaj SC, Singh GP, Gangwar OP, Prasad P, Kumar S (2019) Status of wheat rust research and progress in rust management-Indian context. Agronomy 9(12):892

Bräutigam A, Gowik U (2010) What can next generation sequencing do for you? Next generation sequencing as a valuable tool in plant research. Plant Biol 12(6):831–841

Brenchley R, Spannagl M, Pfeifer M, Barker GL, D'Amore R, Allen AM, McKenzie N, Kramer M, Kerhornou A, Bolser D, Kay S (2012) Analysis of the bread wheat genome using whole-genome shotgun sequencing. Nature 491(7426):705–710

Cao A, Xing L, Wang X, Yang X, Wang W, Sun Y, Qian C, Ni J, Chen Y, Liu D, Wang X (2011) Serine/threonine kinase gene Stpk-V, a key member of powdery mildew resistance gene *Pm21*, confers powdery mildew resistance in wheat. Proc Natl Acad Sci 108(19):7727–7732

Cheng B, Ding Y, Gao X, Cao N, Xin Z, Zhang L (2020) The diversity of powdery mildew resistance gene loci among wheat germplasm in Southwest China. Cereal Res Commun 48(1):65–70

Crossa J, Burgueno J, Dreisigacker S, Vargas M, Herrera-Foessel SA, Lillemo M, Singh RP, Trethowan R, Warburton M, Franco J, Reynolds M (2007) Association analysis of historical bread wheat germplasm using additive genetic covariance of relatives and population structure. Genetics 177(3):1889–1913

Daetwyler H, Hayden M, Bansal U, Bariana H, Hayes B (2013) Genomic selection for disease and morphological traits in diverse wheat landraces. In: Plant and animal genome San-Diego, CA

Deller S, Hammond-Kosack KE, Rudd JJ (2011) The complex interactions between host immunity and non-biotrophic fungal pathogens of wheat leaves. J Plant Physiol 168(1):63–71

Dobon A, Bunting DC, Cabrera-Quio LE, Uauy C, Saunders DG (2016) The host-pathogen interaction between wheat and yellow rust induces temporally coordinated waves of gene expression. BMC Genomics 17(1):1–14

Eberhard FS, Zhang P, Lehmensiek A, Hare RA, Simpfendorfer S, Sutherland MW (2010) Chromosome composition of an F2 *Triticum aestivum*× *T. turgidum* spp. durum cross analysed by DArT markers and MCFISH. Crop Pasture Sci 61(8):619–624

Edae EA, Rouse MN (2019) Bulked segregant analysis RNA-seq (BSR-Seq) validated a stem resistance locus in *Aegilops umbellulata*, a wild relative of wheat. PLoS One 14(9):e0215492

Edae EA, Olivera PD, Jin Y, Poland JA, Rouse MN (2016) Genotype-by-sequencing facilitates genetic mapping of a stem rust resistance locus in *Aegilops umbellulata*, a wild relative of cultivated wheat. BMC Genomics 17(1):1039

Edwards D, Batley J, Snowdon RJ (2013) Accessing complex crop genomes with next-generation sequencing. Theor Appl Genet 126(1):1–11

El Hanafi S, Backhaus A, Bendaou N, Sanchez-Garcia M, Al-Abdallat A, Tadesse W (2021) Genome-wide association study for adult plant resistance to yellow rust in spring bread wheat (*Triticum aestivum* L.). Euphytica 217(5):1–14

Elshire RJ, Glaubitz JC, Sun Q, Poland JA, Kawamoto K, Buckler ES, Mitchell SE (2011) A robust, simple genotyping-by-sequencing (GBS) approach for high diversity species. PLoS One 6(5):e19379

Gaál E, Farkas A, Türkösi E, Kruppa, Szakács É, Szőke-Pázsi K, Kovács P, Kalapos B, Darkó É, Said M, Lampar A (2024) DArTseq genotyping facilitates identification of *Aegilops biuncialis* chromatin introgressed into bread wheat Mv9kr1. Plant Mol Biol 114(6):1–16

Gardiner L-J, Bansept-Basler P, El-Soda M, Hall A, O'Sullivan DM (2020) A framework for gene mapping in wheat demonstrated using the *Yr7* yellow rust resistance gene. PLoS One 15(4):e0231157

González-Neira A (2013) The GoldenGate genotyping assay: custom design, processing, and data analysis. Methods Mol Biol 1015:147–153

Gupta SK, Charpe A, Koul S, Prabhu KV, Haq QMR (2005) Development and validation of molecular markers linked to an *Aegilops umbellulata*–derived leaf-rust-resistance gene, *Lr9*, for marker-assisted selection in bread wheat. Genome 48(5):823–830

Haider MW, Kaur J, Bala R, Singh S, Srivastava P, Sharma A, Singh R, Kumari J (2023) Stripe rust resistance gene (s) postulation in wheat germplasm with the help of differentials and tagged molecular markers. Sci Rep 13(1):9007

He J, Zhao X, Laroche A, Lu Z-X, Liu HK, Li Z (2014) Genotyping-by-sequencing (GBS), an ultimate marker-assisted selection (MAS) tool to accelerate plant breeding. Front Plant Sci 5:484

He H, Liu R, Ma P, Du H, Zhang H, Wu Q, Yang L, Gong S, Liu T, Huo N, Gu YQ (2021) Characterization of *Pm68*, a new powdery mildew resistance gene on chromosome 2BS of Greek durum wheat TRI 1796. Theor Appl Genet 134(1):53–62

Hiebert CW, Moscou MJ, Hewitt T, Steuernagel B, Hernández-Pinzón I, Green P, Pujol V, Zhang P, Rouse MN, Jin Y, McIntosh RA (2020) Stem rust resistance in wheat is suppressed by a subunit of the mediator complex. Nat Commun 11(1):1123

Huang L, Gill BS (2001) An RGA–like marker detects all known *Lr21* leaf rust resistance gene family members in *Aegilops tauschii* and wheat. Theor Appl Genet 103(6):1007–1013

Hurni S, Brunner S, Buchmann G, Herren G, Jordan T, Krukowski P, Wicker T, Yahiaoui N, Mago R, Keller B (2013) Rye *Pm8* and wheat *Pm3* are orthologous genes and show evolutionary conservation of resistance function against powdery mildew. Plant J 76(6):957–969

Jaccoud D, Peng K, Feinstein D, Kilian A (2001) Diversity arrays: a solid state technology for sequence information independent genotyping. Nucleic Acids Res 29(4):e25–e25

Jaiswal S, Sheoran S, Arora V, Angadi UB, Iquebal MA, Raghav N, Aneja B, Kumar D, Singh R, Sharma P, Singh GP (2017) Putative microsatellite DNA marker-based wheat genomic resource for varietal improvement and management. Front Plant Sci 8:2009

Jia J, Zhao S, Kong X, Li Y, Zhao G, He W (2013) *Aegilops tauschii* draft genome sequence reveals a gene repertoire for wheat adaptation. Nature 496(7443):91–95

Jia M, Xu H, Liu C, Mao R, Li H, Liu J, Du W, Wang W, Zhang X, Han R, Wang X (2020) Characterization of the powdery mildew resistance gene in the elite wheat cultivar Jimai 23 and its application in marker-assisted selection. Front Genet 11:241

Jin-Long G, Li-Ping X, Jing-Ping F, Ya-Chun S, Hua-Ying F, You-Xiong Q, Jing-Sheng X (2012) A novel dirigent protein gene with highly stem-specific expression from sugarcane, response to drought, salt and oxidative stresses. Plant Cell Rep 31(10):1801–1812

Klymiuk V, Chawla HS, Wiebe K, Ens J, Fatiukha A, Govta L, Fahima T, Pozniak CJ (2022) Discovery of stripe rust resistance with incomplete dominance in wild emmer wheat using bulked segregant analysis sequencing. Commun Biol 5(1):826

Kumar S, Röder MS, Tripathi SB, Kumar S, Chand R, Joshi AK, Kumar U (2015) Mendelization and fine mapping of a bread wheat spot blotch disease resistance QTL. Mol Breed 35(11). https://doi.org/10.1007/s11032-015-0411-5

Kumar D, Kumar A, Chhokar V, Gangwar OP, Bhardwaj SC, Sivasamy M, Prasad SS, Prakasha TL, Khan H, Singh R, Sharma P (2020) Genome-wide association studies in diverse spring wheat panel for stripe, stem, and leaf rust resistance. Front Plant Sci 11:748

Li X, Zhu C, Yeh CT, Wu W, Takacs EM, Petsch KA, Tian F, Bai G, Buckler ES, Muehlbauer GJ, Timmermans MC (2012) Genic and nongenic contributions to natural variation of quantitative traits in maize. Genome Res 22(12):2436–2444

Li H, Dong Z, Ma C, Xia Q, Tian X, Sehgal S, Koo DH, Friebe B, Ma P, Liu W (2020a) A spontaneous wheat-*Aegilops longissima* translocation carrying *Pm66* confers resistance to powdery mildew. Theor Appl Genet 133(4):1149–1159

Li J, Dundas I, Dong C, Li G, Trethowan R, Yang Z, Hoxha S, Zhang P (2020b) Identification and characterization of a new stripe rust resistance gene *Yr83* on rye chromosome 6R in wheat. Theor Appl Genet 133(4):1095–1107

Li Y, Shi X, Hu J, Wu P, Qiu D, Qu Y, Xie J, Wu Q, Zhang H, Yang L, Liu H (2020c) Identification of a recessive gene *PmQ* conferring resistance to powdery mildew in wheat landrace Qingxinmai using BSR-Seq analysis. Plant Dis 104(3):743–751

Li Y, Wei ZZ, Sela H, Govta L, Klymiuk V, Roychowdhury R, Chawla HS, Ens J, Wiebe K, Bocharova V, Ben-David R (2024) Dissection of a rapidly evolving wheat resistance gene cluster by long-read genome sequencing accelerated the cloning of *Pm69*. Plant Comm 5(1):100646

Lillemo M, Joshi AK, Prasad R, Chand R, Singh RP (2013) QTL for spot blotch resistance in bread wheat line Saar co-locate to the biotrophic disease resistance loci *Lr34* and *Lr46*. Theor Appl Genet 126(3):711–719

Ling H-Q, Zhao S, Liu D, Wang J, Sun H, Zhang C, Fan H, Li D, Dong L, Tao Y, Gao C (2013) Draft genome of the wheat A-genome progenitor *Triticum urartu*. Nature 496:87–90

Liu J, Feng B, Xu Z, Fan X, Jiang F, Jin X, Cao J, Wang F, Liu Q, Yang L, Wang T (2018) A genome-wide association study of wheat yield and quality-related traits in Southwest China. Mol Breed 38(1):1–11

Lu P, Liang Y, Li D, Wang Z, Li W, Wang G, Wang Y, Zhou S, Wu Q, Xie J, Zhang D (2016) Fine genetic mapping of spot blotch resistance gene Sb3 in wheat (Triticum aestivum). Theor Appl Genet 129(3):577–589

Ma J, Qin N, Cai B, Chen G, Ding P, Zhang H, Yang C, Huang L, Mu Y, Tang H, Liu Y (2019) Identification and validation of a novel major QTL for all-stage stripe rust resistance on 1BL in the winter wheat line 20828. Theor Appl Genet 132(5):1363–1373

Marchal C, Zhang J, Zhang P, Fenwick P, Steuernagel B, Adamski NM, Boyd L, McIntosh R, Wulff BB, Berry S, Lagudah E (2018) BED-domain-containing immune receptors confer diverse resistance spectra to yellow rust. Nat Plants 4(9):662–668

Marone D, Laidò G, Saccomanno A, Petruzzino G, Giaretta Azevedo CV, De Vita P, Mastrangelo AM, Gadaleta A, Ammar K, Bassi FM, Wang M (2024) Genome-wide association study of common resistance to rust species in tetraploid wheat. Front Plant Sci 14:1290643

McIntosh RA, Wellings CR, Park RF (1995) Wheat rusts: an atlas of resistance genes. In: McIntosh RA, Wellings CR, Park RF (eds) Wheat rusts and the genetic bases of disease resistance. CSIRO Publishing

McIntosh RA, Dubcovsky J, Rogers WJ, Morris C, Appels R, Xia XC (2010) Catalogue of gene symbols for wheat: 2010 supplement. Ann Wheat Newsl 56:273–282

McIntosh R, Dubcovsky J, Rogers JW, Morris C, Appels R, Xia X (2014) Catalogue of gene symbols for wheat: 2013–14 supplement. Ann Wheat Newsl 58:1–18

Megerssa SH, Ammar K, Acevedo M, Brown-Guedira G, Ward B, Degete AG, Randhawa MS, Sorrells ME (2020) Multiple-race stem rust resistance loci identified in durum wheat using genome-wide association mapping. Front Plant Sci 11:1934

Michelmore RW, Paran I, Kesseli R (1991) Identification of markers linked to disease-resistance genes by bulked segregant analysis: a rapid method to detect markers in specific genomic regions by using segregating populations. Proc Natl Acad Sci 88(21):9828–9832

Millet E, Manisterski J, Ben-Yehuda P, Distelfeld A, Deek J, Wan A, Chen X, Steffenson BJ (2014) Introgression of leaf rust and stripe rust resistance from Sharon goatgrass (Aegilops sharonensis Eig) into bread wheat (*Triticum aestivum* L.). Genome 57(6):309–316

Mourad AM, Sallam A, Belamkar V, Wegulo S, Bowden R, Jin Y, Mahdy E, Bakheit B, El-Wafaa AA, Poland J, Baenziger PS (2018) Genome-wide association study for identification and validation of novel SNP markers for *Sr6* stem rust resistance gene in bread wheat. Front Plant Sci 9:380

Mu J, Liu L, Liu Y, Wang M, See DR, Han D, Chen X (2020) Genome-wide association study and gene specific markers identified 51 genes or QTL for resistance to stripe rust in US winter wheat cultivars and breeding lines. Front Plant Sci 11:998

Narang D, Kaur S, Steuernagel B, Ghosh S, Bansal U, Li J (2020) Discovery and characterisation of a new leaf rust resistance gene introgressed in wheat from wild wheat *Aegilops peregrina*. Sci Rep 10(1):1–9

Neupane A (2022) Genetic mapping of Fusarium head blight resistance, rust resistance, and agronomic traits in winter wheat. University of Manitoba

Nsabiyera V, Baranwal D, Qureshi N, Kay P, Forrest K, Valárik M (2020) Fine mapping of *Lr49* using 90K SNP chip array and flow-sorted chromosome sequencing in wheat. Front Plant Sci 10:1787

Pascual L, Ruiz M, López-Fernández M, Pérez-Peña H, Benavente E, Vázquez JF, Sansaloni C, Giraldo P (2020) Genomic analysis of Spanish wheat landraces reveals their variability and potential for breeding. BMC Genomics 21(1):1–7

Pennisi E (2001) The human genome. AAAS 291(6446):1177–1180

Poland JA, Endelman J, Dawson J, Rutkoski J, Wu S, Manes Y (2012) Genomic selection in wheat breeding using genotyping-by-sequencing. TPG 5(3):103–113

Ramirez-Gonzalez RH, Segovia V, Bird N, Fenwick P, Holdgate S, Berry S, Jack P, Caccamo M, Uauy C (2015) RNA-S eq bulked segregant analysis enables the identification of high-resolution genetic markers for breeding in hexaploid wheat. Plant Biotechnol J 13(5):613–624

Sánchez-Martín J, Steuernagel B, Ghosh S, Herren G, Hurni S, Adamski N (2016) Rapid gene isolation in barley and wheat by mutant chromosome sequencing. Genome Biol 17(1):22.1

Schachermayr GM, Messmer MM, Feuillet C, Winzeler H, Winzeler M, Keller B (1995) Identification of molecular markers linked to the *Agropyron elongatum*-derived leaf rust resistance gene *Lr24* in wheat. Theor Appl Genet 90(7):982–990

Schwessinger B (2017) Fundamental wheat stripe rust research in the 21st century. New Phytol 213(4):1625–1631

Shaheenuzzamn M, Liu T, Shi S, An P, Wu H, Wang Z (2020) Development of sequencing technology and role of next generation sequencing technologies in wheat research: a review. Pak J Bot 52(5):1867–1878

Sharma D, Avni R, Gutierrez-Gonzalez J, Kumar R, Sela H, Prusty MR, Shatil-Cohen A, Molnár I, Holušová K, Said M, Doležel J (2024) A single NLR gene confers resistance to leaf and stripe rust in wheat. Nat Commun 15(1):9925

Shendure J, Ji H (2008) Next-generation DNA sequencing. Nat Biotechnol 26(10):1135–1145

Shewry PR (2009) Wheat. J Exp Bot 60(6):1537–1553

Shiferaw B, Smale M, Braun HJ, Duveiller E, Reynolds M, Muricho G (2013) Crops that feed the world 10. Past successes and future challenges to the role played by wheat in global food security. Food Secur 5(3):291–317

Silva P, Evers B, Kieffaber A, Wang X, Brown R, Gao L, Fritz A, Crain J, Poland J (2022) Applied phenomics and genomics for improving barley yellow dwarf resistance in winter wheat. G3 12(7):p.jkac064

Singh RP, Rajaram S, Saini R, Huerta-Espino J, William M (2004) Genetics of host-pathogen interaction and breeding for durable resistance. In: Jain HK, Kharkwal MC (eds) Plant breeding. Springer, pp 145–166

Smith SM, Maughan PJ (2015) SNP genotyping using KASPar assays. Methods Mol Biol 1245:243–256

Steuernagel B, Vrána J, Karafiátová M, Wulff BB, Doležel J (2017) Rapid gene isolation using MutChromSeq. In: Wheat rust diseases: methods and protocols, pp 231–243

Subramaniam S, Kadirvel P, Periyannan S (2024) Wheat improvement through advances in single nucleotide polymorphism (SNP) detection and genotyping with a special emphasis on rust resistance. Theor Appl Genet Inter J Plant Breed Res 137(10):224

Sun C, Dong Z, Zhao L, Ren Y, Zhang N, Chen F (2020) The Wheat 660K SNP array demonstrates great potential for marker-assisted selection in polyploid wheat. Plant Biotechnol J 18(6):1354–1360

Szőke-Pázsi K, Kruppa K, Tulpová Z, Kalapos B, Türkösi E, Gaál E, Darkó É, Said M, Farkas A, Kovács P, Ivanizs L (2024) DArTseq genotyping facilitates the transfer of "exotic" chromatin from a *Secale cereale*× *S. Strictum* hybrid into wheat. Front Plant Sci 15:1407840

Thind AK, Wicker T, Šimková H, Fossati D, Moullet O, Brabant C, Vrána J, Doležel J, Krattinger SG (2017) Rapid cloning of genes in hexaploid wheat using cultivar-specific long-range chromosome assembly. Nat Biotechnol 35(8):793–796

Thomson MJ, Zhao K, Wright M, McNally KL, Rey J, Tung CW, Reynolds A, Scheffler B, Eizenga G, McClung A, Kim H (2012) High-throughput single nucleotide polymorphism genotyping for breeding applications in rice using the BeadXpress platform. Mol Breed 29(4):875–886

Van Tassell CP, Smith TP, Matukumalli LK, Taylor JF, Schnabel RD, Lawley CT, Haudenschild CD, Moore SS, Warren WC, Sonstegard TS (2008) SNP discovery and allele frequency estimation by deep sequencing of reduced representation libraries. Nat Methods 5(3):247–252

Verges VL, Brown-Guedira GL, Van Sanford DA (2021) Genome-wide association studies combined with genomic selection as a tool to increase Fusarium head blight resistance in wheat. Crop Breed Genet Genom 3(4):e210007

Vikram P, Sehgal D, Sharma A, Bhavani S, Gupta P, Randhawa M, Pardo N, Basandra D, Srivastava P, Singh S, Sood T (2021) Genome-wide association analysis of Mexican bread wheat landraces for resistance to yellow and stem rust. PLoS One 16(1):e0246015

Wang Z, Gerstein M, Snyder M (2009) RNA-Seq: a revolutionary tool for transcriptomics. Nat Rev Genet 10(1):57–63

Wang Y, Cheng X, Shan Q, Zhang Y, Liu J, Gao C, Qiu JL (2014) Simultaneous editing of three homoeoalleles in hexaploid bread wheat confers heritable resistance to powdery mildew. Nat Biotechnol 32(9):947–951

Wenzl P, Carling J, Kudrna D, Jaccoud D, Huttner E, Kleinhofs A, Kilian A (2004) Diversity arrays technology (DArT) for whole-genome profiling of barley. Proc Natl Acad Sci 101(26):9915–9920

Winfield MO, Allen AM, Burridge AJ, Barker GL, Benbow HR, Wilkinson PA, Coghill J, Waterfall C, Davassi A, Scopes G, Pirani A (2016) High-density SNP genotyping array for hexaploid wheat and its secondary and tertiary gene pool. Plant Biotechnol J 14(5):1195–1206

Xu F, Chen S, Zhou S, Yue C, Yang X, Zhang X, Zhan K, He D (2022) Genome-wide association, RNA-seq and iTRAQ analyses identify candidate genes controlling radicle length of wheat. Front Plant Sci 13:939544

Yahiaoui N, Srichumpa P, Dudler R, Keller B (2004) Genome analysis at different ploidy levels allows cloning of the powdery mildew resistance gene *Pm3b* from hexaploid wheat. Plant J 37(4):528–538

You FM, Huo N, Gu YQ, Luo M-c, Ma Y, Hane D, Lazo GR, Dvorak J, Anderson OD (2008) BatchPrimer3: a high throughput web application for PCR and sequencing primer design. BMC Bioinf 9(1):1–13

Yu J, Pressoir G, Briggs WH, Bi IV, Yamasaki M, Doebley JF, McMullen MD, Gaut BS, Nielsen DM, Holland JB, Kresovich S (2006) A unified mixed-model method for association mapping that accounts for multiple levels of relatedness. Nat Genet 38(2):203–208

Yuan M, Ngou BPM, Ding P, Xin XF (2021) PTI-ETI crosstalk: an integrative view of plant immunity. Curr Opin Plant Biol 62:102030

Zeller FJ (1973) 1B/1R wheat-rye chromosome substitutions and translocations. In: Proceedings of the fourth international wheat genetics symposium, Columbia/ MO, pp 209–221

Zhang Z, Lin Z, Xin Z (2009) Research progress in BYDV resistance genes derived from wheat and its wild relatives. J Genet Genom 36(9):567–573

Zhu T, Wu L, He H, Song J, Jia M, Liu L, Wang XL, Han R, Niu LP, Du WX, Zhang X (2020) Bulked Segregant RNA-Seq reveals distinct expression profiling in Chinese wheat cultivar Jimai 23 responding to powdery mildew. Front Genet 11:474

MIX
Papier aus verantwortungsvollen Quellen
Paper from responsible sources
FSC® C105338

If you have any concerns about our products,
you can contact us on
ProductSafety@springernature.com

In case Publisher is established outside the EU,
the EU authorized representative is:
Springer Nature Customer Service Center GmbH
Europaplatz 3, 69115 Heidelberg, Germany

Printed by Libri Plureos GmbH
in Hamburg, Germany